BIODIVERSITY HOTSPOT OF THE COASTAL FORESTS OF EASTERN AFRICA

Cover photo credits:

Top left: *Hildegardia migeodii,* Malvaceae. Photo by Ton Rulkens from Mozambique. Uploaded by Jacopo Werther. https://creativecommons.org/licenses/by-sa/2.0/

Top middle: Black Rhino. Photo courtesy of Working with Wildlife. https://workingwithwildlife.org/ https://creativecommons.org/licenses/by-sa/4.0/

Top right: Mitchell's Reed Frog (*Hyperolius mitchelli*). Photo by John Lyakurwa. https://www.inaturalist.org/photos/101437054. https://creativecommons.org/licenses/by/4.0/

Biodiversity Hotspots of the World

BIODIVERSITY HOTSPOT OF THE COASTAL FORESTS OF EASTERN AFRICA

T. Pullaiah, PhD
Ally J. K. Nkwabi, PhD
Editors

APPLE ACADEMIC PRESS

First edition published 2026

Apple Academic Press Inc.
1265 Goldenrod Circle, NE,
Palm Bay, FL 32905 USA

760 Laurentian Drive, Unit 19,
Burlington, ON L7N 0A4, Canada

CRC Press
2385 NW Executive Center Drive,
Suite 320, Boca Raton FL 33431

4 Park Square, Milton Park,
Abingdon, Oxon, OX14 4RN, UK

Apple Academic Press exclusively co-publishes with CRC Press, an imprint of Taylor & Francis Group, LLC

Library and Archives Canada Cataloguing in Publication

CIP data on file with Canada Library and Archives

Library of Congress Cataloging-in-Publication Data

CIP data on file with US Library of Congress

ISBN: 978-1-998511-08-2 (hbk)
ISBN: 978-1-998511-09-9 (ebk)

DOI: 10.1201/9781998511099

ABOUT THE BOOK SERIES
BIODIVERSITY HOTSPOTS OF THE WORLD

Series Editor: **T. Pullaiah, PhD**
Former Professor, Department of Botany, Sri Krishnadevaraya University,
Andhra Pradesh, India

The term "biodiversity" came into common usage after the 1986 National Forum on Biodiversity, held in Washington, DC, and the publication of selected papers from that event, titled *Biodiversity,* edited by Wilson (1988). Wilson credits Walter G. Rosen for coining the term. In 1988, British ecologist Norman Myers published a seminal paper identifying 10 tropical forest "hotspots." Myers described ten tropical forest "hotspots" on the basis of extraordinary plant endemism and high levels of habitat loss, albeit without quantitative criteria for the designation of "hotspot" status. A subsequent analysis added eight additional hotspots, including four from Mediterranean-type ecosystems (Myers 1990).

A biodiversity hotspot is a biogeographic region that is both a significant reservoir of biodiversity and is threatened with destruction. Conservation International introduced two strict quantitative criteria to qualify as a hotspot: a region had to contain at least 1500 vascular plants as endemics (>0.5% of the world's total), and it had to have 30% or less of its original vegetation (extent of historical habitat cover) remaining. These efforts culminated in an extensive global review (Mittermeier et al., 1999) and scientific publication (Myers et al., 2000) that introduced seven new hotspots on the basis of both the better-defined criteria and new data. In 2005, an additional analysis brought the total number of biodiversity hotspots to 34, based on the work of nearly 400 specialists. In 2011, the Forests of East Australia was identified as the 35th hotspot by a team of researchers from the Commonwealth Scientific and Industrial Research Organisation (CSIRO) working with Conservation International. In February 2016, the North American Coastal Plain was recognized as meeting the criteria and became the Earth's 36th hotspot.

Very few books have been written/edited to document the biodiversity of these hotspots. Hence, these new volumes which aim to be a reference source for all foresters, wildlife biologists, conservationists, botanists, zoologists, policymakers, and others concerned with the conservation of the biodiversity of our planet.

REFERENCES

Mittermeier, R. A., Myers, N., Robles-Gil, P., & Mittermeier, C. G. (1999). *Hotspots: Earth's biologically richest and most endangered terrestrial ecoregions.* CEMEX/Agrupación Sierra Madre.

Myers, N. (1990). The biodiversity challenge: Expanded hot-spots analysis. *Environmentalist, 10,* 243–256.

Myers, N., Mittermeier, R., Mittermeier, C., Fonseca, G. A. B., & Kent, J. (2000). Biodiversity hotspots for conservation priorities. *Nature, 403,* 853–858. https://doi.org/10.1038/35002501 (accessed on 25th June 2025).

Wilson, E. O., & Peter, F. M. (Eds.). (1988). *Biodiversity.* The National Academies Press. https://doi.org/10.17226/989 (accessed on 25th June 2025).

BIODIVERSITY HOTSPOTS OF THE WORLD

North and Central America
- California Floristic Province
- Madrean Pine-Oak woodlands
- Mesoamerica
- North American Coastal Plain

The Carribbean
- Caribbean Islands

South America
- Atlantic Forest
- Cerrado
- Chilean Winter Rainfall-Valdivian Forests
- Tumbes-Choco-Magdalena
- Tropical Andes

Europe
- Mediterranean Basin

Africa
- Cape Floristic Region
- Coastal Forests of Eastern Africa
- Eastern Afromontane
- Guinean Forests of West Africa
- Horn of Africa
- Madagascar and Indian Ocean Islands
- Maputaland-Pondoland-Albany
- Succulent Kaoo

Central Asia
- Mountains of Central Asia

South Asia
- The Himalaya
- Indo-Burma
- Western Ghats–Sri Lanka

Southeast Asia and Asia-Pacific
- East Melanesian Islands
- New Caledonia
- New Zealand
- Philippines
- Polynesia–Micronesia
- Eastern Australian Temperate Forests
- Southwest Australia
- Sundland and Nicobar Islands
- Wallacea

East Asia
- Japan
- Mountains of Southwest China

West Asia
- Caucasus
- Irano-Anatolian

CURRENT BOOKS IN THE SERIES:

- **Biodiversity Hotspot of the Western Ghats and Sri Lanka**
  Editor: T. Pullaiah
- **Biodiversity Hotspot of the Himalaya**
  Editor: T. Pullaiah
- **Biodiversity Hotspot of the Indo-Burma Region**
  Editor: T. Pullaiah
- **Biodiversity Hotspot of the Coastal Forests of Eastern Africa**
  Editors: T. Pullaiah and Ally J. K. Nkawabi
- **Biodiversity Hotspot of the Mediterranean Basin**
  Editors: T. Pullaiah and Loris Galli
- **Biodiversity Hotspot of the Eastern Afromontane**
  Editors: T. Pullaiah and Teshome Soromessa
- **Biodiversity Hotspot of Sundaland**
  Editor: T. Pullaiah
- **Biodiversity Hotspot of Caucasus**
  Editors: T. Pullaiah and Levan Mumladze

FORTHCOMING BOOKS IN THE SERIES:

- **Biodiversity Hotspot of the Chilean Winter Rainfall–Valdivian Forest**
  Editors: T. Pullaiah and Cristian Echeverria
- **Biodiversity Hotspot of Mesoamerica**
  Editors: T. Pullaiah and Irene Pisanty
- **Biodiversity Hotspot of Tumbes-Choco-Magdalena**
  Editors: T. Pullaiah, Diego Tirira, and Pablo Lozano
- **Biodiversity Hotspot of the Irano-Anatolian Region**
  Editor: T. Pullaiah
- **Biodiversity Hotspot of the Mountains of Central Asia**
  Editors: T. Pullaiah and Marc Foggin
- **Biodiversity Hotspot of the Horn of Africa**
  Editors: T. Pullaiah and Teshome Soromessa
- **The Irano-Anatolian Biodiversity Hotspot**
  Editor: T. Pullaiah

- **Biodiversity Hotspot of Cerrado**
  Editor: T. Pullaiah
- **Biodiversity Hotspot of the Tropical Andes**
  Editors: T. Pullaiah and Monica Moraes
- **Biodiversity Hotspot of the Guinean Forests of West Africa**
  Editors: T. Pullaiah and Kouami Kokou
- **Biodiversity Hotspot of the Atlantic Forest**
  Editors: T. Pullaiah and Filipe Nascimento
- **Biodiversity Hotspot of the North American Coastal Plain**
  Editors: T. Pullaiah and Edwin Bridges
- **Biodiversity Hotspot of the Eastern Australian Temperate Forests**
  Editors: T. Pullaiah and Chris Dickman
- **Biodiversity Hotspot of the Caribbean Islands**
  Editors: T. Pullaiah and Chinthapalli Bhaskar Rao
- **Biodiversity Hotspot of the Philippines**
  Editors: T. Pullaiah, Crusty E. Tinio, and Lerma Sj. Maldia
- **Biodiversity Hotspot of Madagascar and Indian Ocean Islands**
  Editors: T. Pullaiah and Zo Lalaina Rakotobe
- **Biodiversity Hotspot of Wallacea**
  Editors: Bernard Michux and T. Pullaiah

ABOUT THE EDITORS

T. Pullaiah, PhD

Department of Botany, Sri Krishnadevaraya University, Anantapur, Andhra Pradesh, India

T. Pullaiah, PhD, is a former professor in the Department of Botany at Sri Krishnadevaraya University in Andhra Pradesh, India, where he taught for more than 35 years. He has held several positions at the university, including Dean of the Faculty of Biosciences, Head of the Department of Botany, Head of the Department of Biotechnology, and Member of the Academic Senate. He served as President of the Indian Botanical Society (2014), President of the Indian Association for Angiosperm Taxonomy (2013), and Fellow of the Andhra Pradesh Akademi of Sciences. He has been awarded the Panchanan Maheshwari Gold Medal, the Prof. P. C. Trivedi Medal, the Dr. G. Panigrahi Memorial Lecture Award of the Indian Botanical Society, the Prof. Y. D. Tyagi Gold Medal of the Indian Association for Angiosperm Taxonomy, and the Best Teacher Award from the Government of Andhra Pradesh.

Under his guidance, 54 students obtained their doctoral degrees. He has authored 65 books, edited 45 books, and published over 340 research papers. His books include *Mangroves: Biodiversity, Livelihoods and Conservation* (Springer); *Ethnobotany of India* (5 volumes, Apple Academic Press); *Global Biodiversity* (4 volumes, Apple Academic Press); *Monograph on Brachystelma and Ceropegia in India* (CRC Press); *Apomixis in Angiosperms* (CRC Press); *Invasive Alien Species* (4 volumes, Wiley International); *Advances in Cell and Molecular Diagnostics* (Elsevier); *Camptothecin and Camptothecin Producing Plants* (Elsevier); *Paclitaxel* (Elsevier); *Red Sanders: Silviculture and Conservation* (Springer); *Genetically Modified Crops* (2 volumes, Springer); *Flora of Andhra Pradesh* (5 volumes); *Flora of Eastern Ghats* (4 volumes); *Flora of Telangana* (3 volumes); *Encyclopedia of World Medicinal Plants* (7 volumes, 2nd edition); and *Encyclopedia of Herbal Antioxidants* (3 volumes).

He was also a member of the Species Survival Commission of the International Union for Conservation of Nature (IUCN). Professor Pullaiah received his PhD from Andhra University, India; attended Moscow State University, Russia; and worked as a Postdoctoral Fellow from 1976 to 1978.

Ally J. K. Nkwabi, PhD

National Coordinator, International Waterbird Census (IWC), Tanzania

Ally J. K. Nkwabi, PhD, is a National Coordinator for the International Waterbird Census (IWC) in Tanzania. His early research focused on the ecology of fauna and flora, human-wildlife conflict as well as assessing biodiversity in different wetlands. After completing his

PhD, Dr. Nkwabi served as Principal Research Officer at the Tanzania Wildlife Research Institute (TAWIRI). He has published 18 papers in peer-reviewed journals, 15 book chapters, and one book: *Birds of Western Tanzania (Gombe, Mahale, Rubondo, Saanane, Malagarasi, Ntakata, Minziro),* Doohyun Publishing Co., National Institute of Biological Resources (NIBR), Korea. Dr. Nkwabi is a member of the AEWA Grey Crowned Crane International Working Group (AEWA-GCCIW). He earned his BSc, MSc, and PhD at the University of Dar es Salaam, Tanzania.

CONTENTS

CONTRIBUTORS

Siri A. Abihudi
Institute of Traditional Medicine, Muhimbili University of Health and Allied Sciences, Dar es Salaam, Tanzania

John K. Bukombe
Tanzania Wildlife Research Institute (TAWIRI), Arusha, Tanzania

Francis Chebby
Department of Biology, College of Natural and Mathematical Sciences, The University of Dodoma, Dodoma, Tanzania

Immaculate K. Constantine
National Herbarium of Tanzania: Tanzania Plant Health and Pesticides Authority (TPHPA), Arusha, Tanzania

Deusdedith B. Fidelis
Tanzania Wildlife Research Institute, Arusha, Tanzania

Kwaslema Malle Hariohay
College of African Wildlife Management, Mweka, Moshi, Tanzania

Rob Harris
World Wildlife Fund for Nature (WWF), Tanzania Country Office, Dar es Salaam, Tanzania

Meshack Haule
Sokoine University of Agriculture, Institute of Pest Management, Morogoro, Tanzania

Denis Ikanda
World Wildlife Fund for Nature (WWF), Tanzania Country Office, Dar es Salaam, Tanzania

Jasson John
Department of Zoology and Wildlife Conservation, College of Natural and Applied Sciences, University of Dar es Salaam, Dar es Salaam, Tanzania

Josephat Kalugasha
Department of Botany, University of Dar es Salaam, Dar es Salaam, Tanzania

Almas Kashindye
World Wildlife Fund for Nature (WWF), Tanzania Country Office, Dar es Salaam, Tanzania

Richard Katondo
World Wildlife Fund for Nature (WWF), Tanzania Country Office, Dar es Salaam, Tanzania

David J. Kavana
Sokoine University of Agriculture, Mizengo Pinda Campus College, Department of Natural Resources Management and Conservation, Mpanda, Katavi, Tanzania

Pius Y. Kavana
Tanzania Wildlife Research Institute (TAWIRI), Arusha, Tanzania

Hamza K. Kija
Tanzania Wildlife Research Institute (TAWIRI), Arusha, Tanzania

William Kindeketa
Tanzania Forest Research Institute, Morogoro, Morogoro, Tanzania

Shadia I. Kilwanila
Department of Zoology, University of Dar es Salaam, Dar es Salaam, Tanzania

Moses Kyando
Department of Biology, College of Natural and Mathematical Sciences, The University of Dodoma, Dodoma, Tanzania

Cecilia Leweri
Tanzania Wildlife Research Institute (TAWIRI), Arusha, Tanzania

Asanterabi Lowassa
Tanzania Wildlife Research Institute (TAWIRI), Arusha, Tanzania

John V. Lyakurwa
Department of Zoology and Wildlife Conservation, University of Dar es Salaam, Tanzania

Richard D. Lyamuya
Tanzania Wildlife Research Institute (TAWIRI), Arusha, Tanzania

Lazaro Mangewa
Department of Wildlife Management, College of Forestry, Wildlife and Tourism, Sokoine University of Agriculture (SUA), Morogoro, Tanzania

Mwita M. Mangora
Institute of Marine Sciences. University of Dar Salaam, Buyu, Zanzibar, Tanzania

Wilfred Marealle
Tanzania Wildlife Research Institute (TAWIRI), Arusha, Tanzania

Teclamageni S. Mayeji
Sokoine University of Agriculture, Institute of Pest Management, Morogoro, Tanzania

Frank Mbago
Department of Botany, University of Dar es Salaam, Dar es Salaam, Tanzania

Nsajigwa Emmanuel Mbije
Sokoine Umiversity of Agriculture, College of Forestry, Wildlife and Tourism, Department of Wildlife.Management, Morogoro Tanzania

Franco Mbise
Department of Biology, College of Natural and Mathematical Sciences, The University of Dodoma, Dodoma, Tanzania

Levina Z. Mcharo
Protect Nature, Save Farmlands Company Limited, Arusha, Tanzania

Naza Mmbaga
Department of Biology, College of Natural and Mathematical Sciences, The University of Dodoma, Dodoma, Tanzania

Neduvoto P. Mollel
National Herbarium of Tanzania: Tanzania Plant Health and Pesticides Authority (TPHPA), Arusha, Tanzania

Elinasi Monga
International Union for Conservation of Nature, Tanzania Country Office, Tanzania

Charles Msagwa
National Herbarium of Tanzania: Tanzania Plant Health and Pesticides Authority (TPHPA), Arusha, Tanzania

Samuel Mtoka
Tanzania Wildlife Research Institute (TAWIRI), Arusha, Tanzania

Fredrick Mugube
Wetlands International, East Africa, Nairobi, Kenya

Emmanuel J. Mwainunu
Tanzania Forestry Research Institute, Morogoro; Tanzania

Machoke Mwita
Tanzania Wildlife Research Institute (TAWIRI), Arusha, Tanzania

Baraka Naftali
Tanzania Wildlife Research Institute (TAWIRI), Arusha, Tanzania

Damian Nguma
Tanzania Wildlife Research Institute (TAWIRI), Arusha, Tanzania

Wilirk Ngalason
Department of Zoology and Wildlife Conservation, University of Dar es Salaam, Dar es Salaam, Tanzania

Kelvin Ngongolo
Department of Biology, College of Natural and Mathematical Sciences, The University of Dodoma, Dodoma, Tanzania

Ally J. K. Nkwabi
Tanzania Wildlife Research Institute (TAWIRI), Arusha, Tanzania

Lilian Nyaega
Wetlands International, East Africa, Nairobi, Kenya

Emmanuel F. Nzunda
Department of Forest Resources Assessment and Management, College of Forestry, Wildlife and Tourism, Sokoine University of Agriculture (SUA), Morogoro, Tanzania

Raymond Okick
Tanzania Wildlife Research Institute (TAWIRI), Arusha, Tanzania

Emmanuel Richard
Tanzania Wildlife Research Institute (TAWIRI), Tanzania

John Sanare
Department of Botany, University of Dar es Salaam, Dar es Salaam, Tanzania

Atuhombye Sigala
Department of Biology, College of Natural and Mathematical Sciences, The University of Dodoma, Dodoma, Tanzania

Carl C. Trettin
USDA Forest Service, Southern Research Station, Asheville, NC, USA

Titus Wamae
Wetlands International, East Africa, Nairobi, Kenya

ABBREVIATIONS

CC	climate change
CF	coastal forests
CFEA	Coastal Forests of Eastern Africa
CR	critically endangered
DD	data deficient
EA	Eastern Africa
EAM	Eastern Arc Mountains
EBA	endemic bird area
EIA	Environmental Impact Assessment
EN	endangered
EVI	enhanced vegetation index
FTEA	Flora of Tropical East Africa
IBA	important bird area
GBIF	Global Biodiversity Information Facility
IUCN	International Union for Conservation of Nature
KBA	key biodiversity area
KIA	kappa index of agreement
LC	least concern
LST	land surface temperature
LUP	land use planning
MEAs	multilateral environmental agreements
NE	not evaluated
NSR	niassa special reserve
NT	near threatened
SNWC	selous-niassa wildlife corridor
TFS	Tanzania Forest Services
VU	vulnerable
WMA	Wildlife Management Area
WWF	World Wildlife Fund

PREFACE

Myers et al. (2000) proposed the concept of Biodiversity Hotspots—areas characterized by a high rate of endemism and a high rate of habitat destruction. Forests and vegetation are being degraded at an alarming rate, resulting in further loss of biodiversity. There is widespread concern over the loss of this precious gift of nature. Hence, there is an urgent need to take stock of the biodiversity in these hotspots to take necessary action for its conservation. The present series is a humble attempt to provide a summary of biodiversity in these hotspots. Very few books have been written or edited to document the biodiversity of these hotspots, and there is currently no book that documents the complete biodiversity of all the hotspots.

Some aspects could not be covered due to two main reasons. First, expertise on certain topics across the entire Coastal Forests of Eastern Africa is not readily available; second, some experts were too busy with their professional activities. I thank all the contributors for their critical review of the biodiversity of the Coastal Forests of Eastern Africa.

The publishers, Apple Academic Press, were very strict regarding plagiarism. They insisted on bringing down the similarity to 5% or below, although for biodiversity reports, where scientific names abound, it is extremely difficult to maintain such a low level of similarity. We must, therefore, thank all the contributors for keeping the similarity to 5% or less. We also thank Rakesh Kumar and Sandra Sickels for their support and encouragement. Their patience and perseverance are very much appreciated.

REFERENCE

Myers, N., Mittermeier, R. A., Mittermeier, C. G., da Fonseca, G. A. B., & Kent, J. Biodiversity hotspots for conservation priorities. *Nature*, **2000,** *403*(6772), 853–858.

CHAPTER 1

GEOGRAPHY, CLIMATE, AND SOIL ALONG THE COASTAL FORESTS OF EAST AFRICA

ALLY J. K. NKWABI

Tanzania Wildlife Research Institute (TAWIRI), Arusha, Tanzania

ABSTRACT

The coastal forest of Eastern African ecoregion extends from Somalia through Kenya, Tanzania to Mozambique, and falls within the Zanzibar-Inhambane regional mosaic. The climate of coastal areas varies from place to place in accordance with geographical location, altitude, relief, and vegetation cover. In general, along the coast, the mean annual rainfall varies from 500 to 2500 mm. Dryness is the most characteristic ecological feature of the forest area along the coastal. During these times, even the forest floor can become quite dry. The soil type of the area is, however, formed from marine sediment deposits.

1.1 INTRODUCTION

1.1.1 GEOGRAPHY

The coastal forest of Eastern African ecoregion extends from Somalia through Kenya, Tanzania to Mozambique and located at an approximately between latitudes 1° North and 25° South, and longitudes 34° and 41° East (Figure 1.1). The coastal forest of Eastern African falls within the Zanzibar-Inhambane regional mosaic (White, 1983; Menegon et al., 2020). They have been defined by World Wildlife Fund (WWF) as the "Eastern Africa Coastal Vegetation Ecoregion" and categorized the coastal vegetation into two WWF Global 200 ecoregions, including the Northern Zanzibar-Inhambane coastal Vegetation mosaic and the Southern Zanzibar-Inhambane Coastal Vegetation Mosaics. The Northern Zanzibar-Inhambane ecoregion is located along the Eastern and Southern coast of Africa, spanning Somalia, Kenya, and Tanzania. The Southern Zanzibar-Inhambane coastal forest mosaic stretches from southern Tanzania to the mouth of the Limpopo River along the Mozambique coast (Figure 1.1).

Biodiversity Hotspot of the Coastal Forests of Eastern Africa. T. Pullaiah and Ally J. K. Nkwabi (Eds.)

DOI: 10.1201/9781998511099-1

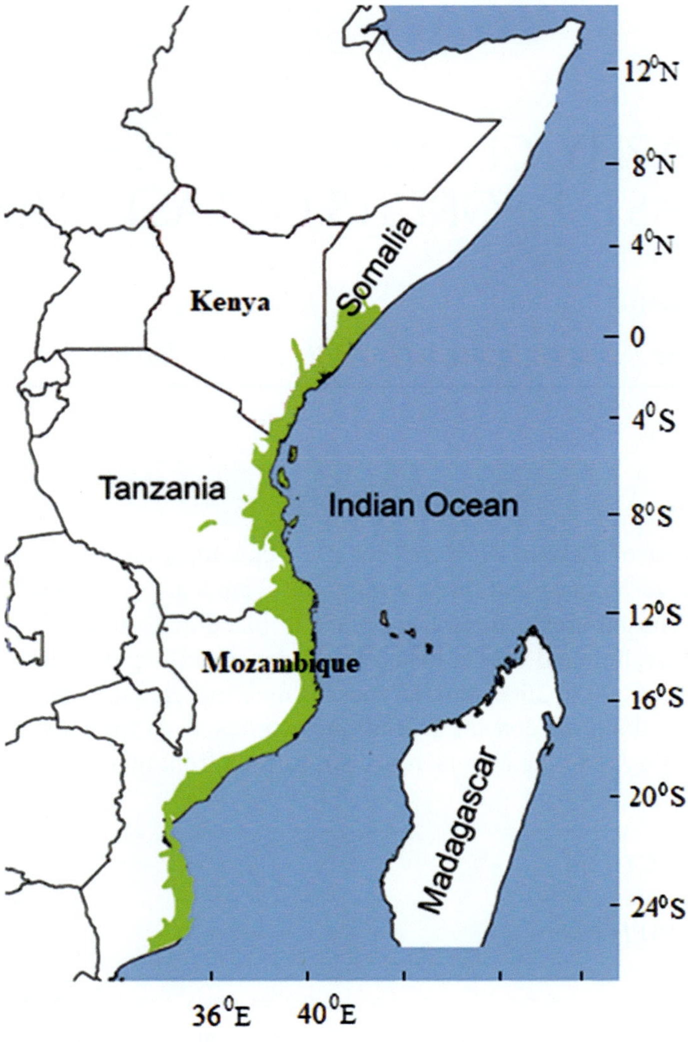

FIGURE 1.1 Coastal ecoregion and areas with coastal forests in Eastern African Countries (green color represents coastal forest).

Lager portion of East Africa Coastal areas located along the Northern Zanzibar-Inhambane ecoregion. The northern margin categorized as isolated vegetation outlier along the Jubba Valley in central coastal Somalia (Madgwick, 1988). The ecoregion observed in southern Somalia extends into northern Kenya, moving inland along the Tana River, before following the narrow coastal strip through central and southern Kenya to the Tanzanian border. In northern Tanzania, the ecoregion extends around the base of the East Usambara Mountains where the habitat intergrades with that of the Eastern Arc Mountains (EAM). It then continues along the coast of northern Tanzania to Dar es Salaam where outliers of the ecoregion extending inland at the base of the Uluguru, Nguru, and Udzungwa EAM. It again narrows around the Rufiji River, continuing along the coast until it meets the border of the next ecoregion around the southern Tanzanian at town of Lindi. The ecoregion includes the large offshore islands of Pemba, Zanzibar, and Mafia and other smaller islets in the Indian Ocean close to the coast. Elevation in this region is mostly below 500 m with some areas being characterized by low hills and plateau that reach as high as 1000 m above sea level (Harper et al., 2010).

1.1.1.1 SOMALIA

Somalia's coastline is considered one of the longest and the easternmost coastline in East Africa with the length of approximately 3025 km (Carbone and Accordi, 2000). This coastline is characterized by tropical, semiarid to arid climatic conditions, with bimodal rainfall influenced by monsoon winds (Carbone and Accordi, 2000; Dua, 2013). The northern portion is the driest with an average mean annual rainfall of 250 mm, whereas the southern portion receives a mean annual rainfall of about 400 mm (Carbone and Accordi, 2000; Dua, 2013). The bimodal rainfall occurs during the months of March to May and from October to November. Diurnal fluctuations in temperatures range between 20°C and 35°C whereby the coolest months are July and August and the hottest ones are March and April (Carbone and Accordi, 2000). The width of the coastal plain in southern Somalia extends inland to about 200 km (Burgess et al., 1998). The relatively dry conditions in Somalia do not offer opportunities for luxuriant forest growth. Only remnants of tiny forests are found in the southern coastal region of Somalia (Rembold et al., 2013).

1.1.1.2 KENYA

The Kenyan coastal region is found south of Somalia immediately south of the equator. It has a total length of about 600 km (Foeken et al., 2022). Coastal Kenya has its inner westernmost extension along Tana River. Different vegetation types occur in the ecoregion including bushland, woodland, forests (dry, swampy, and moist forests), and grasslands with wetlands and mangroves at Tana River delta (Foeken et al., 2022; Ngumbau et al., 2020). Associated with coastal Kenya are the coastal forests distributed within latitude 1°40′ to about 4°40′ south in an elevation of between 0 and 840 m above sea level. Kenyan coastal forests cover an approximately 787 km^2 (Younge et al., 2002) which are home to many species of plants and animals.

1.1.1.3 TANZANIA

The coastal region of Tanzania extends from its northern end at the border with Kenya (4°49'S) to the southern end at its border with Mozambique (10°28′S) with a coastline of over 800 km in length (Yanda and Mung'ong'o, 2019). This region extends from its western border around the base of the EAM of East Africa including the East Usambara, Uluguru, Nguru, and Udzungwa Mountains to the offshore islands of Pemba, Zanzibar, and Mafia up to 100 km on the eastern side (Schipper and Burgess, 2022). The EAM Mountains and the coastal habitats integrate at the foothills to produce a unique vegetation mosaic. The region includes the Kilombero valley that forms an extended lowland floodplain of up to approximately 300 km to the foot hills of the Udzungwa Mountains.

1.1.1.4 MOZAMBIQUE

The Mozambican coastline is the second longest after Somalia in the Eastern African Coast, approximately 2500 km long. It is characterized by diverse habitat types, ecosystems, flora, and fauna. The main vegetation types in this region include lowland forests, thickets, grassland, mangroves, and vegetation on dunes. The region falls within the Zanzibar-Inhambane regional mosaic, which is a phytochorion from the Somali coast to South Africa (White, 1983). It is estimated that Mozambique coastal region contributes two-thirds of the total area covered by coastal forests in the Eastern Africa with the remaining portion in the other coastal Eastern African countries (Albano, 2004). The coastal forests of Mozambique were once continuous, but due to human interferences, they are now fragmented and exist only in smaller patches (Albano, 2004).

1.1.2 CLIMATE

The ecoregion experiences a tropical climate characterized by average temperatures exceeding 25°C, minimal day length fluctuation, and consistently elevated humidity levels. The climate of coastal area varies from place to place in accordance with geographical location, altitude, relief, and vegetation cover. Most rainfall comes in distinctive two major rainfall regimes, first is unimodal (October–April) which experienced in southern part of the coastal area of Tanzania and the second is bimodal occurring from October to December and March to May (Mwandosya et al., 1998; Chang'a et al., 2010), this type of rainfall occurs the east to the coast of East Africa (Chang'a et al., 2010). In general, along the coast, the mean annual rainfall varies from 500 to 2500 mm (Carbone and Accordi, 2000; Dua, 2013), but an average rainfall of between 800 and 1300 mm per year is more typical of the East Africa coastal region (Burgess et al., 1998).

One of the main characteristics of the region is the extreme variability in rainfall from year to year, with dry seasons that sometimes extend into droughts, and when much rain happens in some places during wet seasons, it causes severe flooding in some areas (Burgess and Clarke, 2000). However, it is the dry season that is the most characteristic ecological feature of the forests of the area—during these times even the forest floor can become quite dry. Despite the high annual variability in the rainfall, the long-term climate of the ecoregion has remained relatively stable over millions of years, and potentially as long as 30 million years (Axelrod and Raven, 1978), even during the colder and drier periods of the last Ice Age (Prell et al., 1980).

1.1.3 SOIL

The ecoregion contains a large number of rock types dating from the Precambrian basement rocks through recent unconsolidated sediments at the mouths of rivers and along the coast. The majority of the area is, however, formed from marine sediments deposited since the breakup of Pangea (Burgess and Clarke 2000), with younger sediments closer to the coast. Over the past 30 million years, the Northern Zanzibar-Inhambane Coastal Forest Mosaic has been subject to considerable tectonic activity, which has formed low ridges and swells (generally 100–300 m high). This activity has also caused the shoreline to move in and out from its present position, with the subsequent deposition and erosion of sediments. Faulting associated with the development of the Rift Valley also affected the Eastern African coastal region. One such fault has isolated Pemba Island through the formation of a deep-water "Pemba Trench." Soil and rainfall amount determines vegetation types of the area.

In the Northern Zanzibar-Inhambane ecoregion, the soils are intricate, with the Food and Agriculture Organization of the United Nations having identified at least six major soil units (Moore et al., 2012). There is also considerable variation at the local scale. There are varied soil types in the region mainly sandy, clay, and black cotton influenced by the parent rock type. There are approximately 12 major soil types supporting Coastal Forests in this ecoregion. These range from soils that are "sandy with imperfect drainage," "sandy loams with excessive drainage," "loams with moderately good drainage," "loams with imperfect drainage," and "clays with imperfect drainage" (Burgess and Clarke, 2000). The regular and predictable change in soils down available slopes gave rise to the theory of catenaries succession in soils (Milne, 1935). Such soil variation determines the vegetation types along the east.

The Eastern African coastal area has a large number of rock types resulted from recent unconsolidated sediments. The majority of the area is formed from marine sediments deposited since the breakup of Pangae (Burgess and Clarke, 2000). The coastal region of East Africa is home to several major rivers, such as the Juba River in Somalia, Tana River in Kenya, and various permanent rivers in Tanzania including Ruvu, Pangani, Wami, Rufiji, and Ruvuma. Additionally, the Zambezi River in Mozambique flows into the Indian Ocean. These rivers traverse extensive floodplains that support diverse ecosystems of flora and fauna.

KEYWORDS

- **coastal forests**
- **evolution**
- **location**
- **climate variation**
- **soil type**

REFERENCES

Albano, G. Coastal forests of mozambique: Socio-economic aspects—review, **2004**. coastalforests.tfcg.org/pubs/SocioEco-MZQ.pdf Accessed 26 October 2022.

Axelrod, D. I.; Raven, P. H. Late cretaceous and tertiary vegetation history of Africa. In: Werger, M. J. A. (Ed.), *Biogeography and Ecology of Southern Africa*. Springer: the Netherlands, Dordrecht. **1978,** pp. 77–130.

Burgess, N. D.; Clarke, G. P. (Eds.). *The Coastal Forests of Eastern Africa*. IUCN Forest Conservation Programme, Gland and Cambridge. **2000,** p. 443.

Burgess, N. D.; Clarke, G. P.; Rodgers, W. A. Coastal forests of eastern Africa: Status, endemism patterns and their potential causes. *Biol. J. Linn. Soc.* **1998,** *64*, 337–367.

Carbone, F.; Accordi, G. The Indian Ocean coast of Somalia. *Mar. Pollut. Bull.* **2000**, *41*, 141–159.

Chang'a, L. B.; Yanda, P. Z.; Ngana, J. Spatial and temporal analysis of recent climatological data in Tanzania. *J. Geogr. Reg. Plann.* **2010**, *3* (3), 44.

Dua, J. A sea of trade and a sea of fish: Piracy and protection in the Western Indian Ocean. *J. East. Afr. Stud.* **2013,** *7*(2), 353–370.

Foeken, D.; Hoorweg J.; Obudho, R. A. The Kenya Coast: a regional study, **2022**. https://scholarlypublications.universiteitleiden.nl/access/item%253A2888600/download&ved. Accessed 30 January 2023.

Harper, E. B.; Measey, G. J.; Patrick, D. A.; Menegon, M.; Vonesh, J. R. *Field Guide to the Amphibians of the Eastern Arc Mountains and Coastal Forests of Tanzania and Kenya*. Camerapix Publishers International: Nairobi, Kenya, **2010**.

Madgwick, F. J. Riverine forest in the Jubba Valley: Vegetation analysis and comments on forest conservation. The biogeography of Somalia. *Biogeographia* **1988,** *14*, 67–88.

Menegon, M.; Lyakurwa, J.; Loader, S. The amphibians of the Tanzanian forests Ver 1.1, **2020.** https://www.dropbox.com/sh/010gn9q8syk522r/AACepeVnPj8YkXBQYxbOKbKRa?dl=0. Accessed 14 January 2023.

Milne, G. Some suggested units of classification and mapping, particularly for East African soil. *Soils Res.* **1935**, *4*, 183–198.

Moore, N.; Alagarswamy, G.; Pijanowski, B.; Thornton, P.; Lofgren, B.; Olson, J.; Andresen, J.; Yanda, P.; Qi, J. East African food security as influenced by future climate change and land use change at local to regional scales. *Clim. Change.* **2012**, *110*, 823–844.

Mwandosya, M. J.; Nyenzi, B. S.; Lubanga, M. L. *The Assessment of Vulnerability and Adaptation to Climate Change Impacts in Tanzania*. Centre for Energy, Environment, Science and Technology: Dar es Salaam. **1998.**

Ngumbau, V. M.; Luke, Q.; Nyange, M.; Wanga, V. O.; Watuma, B. M.; Mbuni, Y. M.; Munyao, J. N.; Oulo, M. A., Mkala, E. M.; Kipkoech, S.; Itambo, M.; Hu, G. W.; Wang, Q. F. Annotated checklist of the coastal forests of Kenya, East Africa. *PhytoKeys* **2020**, *147*, 1–191, https://doi.org/10.3897/phytokeys.147.49602.

Prell, W. L., Hutson, W. H..; Williams, D. F., Be, A. W. H.; Geitzenauer, K.; Molfino, B. Surface circulation of the Indian Ocean during the last glacial maximum, approximately 18,000 yr BP. *Quat. Res.* **1980**, *14*(3), 309–336.

Rembold, F.; Oduori, S. M.; Gadain, H.; Toselli, P. Mapping charcoal driven forest degradation during the main period of Al Shabaab control in Southern Somalia. *Energy Sustain. Dev.* **2013**, *17*(5), 510–514.

Schipper, J.; Burgess, N. Eastern Africa: Coastal areas of Kenya, Somalia and Tanzania, **2022.** www.worldwildilife.org. Accessed 22 February 2023.

White, F. *Vegetation of Africa*. UNESCO: Paris, France, **1983.**

Yanda, P. Z.; Mung'ong'o C. G. Researching climate change and socio-ecological systems' vulnerability in the coastal areas of Tanzania: Some theoretical perspectives in climate change and coastal resources. In: Yanda, P. Z.; Bryceson, I.; Mwevura, H.; Mung'ong'o, C. G. (Eds.), *Tanzania: Studies on Socio-Ecological Systems' Vulnerability, Resilience and Governance*. Springer Nature: Switzerland, **2019**, https://doi.org/101007/978-3-030-04897-6_1.

Younge, A.; Negussie, G.; Burgess, N. Eastern African Coastal Forest Programme, WWF –EARPO Regional Workshop Report, Nairobi, Kenya February, **2002**, pp. 4–7.

CHAPTER 2

THE VEGETATION TYPES OF THE COASTAL FORESTS OF EASTERN AFRICA BIODIVERSITY HOTSPOT

RICHARD D. LYAMUYA[1*], WILFRED MAREALLE[1], KWASLEMA M. HARIOHAY[2], ELINASI MONGA[3], and ALLY J. K. NKWABI[1]

[1]Tanzania Wildlife Research Institute, Arusha, Tanzania

[2]College of African Wildlife Management, Mweka, Moshi, Tanzania

[3]International Union for Conservation of Nature, Tanzania Country Office, Tanzania

*Corresponding author

ABSTRACT

This chapter reviews the currently present vegetation types with special emphasis on their distribution, status, and conservation threats along the biodiversity hotspot of the coastal forests of Eastern Africa. Currently, the coastal forests of Eastern Africa are composed of a mosaic of halophytic mangrove forests, lowland mixed rain forests, Brachystegia woodland, swamp and scrub forests, and riverine gallery forests. We found that the coastal forests of Eastern Africa biodiversity hotspot has three major forest categories, including the Eastern Africa closed canopy coastal forests, savanna woodland, and nonforest vegetation belonging to nine major vegetation groups with 11 subgroups of forests represented with several examples. Moreover, the environmental factors (i.e., soil and amount of rainfall) mostly influenced the vegetation types of the area. Again, millions of people living in the area depend very much on these forests for their daily survival as well as economic growth. Despite that, those forests are still found to harbor a variety of life forms, which convinced Conservation International to classify them as biodiversity hotspots. However, poverty is the major root cause for the unstable forest uses, thus causing threats to their survival in the area. These unsustainable forest uses include charcoal production, unsustainable logging, grazing, and conversion to other land uses.

2.1 INTRODUCTION

2.1.1 BACKGROUND

The coastal forests of Eastern Africa, spanning from the southern coast of Mozambique to southern Somalia, were once spread over approximately 291,500 km^2. However, the

Biodiversity Hotspot of the Coastal Forests of Eastern Africa. T. Pullaiah and Ally J. K. Nkwabi (Eds.)

DOI: 10.1201/9781998511099-2

current extent of these forests has significantly reduced to only 29,150 km^2, representing a mere 10% of their original area (Gachanja and Kanyanya, 2004; Githitho, 2004). These forests are composed of the different forest types including the mangrove, mountain and the lowland forest patches (https://coastalforests.tfcg.org/pubs/CFNationalSynthesis-Tz.pdf; Matiku, 2024). While they are primarily found along the eastern coast of the Indian Ocean in the tropics, spanning six countries, including Kenya, Tanzania, and Mozambique (Habel et al., 2017), they are most commonly found in the latter three locations (Gachanja and Kanyanya, 2004; Figure 2.1). Previous studies have revealed that these forest sizes are either increasing or decreasing in some of the Eastern African countries (Burgess et al., 2003; Karyn et al., 2010). For example, in Tanzania, the size of the coastal forest is 692 km^2, but in 1992 the size was estimated at 700 km^2 indicating there is a decline in forest size (Burgess et al., 2003), and 4778 km^2 in Mozambique indicating that the forest size in the country has tremendously increased when compared to 1992 when estimated at 1790 km^2 (Burgess et al., 2003). The main reason for this decrease in coastal forest size was reported to be caused by human activities including agriculture, charcoal making, and grazing by cattle (Karyn et al., 2010).

The coastal strip of Eastern Africa supports diverse and highly heterogeneous vegetation types (Bloesch ans Klötzli, 2004; Annexure 2.1). However, our review came up with three major vegetation types of the area represented first by the Eastern Africa closed canopy and coastal forest, second the woodland, and third the nonforest vegetation with their representative groups (Annexure 2.1).

The Eastern African Closed canopy forest represents only 1% of the area but contains at least 400 endemic plants out of the 3000 plant species identified from coastal forests. The remaining 99% of the area is a heterogeneous mix including savanna woodlands, bushlands, thickets, and farmland that contains more than 500 endemic plant species. However, according to Karyn et al. (2010), these forest habitat occur in small patches in every location found. The survey conducted by Karyn et al. (2010) between 1990 and 2000 showed that around 6820 km^2 of the forest and woodland cover only remained with more forest cover found in Tanzania (66.9%) compared to Kenya (33.1%). Despite of that more forest loss was observed to occur especially in unprotected areas than protected areas and was found to occur more in Tanzania (87.5%) than in Kenya (12.5%; Karyn et al., 2010). Additionally, these forests are biologically distinct from all other forest types in Africa, as they are known to support many unique species of animals and plants (Burgess and Muir, 1994). Therefore, they are currently considered as a global biodiversity hotspot (Karyn et al., 2010).

2.1.2 EAST AFRICA COASTAL FOREST ZONE

The East African coastal forests occupied a total area of around 260,000 km^2 (Burgess et al., 2003). These forests support many of the endemic species, therefore they are termed the hotspot of endemism in Africa (Fungomeli et al., 2020). Currently, the region constitutes one of the 36 global biodiversity hotspots (Habel et al., 2017; Nkwabi et al., 2018).

According to Younge et al. (2002), these forests are composed of a variety of forest types and found to occur in different habitats (Joordens et al., 2019).

FIGURE 2.1 Map of Eastern Africa, showing the distribution of the Coastal Forest of Eastern Africa and Eastern Arc Montane forests.

Source: Joordens et al. (2019)

The coastal forests are distributed and can be found up to higher altitudes above sea level but excluding the mangrove forests, which are referred to as "maritime" areas (Mligo, 2018). According to Gachanja and Kanyanya (2004), currently most of the mangrove forests occur in small patches, fragmented and isolated. This supports what previous studies have reported that the coastal forests of Eastern Africa are one of Africa's foremost biodiversity conservation priorities, as they form a highly threatened center of endemism for plants (Figure 2.3). Moreover, Ngumbai et al. (2020) and Barratt (2017) described the Eastern African coastal forests as the ancient Pan-African Forest.

Today only approximately 3170 km^2 of East Africa's coastal forest still exists and 660 km^2 of its total coverage is restricted to forest patches in Kenya (Habel et al., 2017). What is left is astonishing in its biodiversity—the plant species in two forests just 100 km apart can differ by as much as 80% and brimming with flora scientists are still working to identify, with some species facing the risk of extinction (Woodhouse, 2019).

2.1.3 HUMAN PRESSURE ON COASTAL FOREST

According to Ngumbau et al. (2020), anthropogenic pressures is the main cause of loss for these forests which now occur in small patches. In Tanzania, for instance deforestation rates were calculated and considered to have slowed from 1.0% yr^{-1}, or > 3735 ha yr^{-1} during the 1990s, to 0.4 % yr^{-1}, or > 1233 ha yr^{-1} during 2000–2007 (Godoy et al., 2012). The area, particularly in Kenya and Tanzania, was classified as a biodiversity hotspot by Myers et al. (2000). This designation is based on its inclusion as one of the 36 global biodiversity hotspots, showcasing a significant variety of endemic plant and animal species. This is largely due to the presence of numerous narrowly endemic plants and animals in fragmented forest patches within the region (Ngumbau et al., 2020).

A study by Younge et al. (2002) observed a large portion of the forest being converted and fragmented into small patches as a result of the anthropogenic activities of the area. This millennial use has transformed the indigenous forest cover into small, isolated patches (Burgess et al., 2017). For example, today, only approximately 3170 km^2 of coastal forest remains (about 21% of the total remaining coverage (Habel et al., 2017). In addition, according to Burgess et al. (2017), the overall coastal forest cover declined by over in 2007. The rate of deforestation was found to occur lower in protected areas compared to outside reserves over the past 20-year period (Burgess et al., 2017). For example, the 2010 data analysis showed that patterns of forest loss were concentrated around Dar es Salaam (Burgess et al., 2017). Therefore, threats to the Eastern African coastal forests are and continue to be linked to conversion to farmland, settlement, mining, and grazing land. Additionally, changes in land-use patterns and lack of effective and integrated policies, coupled with inadequate/ineffective implementation of land-use planning (Younge et al., 2002). Current forest destruction trends are affecting a wide range of terrestrial and marine resources in the three countries covered under the program (Younge et al., 2002). For example, the sustainable production potentials and the services they provide to the local livelihoods continue to decline.

2.1.4 FLORA ENDEMIC SPECIES OF EASTERN AFRICAN COASTAL FOREST

The Eastern African coastal forests support the growth of many varieties of plant species, both endemic and near-endemic species (Burgess et al., 2017; Woodhouse, 2019; Ngumbau et al., 2020). The floral endemic species of eastern coastal forests consist of c550 endemic species occurring from northern Kenya to southern Tanzania (Burgess et al., 2017). According to Burgess et al. (2017), many of these forests have a higher number of endemic species which are narrowly distributed.

2.1.5 SOIL AND VEGETATION

The Eastern African coastal zone is predominantly a plain sloping gently upwards away from the sea, with variable plain width. The coastal forest area extends more than 200 km inland at the northern and southern boundaries, including southern Somalia, northern Kenya, and southern Mozambique. However, in certain regions, it narrows to less than 30 km in width. This coastal plain comprises a complex sequence of mainly post-Jurassic (C135 million years old) sediments of both the marine and freshwater sedimentary rocks (sandstones, limestones, marls, shales, mudstones, and unconsolidated sediments) (Kent, 1972).

The ecoregion contains a large number of rock types dating from the Preambrian basement rocks through recent unconsolidated sediments at the mouths of rivers and along the coast. The majority of the area is, however, formed from marine sediments deposited since the breakup of Pangea (Burgess and Clarke, 2000). Over the past 30 million years, the Northern Zanzibar-Inhambane coastal forest Mosaic has been subject to considerable tectonic activity, which has formed low ridges and swells (generally 100–300 m high). This activity has also caused the shoreline to move in and out from its present position, with the subsequent deposition and erosion of sediments. Faulting associated with the development of the Rift Valley also affected the Eastern African coastal region. One such fault has isolated Pemba Island through the formation of a deepwater "Pemba Trench." Soil and rainfall amount determines the vegetation types of the area. The soils in the Northern Zanzibar-Inhambane ecoregion are complex with at least six The Food and Agriculture Organization major soil units. There is also considerable variation at the local scale. The Eastern African coastal forest soils range from soils that are "sandy with imperfect drainage," "sandy loams with excessive drainage," "loams with moderately good drainage," "loams with imperfect drainage," and "clays with imperfect drainage" (Burgess and Clarke, 2000). The regular and predictable change in soils down available slopes gave rise to the theory of catenary succession in soils (Milne, 1947). Such soil variation determines the vegetation types along the east. Different studies have shown that soils varied from one another because of many factors (Milne, 1947; Chaplot et al., 2010; Bai and Zhou, 2020).

2.1.6 THE DESCRIPTION OF VEGETATION TYPES OF EASTERN AFRICAN COASTAL FORESTS

2.1.6.1 EASTERN AFRICA CLOSED CANOPY COASTAL FORESTS

The closed canopy coastal forest of East Africa was classed into Eastern African Coastal Dry Forest, Eastern African Coastal Scrub Forest, Eastern African Coastal Riverine/Groundwater/Swamp Forest, Mixed Forest, and Eastern African Coastal/Afromontane Transition Forest.

2.1.6.1.1 Eastern African Coastal Dry Forest

The coastal dry forest is defined here as essentially dense woody vegetation found within a narrow range of the coast (Timberlake et al., 2011). However, this definition does not include moist forests because moist forests are associated with watercourses such as rivers (Timberlake et al., 2011). The Eastern African coastal dry forests exhibit distinctive qualities, including:

1. Being characterized by semievergreen forest composition.
2. Featuring plants with compact, tough, and thick evergreen leaves.
3. Hosting a unique array of plant species distinct from miombo forests.
4. Having a limited geographic range due to their specialized environmental needs.
5. Harbor a significant number of endemic plant species spanning various families (such as Annonaceae, Fabaceae, Caesalpinioideae, and Rubiaceae)

A good example here of a coastal dry forest is the Arabuko Sokoke dryland coastal forest. This forest has a total forest size of approximately 416 km^2 (Habel et al., 2017). It consists of different forest types composed of different tree species composition due to soil type, for example, Cynometra forest, that is, *Brachylaena huillensis, Cynometra webberi, Euphorbia candelabrum, Manilkara sulcata*, and *Oldfieldia somalensis, Brachystegia* forest, that is, *Brachystegia spiciformis*, mixed forest, that is, *Afzelia quanzensis, Combretum schumannii, Drypetes reticulata, Hymenaea verrucosa*, and *Manilkara sansibarensis* (Habel et al., 2017; Joordens et al., 2019).

2.1.6.1.2 Eastern African Coastal Scrub Forest

This forest is characterized by having a forest canopy height of more than 10 m and bushland or thicket canopy height of less than 10 m (Githitho, 2004). Usually, the present trees do not form a closed canopy. Within this forest, bushes and shrubs contribute at least as much as the trees to the appearance of this vegetation type (Githitho, 2004) while herbs are scarce to absent. This kind of vegetation type resembles very much the "maritime Eastern African coastal scrub forest" that develops on shallow and easily desiccated soils that overly coral rag (Githitho, 2004). Representative examples include the scrub forest near Msambweni and the Zanzibar-Inhambane scrub forest. The main species recorded by Githitho (2004) in this vegetation type were *Adansonia digitata, A. quanzensis, Antiaris toxicaria, Bridelia micrantha, C. schumannii, Dialium orientale, Dichrostachys cinerea, Diospyros mespiliformis, Dodonaea viscosa, Encephalartos hildebrandtii, Euclea racemosa, Euphorbia tirucalli, Ficus sur, Flacourtia indica, Flueggea virosa, M. sulcata, Markhamia zanzibarica, Mimusops obtusifolia, Monodora grandidieri, Ozoroa insignis, Pandanus kirkii, Psydrax schimperiana, Pterocarpus angolensis, Rhodognaphalon schumannianum, Ricinodendron heudelotii, Salvadora persica, Sclerocarya birrea, Sideroxylon inerme, Sorindeia madagascariensis, Sterculia africana, Syzygium cordatum, Tamarindus indica*, and *Xylopia parviflora.* In Kenya and southern Somalia, scrub forests dominated by *Diospyros cornii* and *Manilkara mochisia* form a quasi-continuous narrow band separating the forests of the coastal region from the bushlands of the interior. Similar vegetation extends to southern Tanzania.

2.1.6.1.3 Eastern African Coastal Riverine/Groundwater/Swamp Forest

The Eastern African Coastal Riverine/Groundwater/Swamp Forest is vegetation grown near water sources or waterlogged areas (i.e., along the Tana River). This vegetation occurs in areas along the Jubba River at Bu'ale, Somalia and along the Tana River in Kenya. In Tanzania, for example, this kind of vegetation type occurs in Saadani National Park at Wami Delta. In addition, groundwater forests can be found along Gendagenda Forest, valley bottom areas of the Pugu and Kazimzumbwi forests, swamp forest areas of Mchungu Forest and Kimboza Forest, Tanzania; and riverine forest in the Liwiti village forest reserve, Tanzania (Figure 2.2) and Massenjere Forest Reserve, Malawi.

FIGURE 2.2 A riverine coastal forest in Liwiti village, Kilwa District, Lindi Region (*Photo credit: Mpingo Conservation and Development Initiative*).

Bloesch and Klötzli (2004) reported that the groundwater forests might occur on permeable sandy silt valley bottoms with a high groundwater table. In addition, groundwater forests and gallery forests are in practice difficult to separate, since areas of gallery forests are effectively groundwater forests though they do not meet with river water (Bloesch and Klötzli, 2004). Consequently, many of their species are common and diverse (Bloesch and Klötzli, 2004).

Furthermore, the swamp forests are restricted in occurrence. For example, *Barringtonia racemosa* forest with *Acrostrichum aureum*, *Hibiscus tiliaceus*, *Pandanus* spp., and *Phoenix reclínala* often occurs immediately behind the mangrove zone. In places, *Barringtonia* extends upstream for a considerable distance. On Pemba Island Raphia swamp forest occurs in shallow valleys with very slow drainage. Associates include *Elaeis guineensis, Voacanga thouarsii*, and the aroid *Typhonodorum lindleyanum.*

2.1.6.1.4 Mixed Forest

The mixed forest occupies 7000 ha and occurs along the grey sands soil (https://whc.unesco.org/en/tentativelists/6657). Various tree species, such as *C. schumannii*, can be observed in these forests. The majority of mixed forests are located in Arabuko Sokoke Forest in Kenya, spanning 39,100 hectares. Additionally, similar forests can be found in Pangani, Tanzania, covering approximately 0.5 km^2 at an altitude of 75 m above sea level. It is composed of several types of tree species including *Ficus sycomorus, Lannea welwitschii, Ficus glumosa, Milicia excelsa, Gyrocarpus americanus, Blighia unijugata, Uvaria faulknerae, Combretum chionanthoides, D. cinerea, Euphorbia wakefieldii, Styloaechiton bogneri, C. webberi, E. hildebrandtii,* and *Aristogeitonia monophyla* (Woodhouse, 2019).

2.1.6.1.5 Eastern African Coastal/Afromontane Transition Forest

These are transitional vegetation formation types found in areas where rainfall is higher (Githitho, 2004). Sometimes this forest is found in lowland areas like in Kenya. In other areas, for example, in Tanzania it is found at the summit of Tongwe Hill, in Kimboza Forest (Figure 2.3), and the lowlands of the East Usambara and dry forest on Machemba Hill, Malawi, and Chirinda Forest in Zimbabwe.

2.1.6.2 WOODLAND (EASTERN AFRICAN COASTAL MIOMBO FOREST)

Syampungani et al. (2009) described Miombo as those vegetation dominated by the genera *Brachystegia, Julbernardia*, and/or *Isoberlinia*, belonging to the legume family (Fabaceae, subfamily Caesalpinioideae) mostly found to occur in the Central, Southern, and Eastern African woodlands. At present, *Brachystegia* and *Julbernardia* woodlands are recognized as the two primary types of Miombo Woodland or Miombo Forest.

2.1.6.2.1 Eastern African Coastal Brachystegia Forest

The Eastern African Coastal *Brachystegia* Forest was described by White (1983) as the vegetation dominated by *B. spiciformis* and *Brachystegia microphylla* as seen in Arabuko-Sokoke forest (7700 ha). In addition to the two species of *Brachystegia*, this vegetation type has been reported to have only 19 tree species apart from the two mentioned species, which are considered dominant/common species. Amongst these, only *H. verrucosa* and *Julbernardia magnistipulata* have been frequently recorded as co-dominants with *Brachystegia*. However, according to White, (1983), there are 21 species of *Brachystegia* in miombo woodland.

FIGURE 2.3 Coastal forests in the southern part of Tanzania (*Photo credit: Aklei Albert*).

2.1.6.3 NONFOREST VEGETATION

It is an area covered with grasslands, salt marshes, unmaintained range, clear-cut vegetation, as well as young, replanted forests with trees less than 2 m tall and less than 26% canopy cover (Mwasumbi et al., 1994). Almeida et al. (2022) described nonforest vegetation as consisting of savannahs, shrublands, grasslands, or wetlands. In this section, we will describe the following types of vegetation including edaphic grassland and coastal evergreen bushland.

2.1.6.3.1 Edaphic Grasslands Habitats

The edaphic grasslands habitats are found in areas that experience tropical continental climates. Tropical or subtropical grassland lying on the margin of the trade wind belts.

These areas include lowland parts of the Eastern Africa coastal region. Since the true savanna vegetation has mainly grass and scattered trees. However, the commonest trees found here are *Vachellia* and Miombo, which mostly are scattered.

2.1.6.3.2 Eastern Africa Coastal Evergreen Bushland

Bushland is land of which 40% or more is covered by plants usually ranging from 3 to 7 m tall, but can be smaller or larger (White, 1983). These plants are usually multiple-stemmed and the main axes are frequently 10 cm or more in diameter at the base. Bushy trees are also frequently present in bushland. They too are normally less than 7 m tall and, although there is a main bole, it branches low down so that progress is impeded. Taller trees sometimes occur as emergent but they either are localized in groups or are widely scattered. When they are sufficiently numerous to form a distinct but open canopy, we are dealing with scrub forests. When they are less frequent, the term "wooded bushland" could be used (White, 1983). Grasses are present in most types of bushlands but are physiognomically subordinate. Where bushes occur scattered in a continuous sward of grasses and cover less than 40% of the surface we can talk of "bushed grassland," but the proportion of bushes is usually much less than this (White, 1983). For reasons given elsewhere, bushed grassland is not recognized separately from wooded grassland in this work. Bushland often occurs in rocky or stony places that are unfavorable to grasses, and elsewhere the grasses have been eliminated or much reduced by grazing. Because of the insignificance of grass, it is inappropriate to use the term savanna in such cases although they have frequently been designated as bush- or shrub-savanna in the past. In places where grasses are sparse, the vegetation remains physiognomic-ally bushland even when the cover is much below 40%. Such vegetation can be referred to as open bushland.

White (1983) reported that various types of bushlands and thickets are found where unfavorable soil conditions prevent the development of the forests. Dense thicket occurs on termite mounds in seasonally waterlogged grassland, which occupies parts of the coastal plain, as between Garsen and Lamu in northern Kenya. The thicket of *Capparis, Carissa, Commiphora*, *Euclea natalensis*, *Diospyros consolatae*, *S. inerme* is usually overtopped by emergent trees of *D. cornii*, *Dobera glabra*, *M. mochisia* or *T. indica* (White, 1983). Thus, on Mafia and Pemba Islands *P. mafiensis* forms a loose open canopy at eight of 8 m (White, 1983). Associates include *S. cordatum*, *Uapaca sansibarica, Parinari curatellifolia, M. sansibarensis*, *E. natalensis,* and scattered *Pándanos goetzei*. When *Philippia* is exclusively dominant, the ground is covered with leaf litter or the lichen *Cladonia medusiana*; otherwise, the fern *Phymatodes scolopendria* is conspicuous. According to White (1983), *Philippia, Phylica simii*, forms an open shrubland on poorly drained soils on the coast of Mozambique near Pebane.

2.1.6.3.3 The Littoral Vegetation

Wetland vegetation: Eggers and Reed (2015) described wetland vegetation as the vegetation that grows in water or on soil that at least periodically is deficient in oxygen due to excessive water content. Furthermore, generally, wetland plant communities are organized according to water permanence, depth, and degree of soil saturation from deep-water wetlands to seasonally flooded basins (Eggers and Reed, 2015).

Swampy forests (Mangroves vegetation): Mangroves are woody plant species growing in salt habitats along the shore of the Ocean. The term mangrove may, however, refer to the plants or to the ecosystem in which they are found. The latter is synonymous with tidal forests, mangrove communities (mangrove forest). In Tanzania, 10 mangrove species have been reported to occur (Shunula et al., 2001). Rufiji Delta in Tanzania is one of the unique mangrove areas in East Africa where permanent human settlements occur in the mangrove forest. The area also harbors one-third of the mangrove cover of Tanzania. Threats to mangroves across the coast of East Africa include conversion to other land uses such as agriculture, salt mining, and settlement. Other threats include cutting for poles and timber, erosion due to flooding especially in the deltas, and pollution, especially in forests close to urban areas. Monga et al. (2018) estimated a net loss of 6422 ha of mangroves in Rufiji Delta between 1991 and 2015, implying an annual rate of 0.52% loss. Most of this loss was contributed by the conversion of mangroves into rice farming. This annual loss rate is higher than loss rates reported in other neighboring countries including Kenya where the annual loss rate was estimated at 0.2% (Mbatha et al., 2022). In Mozambique, more mangrove gain than losses have been reported in the last 30 years (Shapiro et al., 2015).

2.2 THE DISTRIBUTION OF BIODIVERSITY HOTSPOT OF THE EASTERN AFRICA COASTAL FORESTS

The Eastern African coastal forest zone described by Joordens et al. (2019) as to be comprised of the narrow (width ~50–100 km) strip vegetation zone distributed along the coastal area from southern Somalia to southern Mozambique (see Figure 2.1). The coastal forest of Eastern Africa is recognized for having high plant endemism, also the coastal forests support high species richness and diversity of other taxonomic groups including mammals, birds, reptiles, amphibians, and invertebrates (Burgess et al., 1998). However, the coastal forests and their rich biological diversity are severely threatened by human impact and predicted future climate change the mentioned factors caused the coastal forest of Eastern Africa to become a "vanishing refuge area."

2.3 THE IMPORTANCE OF BIODIVERSITY HOTSPOTS OF THE EASTERN AFRICA COASTAL FORESTS

The multitude of benefits are derived from the preservation of forests along the Eastern African coast. These involve primarily their uniqueness in terms of habitats that harbor the large number of fauna and flora which are endemic to the area (Burgess et al., 1998). As a result, their status is known as a Biodiversity hotspot of global importance as they are among the 36 global biodiversity hotspots (Habel et al., 2017). In addition, the Eastern African coastal forests are also important because they are a source of medicinal plants;

reduce soil erosion, fuel wood, building materials, and food to rural people (https://coastalforests.tfcg.org/pubs/CFNationalSynthesis-Tz.pdf; Matiku, 2024).

2.4 CHALLENGES AND CONSERVATION EFFORTS OF EASTERN AFRICA COASTAL FORESTS

There are several main threats to the Eastern African coastal forests (Gachanja and Kanyanya, 2004; https://aquadocs.org/handle/1834/8857; Matiku, 2024). Poverty was mentioned as the main threat, which usually contributes to the deforestation of the Eastern African coastal forests through commercial logging (Figure 2.4). This has resulted in causing forest loss and habitat fragmentation for the forest inhabitants. For example, in the Lindi and Tanga Regions, *Swartzia madagascarensis* and *Brachylaena hutchinsii* are the two targeted tree species for commercial harvest.

FIGURE 2.4 Recent tree cutting for timber in Rufiji Delta, Tanzania (*Photo credit: Elinasi Monga*).

The reason behind here is to meet the increasing demand for wood fuel in the city specifically for charcoal production. This threat was supported by the study by Matiku (2024) which revealed that the main threat identified in the 2002 workshop included unsustainable logging for timber; pole wood; charcoal and carving wood. Also, hunting; salt mining (Figure 2.5); agriculture (i.e., cultivation; encroachment; fire; overgrazing); and wildlife-human conflicts were other mentioned threats (https://aquadocs.org/handle/1834/8857; Matiku, 2024). Previous study have found that in Kenya, there are mineral mining activities taking place especially in Mrima Hill (http://coastalforests.tfcg.org/pubs/SocioEco-Ken.pdf, Gachanja and Kanyanya, 2004).

2.5 CONCLUSION AND RECOMMENDATION

This chapter reviewed and presented an overview of the vegetation types of the Eastern African coastal forests with special emphasis on their distribution, status, threats, and conservation efforts. The Eastern African coastal forests support a diverse of vegetation types. In general, the vegetation types in Eastern Africa can be categorized into three main groups: closed canopy coastal forests, woodland (Eastern African coastal Miombo Forest), and nonforest vegetation. These forests are currently undergoing unsustainable harvesting which results in habitat loss and fragmentation forming forest patches. Therefore, the formed forest patches have the highest biodiversity importance per unit area as they harbor many endemics species.

FIGURE 2.5 Showing the establishment of a chemical storage industry near the mangrove forest along the Mtwara coastal area (*Photo by Ally J. K. Nkwabi, January 2017*).

The Eastern African coastal forest is recognized as a biodiversity hotspot facing threats from human activities and natural elements. Extensive biological research conducted in the area uncovered high levels of unique plant and animal species, emphasizing their global conservation importance. Therefore, this baseline data strengthened the need for goal-oriented management of this wilderness and such management systems depend largely on the ecological knowledge of the species, communities, or ecosystems with scientific principles.

KEYWORDS

- **biodiversity hotspots**
- **coastal forests**
- **Eastern Africa**
- **distribution**
- **vegetation types**

REFERENCES

Almeida, C.; Silva, D.; Messias, C.; Soler, L., Gusmão, L.; Maurano, L.; Souza, T.; Soares, I.; Barradas, D.; Cunha, I. P.; Bastos, E.; Belluzo, A. Quadros, C.; Bento, B. P.; Pinheiro, F.; Silva, L. Mapping natural non-forest vegetation removal in the Brazilian Amazon—a pilot project. Int. Arch. Photogramm. Remote Sens. Spatial Inf. Sci. **2022,** *43*, 1341–1348. https://doi.org/10.5194/isprs-archives-XLIII-B3-2022-1341-2022.

Bai, Y.; Zhou, Y. The main factors controlling spatial variability of soil organic carbon in a small karst watershed, Guizhou Province, China. *Geoderma* **2020;** *357*, 113938. https://doi.org/10.1016/j.geoderma.2019.113938

Barratt, C. D. *Biodiversity Patterns and Conservation of the Coastal Forests of Eastern Africa.* University of Basel, **2017.**

Bloesch, U.; Klötzli, F. Coastal forests of the Saadani National Park. In: *Tanzania Wildlife Discussion Paper.* Dar es Salaam, **2004**.

Burgess, N.; Doggart, N.; Doody, K.; Negussie, G.; Sumbi, P.; Perkin, A. J. O. New information on the lowland coastal forests of eastern Africa. *Oryx* **2003,** *37*, 280–281.

Burgess, N. D.; Clarke, G. P. *Coastal Forests of Eastern Africa.* IUCN-The World Conservation Union, Publications Services Unit. **2000.**

Burgess, N. D.; Clarke, G. P.; Rodgers, W. A. Coastal forests of eastern Africa: Status, endemism patterns and their potential causes. *Biol. J. Linn. Soc.* **2008,** *64*(3), 337–367. https://doi.org/10.1111/j.1095-8312.1998.tb00337.x.

Burgess, N. D.; Fjeldså, J.; Botterweg, R. Faunal importance of the Eastern Arc Mountains of Kenya and Tanzania. *J. East African Nat. Hist.* **1998,** *87*(1), 37–58. https://doi.org/10.2982/0012-8317(1998)87[37:FIOTEA]2.0.CO;2.

Burgess, N. D.; Malugu, I.; Sumbi, P.; Kashindye, A.; Kijazi, A.; Tabor, K.; Mbilinyi, B.; Kashaigili, J.; Wright, T. M.; Gereau, R. E. Two decades of change in state, pressure and conservation responses in the coastal forest biodiversity hotspot of Tanzania. *Oryx* **2017.** *51*(1), 77–86. https://doi.org/10.1017/S003060531500099X.

Burgess, N. D.; Muir, C. Coastal forests of Eastern Africa: Biodiversity and conservation. In: *Society for Environmental Exploration.* Royal Society for the Protection of Birds, **1994**, p. 112.

Chaplot, V.; Bouahom, B.; Valentin, C. Soil organic carbon stocks in Laos: Spatial variations and controlling factors. *Glob. Change Biol*, **2010,** *16*(4), 1380–1393. https://doi.org/10.1111/j.1365-2486.2009.02013.x.

Eggers, S. D.; Reed, D. M. *Wetland Plants and Plant Communities of Minnesota and Wisconsin, Version 3.2.* **2015**.

Fungomeli, M.; Githitho, A.; Frascaroli, F.; Chidzinga, S.; Cianciaruso, M.; Chiarucci, A. A new vegetation-plot database for the Coastal forests of Kenya. *Veg. Classif. Surv.* **2020,** *1*, 103–109. https://doi.org/10.3897/VCS/2020/47180.

Gachanja, M.; Kanyanya, E. *Conservation of Kenya's Coastal Forests: Social Economic Resources'*. **2004.**

Githitho, A. *The Coastal Terrestrial Forests of Kenya: A Report on Resources, Threats and Investments*. **2004.**

Godoy, F. L.; Tabor, K.; Burgess, N. D.; Mbilinyi, B. P.; Kashaigili, J. J.; Steininger, M. K. Deforestation and CO_2 emissions in coastal Tanzania from 1990 to 2007. *Environ. Conserv.* **2012,** *39*(1), 62–71. https://doi.org/10.1017/S037689291100035X.

Habel, J. C.; Casanova, I. C. C.; Zamora, C.; Teucher, M.; Hornetz, B.; Shauri, H.; Mulwa, R. K.; Lens, L. East African coastal forest under pressure. *Biodivers. Conserv.* **2017,** *26*(11), 2751–2758. https://doi.org/10.1007/s10531-017-1375-z.

Joordens, J. C. A.; Feibel, C. S.; Vonhof, H. B.; Schulp, A. S.; Kroon, D. Relevance of the eastern African coastal forest for early hominin biogeography. *J. Hum. Evol.* **2019,** *131*, 176–202. https://doi.org/10.1016/j.jhevol.2019.03.012.

Karyn, T.; Neil, D. B.; Boniface, P. M.; Japhet, J. K.; Marc, K. S. Forest and woodland cover and change in Coastal Tanzania and Kenya, 1990 to 2000. *J. East African Nat. Hist.* **2010;** *99*(1), 19–45. https://doi.org/10.2982/028.099.0102.

Kent, P. E. Mesozoic history of the East Coast of Africa. *Nature* **1972,** *238*(5360), 147–148. https://doi.org/10.1038/238147a0.

Matiku, P. The Coastal Forests of Kenya: Forests data, threats, socio-economic issues, values, stakeholders, challenges, strategies, investment and enabling environment—A national synthesis report for the development of the WWF-EARPO Eastern Africa Coastal Forests Ecoregion Programme, **2024**.

Mbatha, A.; Githaiga, M. N.; Kiplagat, K.; Kairo, J.; Mungai, F. How sustainable is mangrove harvesting in Lamu? An analysis of forest structure. *J. Sustain. For.* **2022**, 1–20. https://doi.org/10.1080/10549811.2022.2123357.

Milne, G. A soil reconnaissance journey through parts of Tanganyika Territory December 1935 to February 1936. *J. Ecol.* **1947,** *35*(1/2), 192–265. https://doi.org/10.2307/2256508.

Mligo, C. The plant species composition, diversity and natural regeneration of indigenous trees in the disturbed Ruvu South Forest Reserve, Tanzania. *Tanzania J. Sci.* **2018,** *44*(3), 46–60.

Monga, E.; Mangora, M. M.; Mayunga, J. S. Mangrove cover change detection in the Rufiji Delta in Tanzania. *WIO J. Mar. Sci.* **2018,** *17*(2), 1–10. https://doi.org/10.4314/wiojms.v17i2.1.

Mwasumbi, L. B.; Burgess, N. D.; Clarke, G. P. Vegetation of Pande and Kiono Coastal forests, Tanzania. *Vegetatio* **1994,** *113*(1), 71–81. https://doi.org/10.1007/BF00045465.

Myers, N.; Mittermeier, R. A.; Mittermeier, C. G.; da Fonseca, G. A. B.; Kent, J. Biodiversity hotspots for conservation priorities. *Nature* **2000,** *403*(6772), 853–858. https://doi.org/10.1038/35002501.

Ngumbau, V. M.; Luke, Q.; Nyange, M.; Wanga, V. O.; Watuma, B. M.; Mbuni, Y. M.; Munyao, J. N.; Oulo, M. A.; Mkala, E. M.; Kipkoech, S.; Itambo, M.; Hu, G. W.; Wang, Q. F. An annotated checklist of the coastal forests of Kenya, East Africa. *PhytoKeys* **2020,** *147*, 1–191. https://doi.org/10.3897/phytokeys.147.49602.

Nkwabi, A. K.; Bukombe, J.; Maliti, H.; Liseki, S.; Lesio, N.; Kija, H. An overview of biodiversity in Tanzania and conservation efforts. In: Pullaiah, T. (Ed.), *Global Biodiversity.* Apple Academic Press, Canada, **2018** pp. 295–340.

Shapiro, A. C.; Trettin, C. C.; Küchly, H.; Alavinapanah, S. Bandeira, S. The mangroves of the Zambezi delta: Increase in extent observed via satellite from 1994 to 2013. *Remote Sens*. **2015,** *7*(12), 16504–16518. https://doi.org/10.3390/rs71215838.

Shunula, J. P.; Jumah, M. A.; Mohammed, S. M.; Machano, H.; Mwinyi, A. A. *Inventory of the mangrove forest resources in the Chwaka Bay Conservation Area* Jozani and Chwaka Bay Conservation Project, Issue, **2001**.

Syampungani, S.; Chirwa, P. W.; Akinnifesi, F. K.; Sileshi, G.; Ajayi O. C. The Miombo woodlands at the crossroads: Potential threats, sustainable livelihoods, policy gaps, and challenges. *Nat. Resour. Forum* **2009**, 33, 150–159.

Timberlake, J.; Goyder, D.; Crawford, F.; Burrows, J.; Clarke, G. P.; Luke, Q.; Matimele, H.; Müller, T.; Pascal, O.; de Sousa, C.; Alves, T. Coastal dry forests in northern Mozambique. *Plant Ecol. Evol.* **2011,** *144*(2), 126–137. https://doi.org/10.5091/plecevo.2011.539.

White, F. The vegetation of Africa. A descriptive memoir to accompany the UNESCO/AETFAT/UNSO vegetation map of Africa. *Revue d'Écol.* **1983**, *39*(2), 247–247.

Woodhouse, G. *The Importance of coastal Kenyan Isolated Hill Forest Outliers and Limestone Outcrops to Botanical Conservation in East Africa*. **2019**.

Younge, A.; Negussie, G.; Burgess, N. Eastern Africa Coastal Forest Programme: Regional workshop report. Eastern African coastal forest programme regional workshop, Nairobi, 4–7 February, 2002.

ANNEXURE 2.1 Showing the Vegetation Types of Coastal Forests of Eastern Africa Biodiversity Hotspot

Vegetation Categories	Major Vegetation Group	Subgroup	Representative Examples
i. Eastern Africa Closed Canopy Coastal Forests	Eastern African Coastal Dry Forest	Low-coastal forest High-coastal forest	*"Cynometra webberi-Manilkara sulcata"* community of the Arabuko-Sokoke forest and Shimba hills forests of Kenya; the forests on Gendagenda Hill, Tanzania; Rondo Plateau, Saadani legume-dominated dry evergreen forests and mixed dry forest, Tanzania, Inhansato and Inhamitanga forests, Cheringoma, Mozambique; Matandwe forest, Malawi Hills; Haroni and Rusitu forests, Zimbabwe
	Eastern African Coastal Scrub Forest	i. Southern tropical semi evergreen forests ii. The Saadani forest-savanna mosaic	Scrub forest near Raas Kaamboni, Somalia; at Msambweni, Kenya; on the northern slopes of the Western Usambara Mountains, Saadani mixed scrub forest and maritime scrub forest, Msua Thickets and on Mbudya Island near Dar es Salaam, Tanzania; also, on Mefunvo Island, Quirimba Islands, Mozambique
	Eastern African Coastal Riverine/Groundwater/ Swamp Forest	Riverine forest	Riparian forest areas along the Jubba River at Bu'ale, Somalia; along the Tana River, Kenya; Saadani (Delta of Wami River?) ground water forest, Gendagenda Forest, valley bottom areas of the Pugu and Kazimzumbwi forests, swamp forest areas of Mchungu Forest and Kimboza Forest, Tanzania; and riverine forest in the Massenjere Forest Reserve, Malawi.
	Mixed forest	Mixed forest (7000 ha) in the east, on grey sands.	Characteristic trees include *Combretum schumannii, Drypetes reticulata, Afzelia quanzensis, Dialium orientale, Hymenaea verrucosa* and *Manilkara sansibarensis.*
	Eastern African Coastal/Afromontane Transition Forest		Forest on the summit of the Shimba Hills, Kenya; on the summit of Tongwe Hill, in Kimboza Forest, Tanzania, and in the lowlands of the East Usambaras; dry forest on Machemba Hill, Malawi and Chirinda Forest, Zimbabwe.
ii. Woodland	Eastern African Coastal Miombo Forest	Brachystegia woodland	i. *Brachystegia* woodland, e.g., Rondo Plateau
		Julbernardia woodland	ii. Julbernardia Woodland ii. Woodland (0.1%) occurrence in Kenya coastal forest dominated by *Brachystegia spiciformis*, Chiniziua Forest, Cheringoma, Mozambique; parts of Arabuko-Sokoke Forest, Kenya and parts of Tong'omba forest and Rondo Forest, Tanzania. Dominated by *Julbernardia* sp
iii. Nonforest vegetation	i. Moist Forest, edaphic grasslands habitats	i. Palm Forest	i. Palm Forest and savanna grassland of coastal region of Kenya

ANNEXURE 2.1 *(Continued)*

Vegetation Categories	Major Vegetation Group	Subgroup	Representative Examples
	ii. Bushland	ii. Savanna grasslands	ii. Coastal evergreen bushland (0.4%) in Kenya
	iii. The littoral vegetation	iii. Coastal evergreen bushland	iii. Coastal evergreen bushland (1%) in Tanzania?
		iv. Wetland vegetation, swampy forests, i.e., Mangroves vegetation, ten species of mangroves reported (Shunula et al. 2001; Shunula 2002).	iv. *Bruguiera gymnoriza*,
			v. *Ceriops tagal*,
			vi. *Rhizophora mucronata*,
			vii. *Sonneratia alba*,
			viii. *Avicennia marina*,
			ix. *Xylocarpus granatum*,
			x. *Xylocarpus moluccensis*,
			xi. *Heritiera littoralis*,
			xii. *Lumnitzera racemosa*
			xiii. *Pemphis acidula*. However, other authors do not consider *P. acidula* as a mangrove tree but regard it as mangrove associate (e.g., Banyikwa and Semesi, 1986).

CHAPTER 3

MANGROVE DIVERSITY, CONSERVATION, AND CHALLENGES ALONG THE COASTAL BIODIVERSITY HOTSPOT OF EASTERN AFRICA

EMMANUEL J. MWAINUNU[1*], MWITA M. MANGORA[2], CARL C. TRETTIN[3], TITUS WAMAE[4], FREDRICK MUGUBE[4], and LILIAN NYAEGA[4]

[1,4]*Tanzania Forestry Research Institute, Morogoro, Tanzania*

[2]*Institute of Marine Sciences. University of Dar Salaam, Buyu, Zanzibar, Tanzania*

[3]*USDA Forest Service. Southern Research Station, Asheville, NC, USA*

[4]*Wetlands International, East Africa, Nairobi, Kenya*

*Corresponding author

ABSTRACT

Mangrove ecosystems in Eastern Africa (EA) serve as breeding grounds and provide habitat for abundant wildlife. However, Mangroves have experienced increased deforestation and degradation, affecting their ecological integrity. This chapter reviewed and presented an overview of the status of mangroves, diversity, distribution, conservation efforts, and challenges along the coastal biodiversity hotspots of the EA region. The region covered 9298.53 km^2, whereas 98.4% of mangroves are located in four countries, such as Mozambique, Tanzania, Kenya, and Somalia. Ten species of mangrove tree species occur in EA. *Rhizophora mucronata* and *B. gymnorrhiza* are the most distributed mangrove tree species along the coastal forest of EA, while *Sonneratia alba* and *Heritiera littorals* are the least. Crustaceans, mollusks, polychaetes, fish, and terrestrial animals are the main taxonomic fauna in mangroves. The region uses International and National policies and legal frameworks to conserve mangroves. Lack of substantive policy and legislative frameworks for only mangroves and conflicts of interest in using mangroves have been contributing challenges to sustaining mangrove conservation. This review demonstrated that the region is highly diverse in flora and fauna. The existing conservation measures, including policies and legal framework, are not strong enough to sustain the ecological integrity of the mangroves. Therefore, more is to be done.

Biodiversity Hotspot of the Coastal Forests of Eastern Africa. T. Pullaiah and Ally J. K. Nkwabi (Eds.)

DOI: 10.1201/9781998511099-3

3.1 INTRODUCTION

Eastern Africa (EA) is the Western Indian Ocean region comprising 10 countries, including five mainland continental states (Somalia, Kenya, Tanzania, Mozambique, and South Africa) and 5 Islands states such as Mauritius, Seychelles, Madagascar, France, and Comoros. The region has about 1 million ha of mangroves (Bosire et al., 2016a) and is considered one of the Coastal Biodiversity Hotspots of EA. Biodiversity hotspots are regions with rich species (Figure 3.1a and b), endemic species and facing strong habitat loss (Myers et al., 2000; Reid, 1998).

FIGURE 3.1a *Ciconia episcopus* a vulnerable bird species in a developing mangrove stand of *Avicenia marina* species.

Mangrove vegetation includes a range of functional forms, including tree shrubs and ground ferns, generally exceeding 0.5 m in height (Duke et al., 1998). They are the only woody plants that grow at the interface between land and sea in tropical and subtropical latitudes where they exist in conditions of high salinity, extreme tides, strong winds, high temperatures, muddy and anaerobic soils (Kathiresan and Bingham, 2001; Alongi, 2012). Mangroves form one of the world's most ecologically and economically valuable ecosystems (Nagelkerken et al., 2014).

They are confined to the tropics and subtropics, developing and growing best where low-wave energy and shelter enable the deposition of fine sediments (Alongi, 2009; Hogarth, 2015). Due to their high productivity, mangrove ecosystems provide important goods and services that support human well-being and coastal and marine ecosystems (Bosire et al., 2016a). The ecosystem goods include food, timber, fuel, and medicine to people, whereas services include soil stabilization and nutrient cycling, habitats for birds, fish, crustaceans, reptiles, and mammals (Hogarth, 2015; Mangora et al., 2016a). Mangroves connect the coastal and terrestrial ecosystems and protect other immediate coastal ecosystems such as seagrass beds and coral reefs. According to Bosire et al. (2016a), mangroves help mitigate eutrophication by stripping off nutrients to enter into coastal habitats such as seagrass beds and coral reefs. Mangroves are increasingly verified to be one of the largest carbon sinks and have therefore gained attention in the debates on their role in climate change mitigation (Donato et al., 2011; Sanderman et al., 2018; Adame et al., 2021).

FIGURE 3.1b *Intermediate Egret* in *Rhizophora mucronata* stand of mangrove of the Rufiji Delta.

Eastern African mangroves provide important habitats for significant biodiversity, ranging from benthic invertebrates, fishes, insects, and birds to larger wildlife such as buffaloes, hippos, crocodiles, and primates. Despite their importance, mangrove forests are at high risk due to threats caused by human populations (Lang'at, 2013). Such threats include tree felling for fuelwood and building materials, clearing mangrove areas for agriculture, aquaculture and salt works, urban and tourist development, and human settlement (Mangora et al., 2016a).

This chapter reviewed and presented an overview of the status of mangroves, diversity, distribution, conservation efforts, and challenges along the coastal biodiversity hotspots of the EA region. The information is important to managers and decision-makers to properly manage mangroves and biodiversity in EA for their sustainability of the resources. Further, the chapter sets a benchmark and foundation for researchers, academicians and university students to build their future studies in the mangrove forest of EA.

3.2 STATUS OF MANGROVES IN EA

3.2.1 OCCURRENCE AND COVERAGE

In EA, Mangroves occur in enclosed lagoons, deltas, estuaries, and bays, where they are protected from wave action (Bosire et al., 2016a). More than 90% of the mangroves occur in coastal areas of EA, which are Mozambique, Tanzania and Kenya (Bosire et al., 2016b; Mangora et al., 2016a; Macamo et al., 2016; Ratsimbazafya et al., 2016). The most extensive mangroves occur in deltas and estuaries of Rufiji and Zambezi Deltas (Bosire et al., 2016a).

The total area of mangroves in EA is 9298.53 km^2, 10.94% of the total mangrove area (84,984.56 km^2) across the Indian Ocean. Also, it is 6.75% of the world's mangrove area (137,760.00 km^2) (Kathiresan and Rajendran, 2005). The region's largest (42.24%) were found in Mozambique, followed by Madagascar (35.01%), Tanzania (15.4%) and Kenya (5.7%) (Figure 3.2a–c). Other countries with much less coverage of mangroves include Comoros, Mauritius, Somalia, South Africa, and Seychelles (Bosire et al., 2016a). The mangrove forest formations in the region occur inside lagoons, deltas, estuaries, and bays, where they are protected from wave action (Mangora, 2011). Rufiji Delta in Tanzania and the Zambezi River Delta in Mozambique have the largest mangrove areas in EA (Bosire et al., 2016a).

3.2.2 DIVERSITY AND DISTRIBUTION OF MANGROVE FLORA

A total of 10 species of mangroves exist in EA. This is 18% of the total of 55 mangrove species reported to occur in the entire Indian Ocean region and 14% of the total 70 mangrove species that occur globally (Kathiresan and Rajendran, 2005), indicating higher diversity of mangrove tree species in the region of EA compared to West Africa with 10% of mangrove species occurring globally (Japhet, 2021). Tanzania, Kenya and Mozambique have many number of mangrove tree species than other countries in the region (Table 3.1). *Rhizophora mucronata* and *Bruguiera gymnorrhiza* are highly distributed as they are found in all countries of EA. *Pemphis acidula* is only reported in Tanzania, and it is still debated in other countries whether it is a true or an associated mangrove species (Mangora et al., 2016a). Gazi Bay in Kenya and Rufiji Delta in Tanzania are among the mangrove stands with the most diverse mangrove tree species. Nine (9) species occur in the Gazi Bay (Githaiga, 2013), while Rufiji Delta has eight (8) species (Lupembe, 2014; Wagner and Sallema-Mtui, 2016; Mangora et al., 2016b).

Mangrove species exhibit zonation (Okello, 2016). The distribution of species across zones depends on the environmental gradient, as mangrove forest tree species respond differently to different environmental gradients (Njana et al., 2015). Environmental gradients include temperature, rainfall, waves, salinity, and soil properties across the mangrove stand. A common pattern of zonation of mangroves in EA (Figure 3.3) is that *Sonneratia alba* occurs in the lower tidal zone on the seaward margin. In the middle are *R. mucronata*, *Ceriops tagal*, *Xylocarpus granatum*, *B. gymnorrhiza*, and *Avicennia marina*, while *Heritiera littoralis* is located in a higher tidal zone.

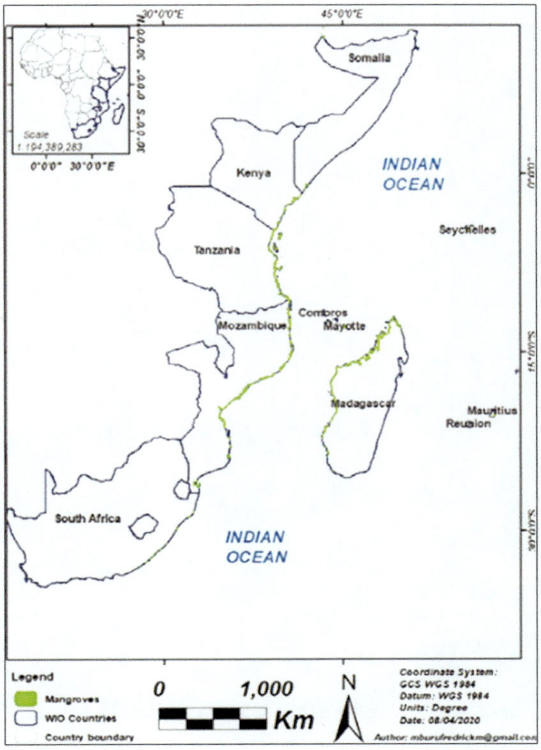

FIGURE 3.2a Map of the coastal forest of Eastern Africa showing the occurrence of mangrove forests
Source: UNEP-Nairobi Convention/USAID/WIOMSA, 2020.

FIGURE 3.2b A *Rhizophora mucronata* stand at Rufiji Delta, Tanzania (*Photo credit: Genovefa Feksi*).

FIGURE 3.2c A *Soneratia Alba* stand at Rufiji Delta, Tanzania (*Photo credit: Genovefa Feksi*).

TABLE 3.1 Mangrove Species Composition and Distribution in Eastern Africa

Species	**Countries**								
	Kenya	**Tanzania**	**Mozambique**	**Madagascar**	**Comoros**	**Seychelles**	**Somalia**	**South Africa**	**Mauritius**
Rhizophora mucronata	✓	✓	✓	✓	✓	✓	✓	✓	✓
Bruguiera gymnorrhiza	✓	✓	✓	✓	✓	✓	✓	✓	✓
Ceriops tagal	✓	✓	✓	✓	✓	✓	✓	✓	
Avicennia marina	✓	✓	✓	✓	✓	✓	✓	✓	
Lumnitzera racemosa	✓	✓	✓	✓	✓	✓	✓	✓	
Sonneratia alba	✓	✓	✓	✓	✓	✓	✓		
Xylocarpus granatum	✓	✓	✓	✓		✓		✓	
Heritiera littorals	✓	✓	✓	✓	✓				
Xylocarpus mollucensis	✓	✓	✓						
Pemphis acidula		✓							
Total	9	10	9	8	7	7	6	6	2

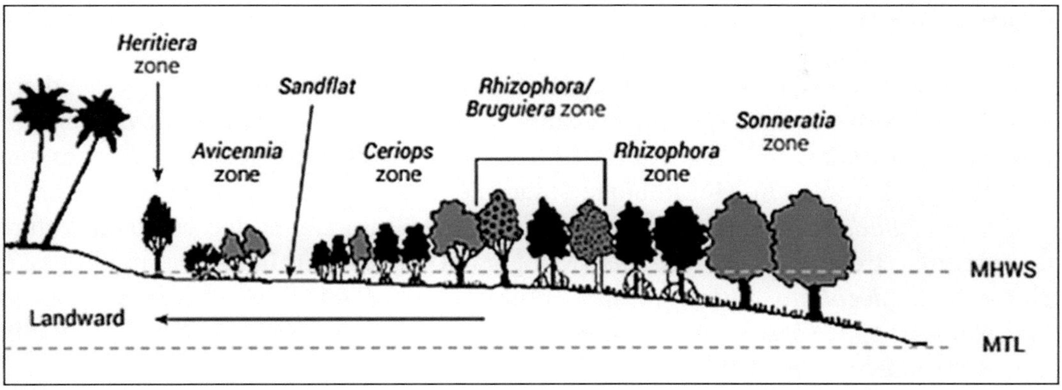

FIGURE 3.3 A common pattern of zonation of mangroves in Eastern Africa.
Source: UNEP-Nairobi Convention/USAID/WIOMSA, 2020.

Mangrove tree species distributed in the EA coastal have varying ecological ranges. *R. mucronata* and *B. gymnorrhiza* are present in all East African countries, indicating a broad ecological distribution. In contrast, *H. littoralis* is limited to only five countries, suggesting a more restricted ecological range. Faridah-Hanum et al. (2012) associate such distribution of mangrove species to local physiological tolerances of individual species to physical conditions such as tidal inundation and high salinity, but may also be influenced by other factors such as predation of plant seedlings. The patterns of mangrove distribution might be due to the variation in the physiological tolerance of individual species to salinity. Shan et al. (2008) suggested that differences in salt tolerance of different species define their distinct positions along the intertidal zones. For example, *H. littoralis* was encountered only in the Northern zone along the freshwater upper fringes of the delta (Japhet, 2021).

3.2.3 *FAUNA DIVERSITY AND DISTRIBUTION IN MANGROVE FOREST*

Mangrove ecosystems in EA serve as breeding grounds and provide a habitat for abundant animals (Ratsimbazafy et al., 2016). Fauna in mangroves is found in the sediment, on the forest floor and in the plant's aerial roots, stem, branches, leaves, and canopy (Bosire et al., 2016b). For several faunal species, mangroves are permanent homes for migratory birds, for example, *Terpsiphone viridis* (intra-African migrants), *Numenius phaeopus* (Palaearctic-African migrant), and *Ardea cinerea* (Palearctic-African migrant), and other species live in the mangroves only at their juvenile stage. Therefore, mangrove act as breeding grounds for some species of animals (Ratsimbazafy et al., 2016). The main taxonomic groups common among mangroves include marine (crustaceans, mollusks, polychaetes, fish) and terrestrial animals (Macamo et al., 2016).

The crustacea are the most abundant epibenthic group in mangroves and are well-represented by crabs, shrimps, barnacles, and prawns (Macamo et al., 2016). Crabs are the most conspicuous invertebrates in the mangroves and have been distributed in all countries of EA (Bosire et al., 2016a). Further, the prawn species that exist in EA include

Penaeus indicus, P. semisulcatus, Metapenaeus monoceros, Fenneropenaeus indicus, Penaeus monodon, and *Penaeus japonicas* (Bosire et al. 2004, 2016b; Mangora et al., 2016a). For the mollusks group, the gastropods are well represented in the southern mangroves, with common species including *Littoraria scabra, Littoraria intermedia, Littoraria subvittata*, and *Littoraria pallescens* which dominate the seaward edges and the *Rhizophora* areas in Mozambique (Macamo et al., 2016. Gastropods such as *L. scabra, Cerithidea decollate*, and *Terebralia palustris* are among the common feature on mangrove tree trunks in Tanzania (Mangora et al., 2016a). In Kenya, molluscs (*T. palustris, C. decollata, Littoraria* spp. (Bosire et al., 2004, 2016b). Oysters such as *Brachydontes* spp., *Crassostrea cuculata* and cockles, *Donax* spp.) (Bosire at al., 2016b) are also common in the mangroves of EA.

A higher abundance of polychaetes, which consist of a group of worms and mice, was specifically documented in the northern mangroves of Mozambique, with a density of 612 individuals/m^3 (Macamo et al., 2016). In the Mangroves, a variety of fish species, which are resident or visit mangroves for shelter, feeding and breeding, include juveniles of commercially important fish groups such as snappers, emperors, groupers, milkfish, and mullets (Mchenga and Ali 2013; Macamo et al., 2016; Mangora et al., 2016a). Various fish species use the mangroves and estuarine habitats (Rajkaran and Adams, 2016). There is particularly high diversity among the fish populations that inhabit mangrove forests and creeks during high tides, the families of which include Mugelidae, Serranidae, Carangidae, Gerridae, Hemiramphidae, Plectrorhynchidae, and Elopidae (Ratsimbazafy et al., 2016).

Terrestrial fauna includes many species of birds, reptiles including crocodiles, mammals (pigs and monkeys), and insects (Bosire et al., 2016b). Crocodiles are common in the Zambezi River delta (Macamo et al., 2016). In Rufiji Delta, Tanzania, Japhet (2018) reported 20 reptiles from four groups, including the Nile crocodile, 93 birds (Figure 3.4), most being migratory species. Notable mammals in EA mangroves, particularly Kenya, Tanzania, Mozambique, and South Africa, include Hippopotamus and Nile crocodiles, baboons, duikers, rodents, and fruit bats. Some sea turtles and Dugongs (*Dugong dugon*) are occasionally found in the mangrove of EA (Ratsimbazafy et al., 2016).

3.4 THREATS TO MANGROVES ACROSS EAST AFRICAN COUNTRIES

Mangrove forests in the EA region face several common threats (Table 3.2), leading to changes in mangrove ecosystem qualities (Macamo et al., 2016). However, certain threats are limited to specific areas, whereas others such as oil spills, mining, and minimal freshwater supplies, were documented in Kenya, Madagascar, and Comoros, respectively. Footpaths, trampling, mud disturbances, forest fires, invasive species, and bush encroachment were solely reported in South Africa (Appadoo et al., 2016; Rajkaran and Adams, 2016; Ratsimbazafy et al., 2016). This notwithstanding mangroves across the countries are threatened due to the high demand for mangrove products such as firewood, charcoal, building, and boat making. In addition to commercial cutting and over-harvesting, coral burning, lime production, salt making, clear-cutting for building sites, solar salt pans, commercial projects, and clearance for agriculture (ASCLME, 2012) as well as clearance for coastal development, coral stone mining and boat making. Natural factors such as flooding and coastal erosion have also contributed to declines in area coverage in key areas. In a study conducted in Rufiji, researchers observed a decrease in mangrove cover from 51,941 to 45,519 ha between 1989 and 2015 (Monga et al., 2018). The result has been attributed to the continued destruction of mangroves due to rice farming and the demand for timber in the Rufiji Delta.

FIGURE 3.4 Mangrove King Fisher, a common bird species in the Eastern African Mangroves (*Photo credit: Frank Sima*).

TABLE 3.2 Threats to Mangroves Diversity in Eastern Africa

Threats	Countries							
	Kenya	Tanzania	Mozambique	South Africa	Madagascar	Comoros	Seychelles	Mauritius
Anthropogenic								
Overexploitation of wood products	✓	✓	✓	✓	✓	✓		
Salt pans	✓	✓	✓	✓				
Rice farming/agriculture		✓		✓	✓		✓	
Mariculture ponds		✓	✓	✓	✓			
Infrastructure development/settlement	✓	✓	✓	✓		✓	✓	✓
Pollution			✓	✓		✓	✓	✓
Charcoal production			✓	✓	✓			
Oil spills	✓							
Damming/dredging of rivers	✓		✓	✓				
Cattle grazing				✓	✓			
Footpaths and trampling				✓				
Mud disturbances				✓				
Forest fire				✓				
Mining					✓			
Natural Threats								
Excessive flooding	✓	✓	✓	✓				
Pest infection	✓	✓						
Sea level rise/cyclones	✓	✓		✓		✓	✓	✓
Sand movement/deposition	✓	✓				✓	✓	✓
Riverbank erosion	✓	✓		✓				
Alien invasion and bush encroachment				✓				
Low freshwater inputs						✓		

Source: Bosire et al. (2016a).

3.5 CONSERVATION INITIATIVE FOR MANGROVE RESOURCES

The Eastern African countries have set various governances for conserving mangrove resources. Mangrove governance in the EA region is recognized in various Multilateral Environmental Agreements (MEAs) such as the Ramsar Convention on Wetlands of International Importance Especially as Waterfowl Habitat, 1971; the Amended Nairobi

Convention for the Protection, Management and Development of the Marine and Coastal Environment of the Western Indian Ocean (Amended Nairobi Convention), 2015; the Sendai Framework for Disaster Risk Reduction (2015–2030); the Convention on Biological Diversity (CBD), 1992; the Convention on International Trade in Endangered Species, 1973; the U.N. Framework Convention on Climate Change, 1992; the Paris Agreement, 2015; and the Sustainable Development Goals, 2015.

Most of the East African countries have either ratified or acceded to these agreements, and they have progressively incorporated mangroves in key global processes, including the Nationally Determined Contributions, the National Biodiversity Action Plans, and Voluntary Commitments on Ocean Action. Other relevant MEAs include the Convention Concerning the Protection of the World Cultural and Natural Heritage (World Heritage Convention); the International Convention for the Prevention of Pollution from Ships (MARPOL), 1973.

At a national level, Kenya has developed the National Mangrove Ecosystem Management Plan (2017–2027) and established the National Management Committee for managing mangroves (GOK, 2017). In Madagascar, coral reefs and islands are classified as sensitive zones under the Environmental Impact Assessment (EIA) legislation. Consequently, EIAs are necessary for projects that could potentially impact mangrove ecosystems. The National Strategy and Action Plan for Mangroves Management (2020–2024) of Mozambique provides for the conservation and restoration of mangrove forests to maintain the ecological processes and functions of mangrove ecosystems (RFM, 2020). In Tanzania, mangrove forests are designated as state forests, with the Tanzania Forest Services (TFS) agency appointed as the main caretaker of these ecosystems (URT, 1991). The country formulated the National Mangrove Management Plan in 1991 to regulate the use and monitor mangrove habitats based on zonation and delineation for different uses. In Seychelles, the Environment Protection Act, 1994 provides integrated coastal zone management, including mangroves and atmosphere protection (R.S., 1994). The Integrated Coastal Management Act of 2008 in South Africa defines coastal wetlands to include adjacent mangroves. Additionally, under the National Forests Act of 1998, protection is provided to four out of the six species of mangroves found in South Africa (RSA, 1998).

3.6 CHALLENGES FOR CONSERVING MANGROVE DIVERSITY

3.6.1 POLICY AND LEGISLATIVE FRAMEWORKS

In EA, none of the countries has substantive policy and legislative frameworks for managing mangroves. The existing policy and legal frameworks are highly fragmented, leading to conflicts and overlapping mandates that compromise mangrove governance. Poor coordination across sectors, weak or dysfunctional state institutions, weak participatory awareness and inadequate financial and technical resources are the main challenges toward mangrove governance in EA.

3.6.2 CONFLICTS OF INTERESTS

In Kenya, the sectoral approach to development planning and management, combined with population pressure and the intensity and complexity of human activities, has resulted in resource use conflicts and adverse effects on mangrove resources (Bosire et al., 2016b). In Tanzania, the conflicts of interest are due to overlapping jurisdiction between sectors operating in mangroves. While the principal forest management jurisdiction is mandated to the TFS under the Ministry of Natural Resources and Tourism, issuance of mining permits for salt evaporation pans (Figure 3.5) is mandated to the Ministry of Energy and Minerals, while fisheries activities in the mangroves, including Mariculture, are jurisdictional responsibility of the Ministry of Livestock and Fisheries (Mangora et al., 2016a). In Mozambique, a challenge for mangrove conservation comes mainly from infrastructure development and offshore gas exploration. For example, in Western Nacala Bay, Beira, and Maputo, where ports and other infrastructure are being built or expanded, leads to clearing significant mangrove areas. Further, offshore gas exploration, such as for port facilities at Palma near Pemba, can potentially degrade mangrove forests (Macamo et al., 2016). In Madagascar, mangrove clearance often occurs in localities where rice is produced (Ratsimbazafy et al., 2016).

FIGURE 3.5 A salt evaporation pan constructed behind a mangrove area partly cleared in Kilwa District, Tanzania.

3.5 CONCLUSIONS AND RECOMMENDATION

The reviewed information demonstrates that Mozambique, Madagascar, Tanzania, and Kenya are the major sites of mangroves in EA, occupying 98.4%. The region has 10 species of mangrove trees, more diverse than Western Africa. All mangrove species of the region exist in Tanzania, implying the country has a variety of environments for harboring the mangrove species. *R. mucronata* and *B. gymnorrhiza* are the most distributed mangrove tree species, implying a wide ecological range. Fauna comprises crustaceans, mollusks, polychaetes, fish, mammals, reptiles, and avifauna taxonomic groups. Crustacea comprise crabs, shrimps, barnacles, and prawns, and mollusks consist of gastropods, while Polychaetes comprise worms and sea mice groups. More than 90 birds, including migratory ones, reptiles (crocodiles and sea turtles), mammals (monkeys and dugongs), and hippos are found in the WI region. International and National policies and legal frameworks are used to conserve mangroves in the region. The existing policy and legal frameworks are highly fragmented, leading to conflicts and overlapping mandates that compromise mangrove governance. Lack of substantive policy and legislative frameworks for only mangroves and conflicts of interest in using mangroves have been contributing challenges to sustaining mangrove conservation. This review demonstrated that the region is highly diverse in flora and fauna. Nonetheless, the existing conservation measures, including policies and legal framework, are not strong enough to sustain the ecological integrity of the mangroves. Therefore, research to explore mangrove diversity and conservation measures is recommended to sustain the mangrove ecosystem.

ACKNOWLEDGMENTS

We are grateful for the assistance received from the Tanzania Forestry Research Institute and Wetlands International. Dr. A. Mkwabi from the Tanzania Wildlife Research Institute provided valuable guidance that was instrumental in the successful completion of the chapter. Ms. Genovefa Fecks of Wetlands International and Frank Sima of Tanzania Forest Services agencies provided crucial photos.

KEYWORDS

- **distribution**
- **flora**
- **fauna**
- **mangrove ecosystem**
- **taxonomic groups**
- **epibenthic group**

REFERENCES

Adame, M. F.; Reef, R.; Santini, N. S.; Najera, E.; Turschwell, M. P.; Hayes, M. A.; Masque, P.; Lovelock, C. E. Mangroves in arid regions: Ecology, threats, and opportunities. *Estuar. Coast. Shelf Sci.* **2021,** *248*, 106796. https://doi.org/10.1016/j.ecss.2020.106796.

Alongi, D. *The Energetics of Mangrove Forests*. Springer Science & Business Media. **2009.**

Alongi, D. M. Carbon sequestration in mangrove forests. *Carbon Manag.* **2012**, *3*(3), 313–322.

Appadoo, C.; Rabi, F. A.; Antha, S.; Bastienne, L. J.; Hurbungs, M.; Mougal, J.; Ragoonaden, R.; Soogun, N.; Vel, T. Mangroves of small islands. In: Bosire J. O., Mangora, M. M., Bandeira S., Rajkaran A., Ratsimbazafy R., Appadoo., Kairo J. G. (Eds.), *Mangroves of the Western Indian Ocean: Status and Management*. WIOMSA: Zanzibar Town, **2016,** p. 161.

ASCLME. *National marine ecosystem diagnostic analysis, Tanzania*. Contribution to the Agulhas and Somali Current Large Marine Ecosystems (ASCLME) Project (supported by UNDP with GEF grant financing). **2012,** p. 92.

Bosire, J. O.; Dahdouh-Guebas, F.; Kairo, J. G.; Cannicci, S.; Koedam, N. Spatial variations in macrobenthic fauna recolonisation in a tropical mangrove bay. *Biodiv. Conserv.* **2004,** *13*, 1059–1074.

Bosire, J. O.; Mangora, M. M.; Bandeira S.; Rajkaran A.; Ratsimbazafy R.; Appadoo; Kairo J. G. *Mangroves of the Western Indian Ocean: Status and management*. WIOMSA: Zanzibar Town, **2016a**, p. 161.

Bosire, J. O.; Sigi Lang'ati, J. K.; Kirui, B.; Kairo, J. G.; Mungi, L. M.; Hamza, A. J. Mangrove of Kenya. In: Bosire J. O., Mangora, M. M., Bandeira S., Rajkaran A., Ratsimbazafy R., Appadoo., Kairo J. G. (Eds.), *Mangroves of the Western Indian Ocean: Status and management.* WIOMSA: Zanzibar Town, **2016b,** p 161.

Donato, D.C.; Kauffman, J. B.; Murdiyarso, D.; Kurnianto, S.; Stidham, M.; Kanninen, M. Mangroves among the most carbon-rich forests in the tropics. *Nat. Geosci.* **2011,** *4* (5), 293–297. https://doi.org/10.1038/ngeo1123.

Duke, N. C. Phenological trends with latitude in the mangrove tree *Avicennia marina*. *J. Ecol.* **1990,** *78* (1), 113. https://doi.org/10.2307/2261040.

Faridah-Hanum, I.; Kudus, K. A.; Saari, N. S. Plant diversity and biomass of Marudu Bay mangroves in Malaysia. *Pak. J. Bot.* **2012,** *44* (2): 151–156.

GOK. *National Mangrove Ecosystem Management Plan (2017–2027)*. Kenya Forest Service: Nairobi, Kenya. **2017**.

Githaiga, N. M. *Structure and Biomass Accumulation of Natural Mangrove Forest at Gazi Bay, Kenya.* Department of Plant and Microbial Sciences, Kenyatta University: Nairobi, **2013,** p. 69

Hogarth, P. J. *The Biology of Mangroves and Seagrasses*. Oxford University Press, **2015.**

Japhet. *The Status of Ecological Resources in the Rufiji Delta, Tanzani*a. A report submitted to Wetlands International Eastern Africa. **2018**.

Japhet, E. Succession and productivity of mangroves. In: *The Rufiji Delta, Tanzania*. PhD thesis, University of Dar es Salaam, Tanzania, **2021.**

Kathiresan, K.; Bingham, B. L. Biology of mangroves and mangrove ecosystems. *Adv. Marin. Biol.* **2001,** *40*, 81–251.

Kathiresan, K.; Rajendran, N. Coastal mangrove forests mitigated tsunami. *Estuar. Coast. Shelf Sci.* **2005,** *65*(3), 601–606. https://doi.org/10.1016/j.ecss.2005.06.022.

Lang'at, J. K. S. *Impacts of Tree Harvesting on the Carbon Balance and Functioning in Mangrove Forests*. Doctoral dissertation. **2013**. Edimburgh Napier University, UK.

Lupembe, I. B. *Carbon Stocks in the Mangrove Ecosystem of Rufiji*. Sokoine University of Agriculture: Morogoro, Tanzania, **2014,** p. 59.

Macamo, C.; Blandeira, S.; Muando, S.; Abreu, D.; Mabilana, H. Mangrove of Mozambique. In: Bosire J. O.; Mangora, M. M.; Bandeira S.; Rajkaran, A.; Ratsimbazafy, R.; Appadoo; Kairo J. G. (Eds.), *Mangroves of the Western Indian Ocean: Status and Management.* WIOMSA: Zanzibar Town, **2016,** p. 161.

Mangora, M. M. Poverty and institutional management stand-off: a restoration and conservation dilemma for mangrove forests of Tanzania. *Wetl. Ecol. Manag.* **2011,** *19*, 533–543. https://doi.org/10.1007/s11273-011-9234-2.

Mangora, M. M.; Lugendo, B. R.; Shali, M. S.; Semesi, S. Mangrove of Tanzania. In: Bosire J. O.; Mangora, M. M.; Bandeira, S.; Rajkaran, A.; Ratsimbazafy, R.; Appadoo; Kairo J. G. (Eds.), *Mangroves of the Western Indian Ocean: Status and management*. WIOMSA: Zanzibar Town, **2016a,** p. 161.

Mangora M. M.; Trettin C. C.; Shalli M. S. *Mangrove Research and Training Forest, Rufiji Delta – Tanzania. Project Progress Report*. Institute of Marine Sciences, US Forest Service, Tanzania Forest Services Agency and USAID. **2016b,** p. 35.

Mchenga, I. S. S.; Ali, A. I. Mangrove litter production and seasonality of dominant species in Zanzibar, Tanzania. *J. East Afr. Nat. Hist.* **2017,** *106* (1): 5–18. https://doi.org/10.2982/028.106.0103.

Monga, E.; Mangora, M. M.; Mayunga, J. S. Mangrove cover change detection in the Rufiji Delta in Tanzania. *WIO J. Mar. Sci.* **2018,** *17* (2): 1. https://doi.org/10.4314/wiojms.v17i2.1.

Myers, N.; Mittermeier, R. A.; Mittermeier, C. G.; Da Fonseca, G. A.; Kent, J. Biodiversity hotspots for conservation priorities. *Nature* **2000,** *403*(6772), 853–858.

Nagelkerken, I.; Blaber, S. J. M.; Bouillon, S.; Green, P.; Haywood, M.; Kirton, L. G.; Meynecke, J. O.; Pawlik, J.; Penrose, H. M.; Sasekumar, A. The habitat function of mangroves for terrestrial and marine fauna: A review. *Aquat. Bot.* **2008,** *89,* 155–185. https://doi.org/10.1016/j.aquabot.2007.12.007.

Njana, M. A.; Eid, T.; Zahabu, E.; Malimbwi, R. Procedures for quantification of belowground biomass of three mangrove tree species. *Wetl. Ecol. Manag.* **2015,** *23* (4), 749–764. https://doi.org/10.1007/s11273-015-9417-3.

Okello, J. A. *The Role of Disturbances in Mangrove Wood Formation and Forest Structure Effect of Large Sedimentation Events*. Vrije Universiteit Brussel and Université Libre de Bruxelles. **2016**.

RFM (Republic of Mozambique). *National Strategy and Action Plan for Mangroves Management (2020–2024)*. **2020**.

R. S. (Republic of Seychelles). Environment Protection Act of **1994**.

RSA (Republic of South Africa) National Forests Act, **1998**.

Rajkaran, A.; Adams, J. Mangrove of South Africa. In: Bosire, J. O.; Mangora, M. M.; Bandeira, S.; Rajkaran A.; Ratsimbazafy, R.; Appadoo; Kairo, J. G. (Eds.), *Mangroves of the Western Indian Ocean: Status and Management.* WIOMSA: Zanzibar Town, **2016,** p. 161.

Ratsimbazafy, R.; Randriaamanantena, D.; Rakotondrazafy, J.; Rakotomalala, H.; Ramahery, V.; Edmond, R.; Razakaniria, G.; Rabarison, H.; Lavitra, T.; Mahafiana, J.; Ravaoarinorotsihoarana, L.; Cripps, G.; England, K.; Carro, A.; Jones, T. G.; Glass, L.; Taylor, B.; Danhaive. Mangrove of Madagascar. In: Bosire, J. O.; Mangora, M. M.; Bandeira, S.; Rajkaran, A.; Ratsimbazafy, R.; Appadoo; Kairo, J. G. (Eds.), *Mangroves of the Western Indian Ocean: Status and managemen*t. WIOMSA: Zanzibar Town, **2016,** p. 161.

Reid, W. V. Biodiversity hotspots. *Trends Ecol. Evol.* **1998,** *13*(7), 275–280.

Sanderman, J.; Hengl, T.; Fiske, G.; Solvik, K.; Fernanda, M.; Benson, L.; Bukoski, J. J.; Carnell, P.; Cifuentes-Jara, M.; Donato, D.; Duncan, C.; Eid, M. E.; Ermgassen, P.; Lewis, E. C.; Macreadie, P. I.; Glass, L.; Gress, S.; Jardine, L. S.; Jones, G. T.; Nsombo, N. E.; Rahman, M.; Sanders, J. C.; Spalding, M.; Landis, E. A global map of mangrove forest soil carbon at 30 m spatial resolution. *Environ. Res. Lett.* **2018,** *13* (5) 055002. https://doi.org/10.1088/1748-9326/aabe1c.

Shan, L.; Renchao, Z.; SuiSui, D.; Suhua, S. Adaptation to salinity in mangroves: Implication on the evolution of salt-tolerance. *Chinese Sci. Bull.* **2008,** *53,* 1708–1715.

UNEP-Nairobi Convention/USAID/WIOMSA. *Guidelines on Mangrove Ecosystem Restoration for the Western Indian Ocean Region*. UNEP, Nairobi, **2020,** p. 71.

URT (United Republic of Tanzania). National Mangrove Management Plan in **1991**.

Wagner, G. M.; Sallema-Mtui, R. *The Rufiji Estuary: Climate Change, Anthropogenic Pressures, Vulnerability Assessment and Adaptive Management Strategies*. **2016,** pp. 183–207. https://doi.org/10.1007/978-3-319-25370-1_12.

CHAPTER 4

PTERIDOPHYTES RICHNESS ALONG THE COASTAL FORESTS OF EASTERN AFRICA

IMMACULATE K. CONSTANTINE[1*], ALLY J. K. NKWABI[2], SIRI A. ABIHUDI[3], CHARLES MSAGWA[1], and NEDUVOTO P. MOLLEL[1]

[1]*National Herbarium of Tanzania, Tanzania Plant Health and Pesticides Authority (TPHPA), Arusha, Tanzania*

[2]*Tanzania Wildlife Research Institute, Arusha, Tanzania*

[3]*Institute of Traditional Medicine, Muhimbili University of Health and Allied Sciences, Dar es Salaam, Tanzania*

*Corresponding author

ABSTRACT

Pteridophytes are one of the very first plants to adapt to life on land, before the flowering plants. Globally, there are about 12,000 species of pteridophytes: 97% represent ferns, while 3% are fern allies. The East African Plant database and the National Herbarium of Tanzania specimens' data were used to filter Coastal Eastern Africa forest species of pteridophytes that are analyzed in this chapter. Two hundred sixty-nine (269) species of pteridophytes of the Polypodiopsida class in 10 orders, namely Cyatheales, Gleicheniales, Hymenophyllales, Isoetales, Ophioglossales, Osmundales, Polypodiales, Psilotales, Salviniales, and Schizaeales and 14 families were studied for their distribution and conservation status. Most of these were found to be of least concern in the IUCN categories of threats due to their wider distribution.

4.1 INTRODUCTION

4.1.1 BRIEF HISTORY OF PTERIDOPHYTES DOCUMENTATION IN THE COAST FOREST

Pteridophytes in coastal forests of Eastern Africa have been a subject of interest for botanists and ecologists for many years. These ancient plants are one of the older groups of plants present in the plant kingdom, commonly called true ferns and fern allies. Ferns and

Biodiversity Hotspot of the Coastal Forests of Eastern Africa. T. Pullaiah and Ally J. K. Nkwabi (Eds.)

DOI: 10.1201/9781998511099-4

their allies are found everywhere, and they are some of the most prevalent plants in some ecosystems. Additionally, many ferns are grown as ornamentals because of their beautiful foliage, as a result adding to the number of species for pteridophytes (Oloyede, 2012).

The coastal forests of Eastern Africa encompass a diverse range of vegetation types, shaped by the region's unique climatic and ecological factors. While there are variations within this region, one of prominent vegetation types includes lowland coastal forests. These forests are found in the low-lying areas near the coast and are characterized by a mix of evergreen and semideciduous tree species. They thrive in the humid tropical climate and include a high number of endemic species (Burgess et al., 1998). Other vegetation is dry coastal forests, mangrove forests, coral rag forests, and scrub and grasslands. Pteridophytes distribution along coastal forest of Eastern Africa was observed to be influenced by ecological factors, including climate, moisture availability, temperature, and soil conditions. It is important to note that within each of these broad categories, there is a high degree of regional variation based on specific ecological niches and local environmental conditions.

The documentation of pteridophytes in coastal forests can be traced back to the early botanical explorations in the 18th and 19th centuries. Early botanists like Carl Linnaeus and Carl Thunberg made observations and collections of ferns during their expeditions (Jeremy and Schelpe, 1982; Hemp and Crouch, 2018; Watuma et al., 2022; Du et al., 2023). Botanical journals and publications of the 19th century often featured accounts of pteridophytes found in coastal regions. One notable example is the "American Fern Journal," which began publishing in the late 1800s and included numerous reports on coastal ferns. As botanical knowledge advanced, regional surveys and floras started including comprehensive lists of pteridophytes in coastal ecosystems. For example, the "Flora of Tropical East Africa" and "Flora Zambesiaca" began to provide detailed information on fern species in coastal regions. In the mid to late 20th century, ecological studies started focusing on the role of pteridophytes in coastal ecosystems. These studies explored their ecological significance, such as their role in stabilizing coastal dunes and providing habitat for wildlife (Lovett and Thomas, 1986; Jacobsen and Jacobsen, 1989; Kornas, 1993; Moran, 1995; Roux, 2001, 2004, 2009, 2011, 2012; Aldasoro et al., 2004; Oloyede, 2012; Liu et al., 2016).

Taxonomic advancements in pteridology, including the use of molecular techniques, have led to the identification of new species and the reclassification of existing ones. These findings have contributed to a more accurate understanding of the diversity of pteridophytes in coastal areas of East Africa. In accounting for various reports of new species or occurrences of pteridophytes species published, scientific explorations have yielded a number of published taxonomics, which are new species descriptions in science or new records (Anthony and Schelpe, 1982; Fischer et al., 2010; Roux, 2011, 2012; Hemp, 2012; Dubuisson et al., 2013, 2016; Schwartsburd and Prado, 2014). *Tectaria puberula* was described by Luke (2005), the species was observed to be restricted to Kenyan coastal forest of Shimba Hill and Tanzanian mountain forests (including Zanzibar and Pemba) in coastal forest of Tanzania (White, 1983; Luke, 2005). In other part of Tanzania, the species was also observed and recorded along the Eastern Province of Morogoro District in Uluguru Mountains, Mwere Valley, the wet evergreen forest along the stream, and Hemp and Crouch (2018) also reported *Asplenium arcumontanum* (Aspleniaceae) to be a new species from the Eastern Arc Mountains of Tanzania. *Asplenium* were encountered on the floor of montane forests of the North and South Pare and West Usambara Mts. Due to few studies of this division these species might occur in Eastern African coastal forests and therefore we encourage more studies to be conducted in the coastal forest of Eastern Africa. However, a new record within the coastal forest of East Africa and Africa in general shows that these forests are a biodiversity reservoir and hotspot with more potential for scientific exploration.

4.1.2 PTERIDOPHYTES OF THE COASTAL FOREST OF EASTERN AFRICA

This chapter examined the distribution patterns of pteridophytes of the coastal forest of Eastern Africa. Coastal forests often support a high diversity of Pteridophyte species due to the relatively stable and humid conditions found in these ecosystems. The combination of moist air from the sea and abundant rainfall creates suitable habitats for ferns and other fern allies. Here, we found 269 species belonging to one class, Polypodiopsida (Table 4.1), 10 orders, 25 families, and 85 genera. According to the updated catalogue of life (https://www.catalogueoflife.org/ 2022), pteridophytes fall under one class, Polypodiopsida. For the species of the East African Coast, the class Polypodiopsida had 10 orders. These are Cyatheales, Gleicheniales, Hymenophyllales, Isoetales, Ophioglossales, Osmundales, Polypodiales, Psilotales, Salviniales and Schizaeales. The order Polypodiales ranked highest with 14 families, 66 genera, and 206 species. Other orders had one to two families and relatively fewer species (Table 4.1).

TABLE 4.1 Classification Diversity of Pteridophytes of the Coastal Forest of Eastern Africa

Class	Orders	No. of families	No. of Genera	No. of Species
Polypodiopsida	Osmundales	1	1	1
Polypodiopsida	Psilotales	1	1	1
Polypodiopsida	Gleicheniales	1	2	2
Polypodiopsida	Isoetales	1	1	3
Polypodiopsida	Schizaeales	2	3	4
Polypodiopsida	Salviniales	2	2	7
Polypodiopsida	Ophioglossales	1	1	8
Polypodiopsida	Cyatheales	1	2	9
Polypodiopsida	Hymenophyllales	1	6	28
Polypodiopsida	Polypodiales	14	66	206
	Total	**25**	**85**	**269**

4.1.3 DISTRIBUTION OF PTERIDOPHYTES

The distribution of pteridophytes can vary depending on the specific vegetation types and environmental conditions. The poor species number and diversity in Africa have been reported by Kornas (1993) and explained to be caused by the low amount of the rain forest on that continent. This low amount of rain resulted in many important families and genera of rain forest pteridophytes to poorly germinate in tropical forests in Africa (Kornas, 1993). There are about 12,000 species of pteridophytes globally, of which about 97% are ferns and 3% the fern allies or Lycophytes. Africa is poorly represented in pteridophytes as compared to other areas around the world, about 600 species of pteridophytes are found in Tropical areas. Other authors have documented the absence of extensive mountain ranges in Africa as the main reason for low fern diversity (Kornas, 1993; Moran, 1995; Aldasoro et al., 2004; Liu et al., 2016), for example, in tropical America there is a higher number of the filmy ferns Hymenophyllaceae, or the Cyatheaceous tree ferns having 197 species, in Malaysia, including Malay Peninsula there are 186 species, in Africa only 15 species of Hymenophyllaceae have been recorded and identified (Kornas, 1993).

In coastal forests of Eastern Africa, the distribution of pteridophytes can vary greatly depending on the specific geography and microclimates of the region. Pteridophytes are less common in mangrove forests due to the saline and waterlogged conditions, which are not favorable for many fern species. However, some salt-tolerant ferns, like *Acrostichum aureum* (golden leather fern), may be found on the fringes of mangrove forests of Eastern Africa. In regions with tropical rainforests along the coast, pteridophytes can be abundant. They thrive in the moist, shaded understory and may form a significant component of the ground vegetation. Pteridophytes may be sparser in coastal scrub habitats, which often experience harsh, sandy, and windy conditions. Species like *Pteridium aquilinum* (bracken fern) can tolerate these conditions and may be present. Pteridophytes are abundant in the understory of lowland coastal forests, where they thrive in the shade of towering trees. Ferns, in particular, are commonly found in these forests, including species like the tree ferns (*Cyathea* spp.) and epiphytic ferns such as *Asplenium* and *Platycerium*. Along freshwater streams and rivers within coastal forests, you can find a variety of fern species that thrive in the consistently moist conditions, including *Osmunda regalis* (royal fern) and *Thelypteris palustris* (marsh fern). Another example, *Nephrolepis biserrata* (giant sword fern), is a very common lowland fern occurring in wet boggy ground, grassland, plantations, secondary forest, or on forest-grassland margins. Very few pteridophytes are able to grow in the zonal vegetation in miombo woodland because of the competition of tall grasses and forbs (Kornas, 1985). Additionally, disturbances such as fires, hurricanes, and human activities can also impact the composition and distribution of pteridophytes within these ecosystems.

4.1.4 PTERIDOPHYTES ALONG ALTITUDINAL GRADIENTS

Altitudinal gradients refer to the change in environmental conditions (such as temperature, precipitation, and vegetation) as one move from low elevations near the coast to higher elevations further inland. Coastal forests in Eastern Africa often exhibit altitudinal gradients as they extend from sea level up into the interior regions. Pteridophytes are generally distributed along an altitudinal gradient, with the highest diversity in the tropics along mountainous areas in the climatic climax of rain forests of coastal areas in East Africa (Moran, 1995; Mehltreter, 2022).

4.2 PTERIDOPHYTES OF THE COASTAL FORESTS OF EASTERN AFRICA

The degree to which Pteridophyte species were found in coastal forests of Eastern Africa in each region (Somalia, Kenya, Tanzania, Mozambique, and Tanzania) was observed using the available data from the Global Biodiversity Information Facility (GBIF). Tanzania has the highest number of fern flora when compared to other countries; we obtained overall 396 species (Roux, 2001; Liu et al., 2016), and out of this total only 268 species were found in the coastal forests of Tanzania. Mozambique had an overall 301 species, and when we consider coastal forest of Mozambique, the area accommodates 125 species. Kenya had overall 278 species with 163 in coast forest and least in Somalia with overall 38 species (Roux, 2001; Liu et al., 2016), with 18 species in the coastal forest as presented in Figure 4.1. The few species presented in coast of Somalia can potentially be attributed to the prevailing political instability within the country, which poses significant obstacles to conducting comprehensive research in the area, furthermore, it may be contributed to the coastline of Somalia, which is semiarid to arid climatic conditions, hence receiving a low amount of rainfall, whereby the northern portion is the driest, with an average mean annual rainfall of 250 mm, whereas the southern portion receives a mean annual rainfall of about 400 mm. For that reason, wet-tropical, mountain habitats are lacking in Somalia, which may hold the highest number of species of pteridophytes. The higher species number of pteridophytes in Tanzania is probably due to the presence of moist mountain forests forming part of the Afromontane Region. These regions include the coastal forests, also termed as the Eastern Arc Mountain Forests, and the volcanic mountain forests.

4.2.1 *DISTRIBUTION OF PTERIDOPHYTES IN THE REGION*

When we consider pteridophytes distribution in coastal areas of East Africa, the species are highly diversified in the eastern mountain arc of the coastal forest of Eastern Africa and somewhat less common in the western mountains of Africa (Roux, 2004; Mehltreter, 2022). The coastal forests of Tanzania and Kenya are renowned as an area of global importance and have been remarked as one of the 36 world's hotspots of biodiversity due to the higher concentration of many narrowly endemic flora species, including pteridophytes (Myers et al., 2000).

The pteridophytes diversity across each region in coastal forests of Eastern Africa has been observed to be richer in moist areas than dry areas. Using our data, we found 47% in Tanzania, 28% in Kenya, 22% in Mozambique, and only 3% in Somalia of pteridophytes species along the coastal forests of Eastern Africa. For instance, *Christella dentata* and *Nephrolepsis undulata* are found along the coastal forest and extend to the southern highlands of Tanzania in Kitulo National Park and further to northern Mozambique

The data for the map came from the GBIF, which may be found online at https://www.gbif.org/. Pteridaceae (26) had the highest number of species recorded within the Coastal Forest of Eastern Africa (CFEA) ecosystem (Figure 4.2), followed by Aspleniaceae which had 16 species (see Figure 4.1).

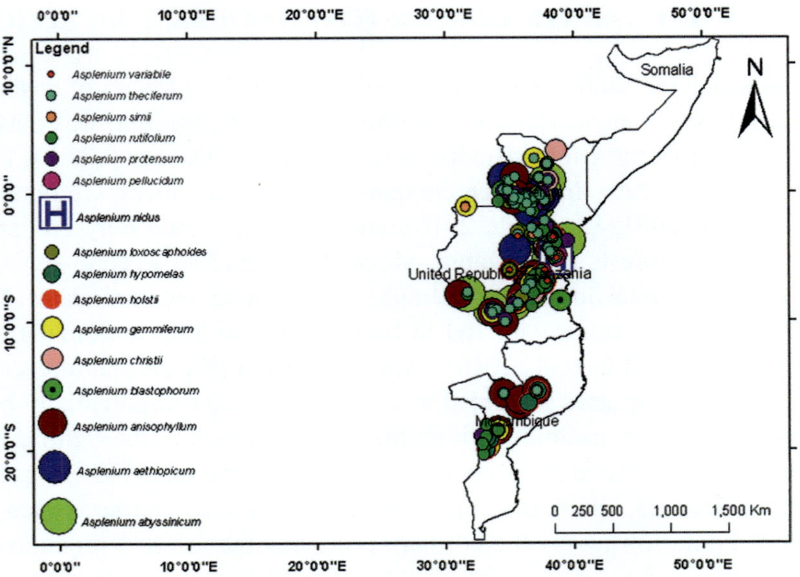

FIGURE 4.1 Combined Map of Somalia, Kenya, Tanzania, and Mozambique presenting the distribution of the family Aspleniaceae, in the coastal forest of East Africa.

Source: Map by Dr. Siri A. Abihudi, November 2023, GIS Office; Muhimbili University of Health and Allied Sciences, Dar es Salaam.

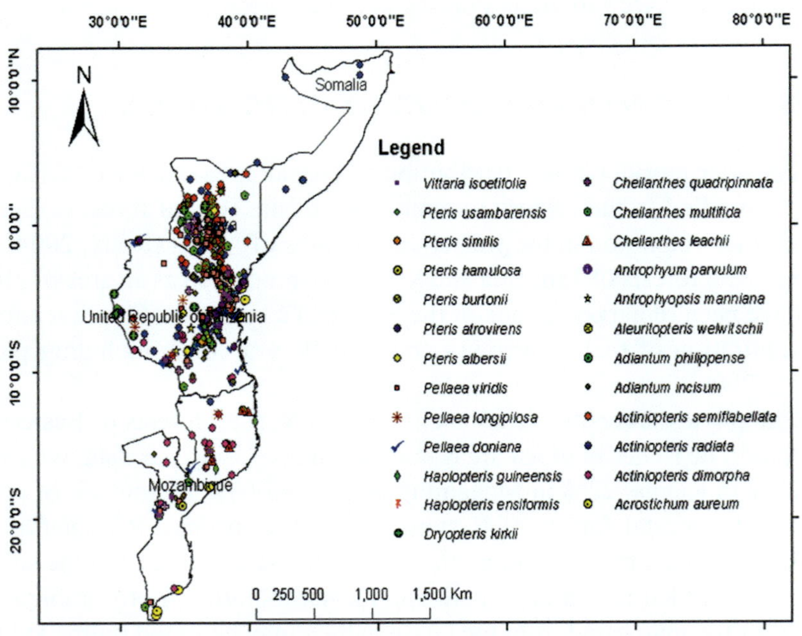

FIGURE 4.2 Map of East African country (Somalia, Kenya, Tanzania, and Mozambique) showing the distribution of members of the family Pteridaceae in the coastal forest of East Africa.

Source: Map by Dr. Siri A. Abihudi, November 2023, GIS Office; Muhimbili University of Health and Allied Sciences, Dar es Salaam.

Other families that were observed to be found along the CFEA included Polypodiaceae with six species (Figure 4.3), Thelypteridaceae with nine species (Figure 4.4), Hymenophyllaceae (7 spp) (Figure 4.5), Athyriaceae (2 spp), Ophioglossaceae (2 spp), as well as one species each from Azollaceae, Cyatheaceae, Oleandraceae, and Marsileaceae (Figure 4.6). It has been observed that Tanzania has the greatest abundance of pteridophytes (61 spp) along the CFEA, followed by Mozambique (17 spp) and Kenya (15 spp), with no species having been recorded in Somalia. This lack of observation of pteridophytes along the coast of Somalia can potentially be attributed to a lack of information recorded by GBIF, which might have contributed to the prevalent instability and lack of peace within the country, which poses significant obstacles to conducting comprehensive research in the area.

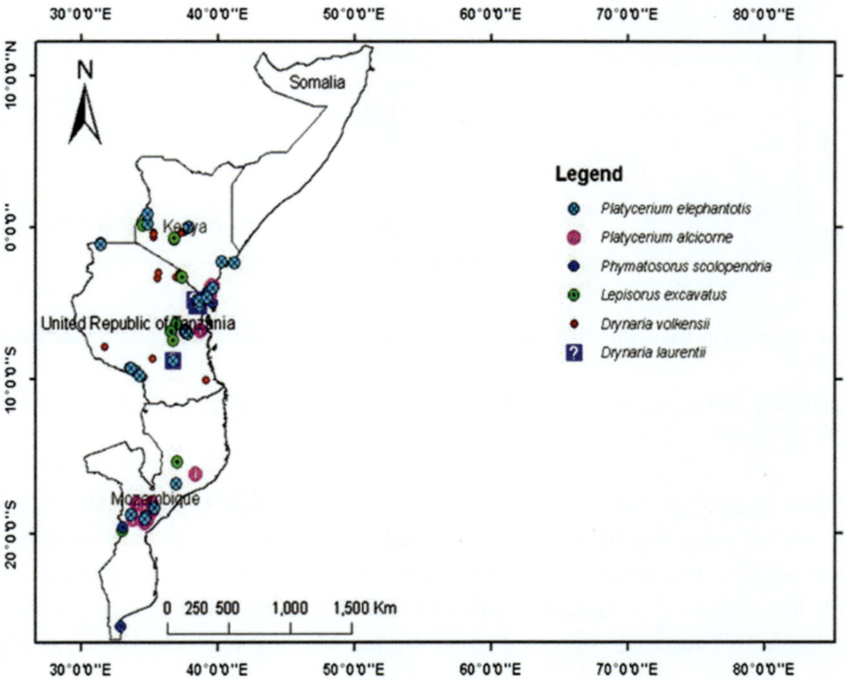

FIGURE 4.3 Map distribution of the family the Polypodiaceae in Somalia, Kenya, Tanzania, and Mozambique coastal forest.

Source: Map by Dr. Siri A. Abihudi, November 2023, GIS Office; Muhimbili University of Health and Allied Sciences, Dar es Salaam.

4.2.2 SPECIES ENDEMISM IN THE EASTERN AFRICA

Coastal forests of Eastern Africa harbor a remarkable diversity of flora and fauna species, many of which are endemic (Burgess et al., 1998). This level of endemism is often attributed to unique environmental conditions, historical factors, and long-term isolation from other ecosystems (Lovett and Wasser, 1993). However, endemicity in pteridophytes is found to

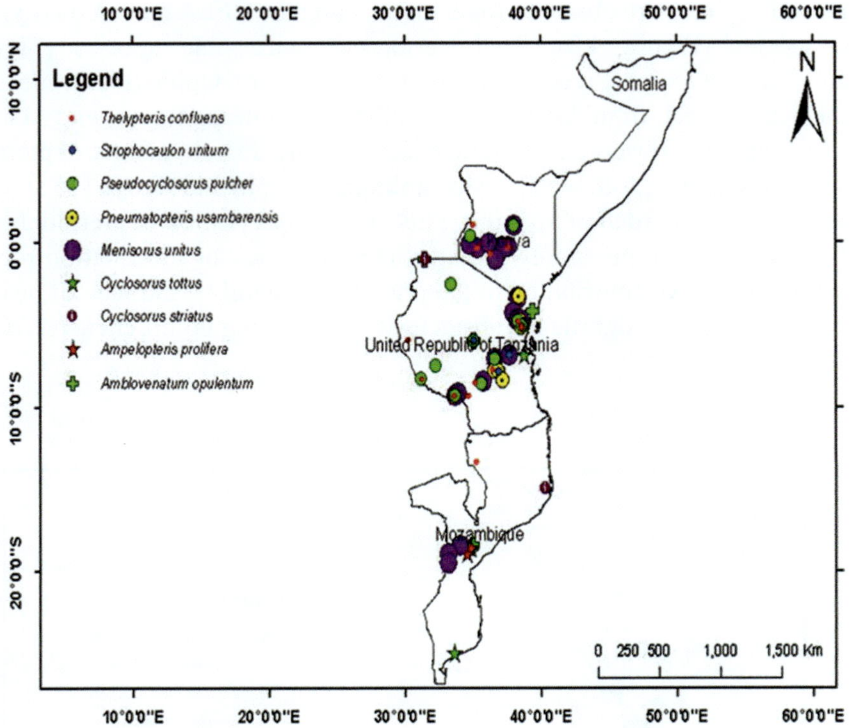

FIGURE 4.4 Species distribution of the family Thelypteridaceae in Somalia, Kenya, Tanzania, and Mozambique coastal forest.

Source: Map by Dr. Siri A. Abihudi, November 2023, GIS Office; Muhimbili University of Health and Allied Sciences, Dar es Salaam.

be lower than in seed plants (Aldasoro et al., 2004). Out of 269 species studied, 18 species of pteridophytes were found to have restricted distribution in the coastal region. Out of which four (4) species are endemic to the coastal forests of Tanzania and Kenya. Seven (7) species are endemic to Tanzania alone (Table 4.2). The other seven (7) species out of 18 are endemic to Tanzania, Kenya, Mozambique, Madagascar, Comoros, and Seychelles (Table 4.2).

4.3 THREATS TO THE PTERIDOPHYTES—PRESENT AND FUTURE SITUATION

The evolutionary adaptations collectively enable pteridophytes to occupy specific ecological niches within the coastal forests of Eastern Africa. They have evolved to thrive in the challenging and ever-changing conditions of this region, contributing to the overall biodiversity and ecosystem stability of these unique coastal ecosystems. While pteridophytes have adapted to various environments over millions of years, typically they reproduce via spores that require water for dispersal and fertilization. Coastal environments in Eastern Africa can have variable moisture levels, including periods of drought. This reliance on water for reproduction can be a limitation when water availability is inconsistent.

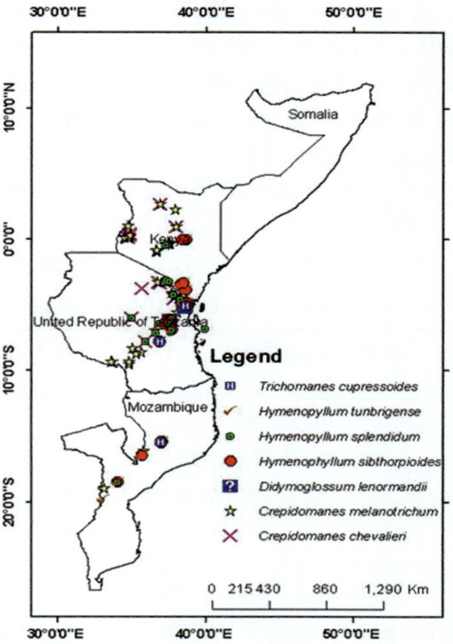

FIGURE 4.5 Species distribution of the family Hymenophyllaceae in Somalia, Kenya, Tanzania, and Mozambique coastal forest.

Source: Map by Dr. Siri A. Abihudi, November 2023, GIS Office; Muhimbili University of Health and Allied Sciences, Dar es Salaam.

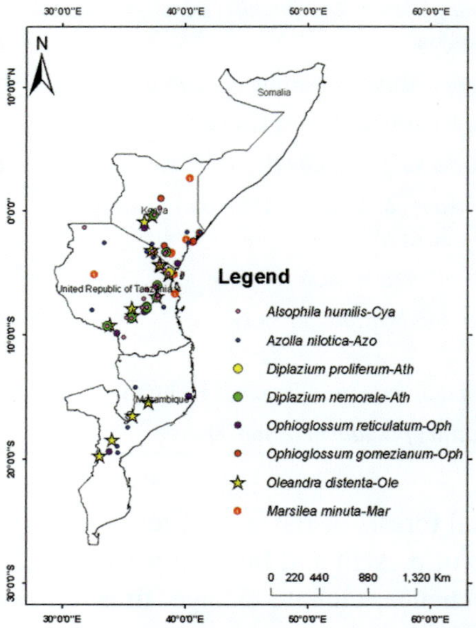

FIGURE 4.6 Species distribution of the families; Ath = Athyriaceae, Oph = Ophioglossaceae, Azo = Azollaceae, Cya = Cyatheaceae, Ole = Oleandraceae, and Mar = Marsileaceae in Somalia, Kenya, Tanzania, and Mozambique coastal forest.

Source: Map by Dr. Siri A. Abihudi, November 2023, GIS Office; Muhimbili University of Health and Allied Sciences, Dar es Salaam).

Pteridophytes generally lack some of the adaptive traits found in angiosperms, such as the ability to produce seeds with protective coverings, more efficient vascular systems, and specialized mechanisms for nutrient uptake and resource allocation. While some pteridophytes can disperse spores over short distances, they lack the advanced dispersal mechanisms seen in some angiosperms, like fruits, and specialized adaptations for animal dispersal. This limitation can hinder their ability to colonize new areas within the coastal forest environment.

TABLE 4.2 Endemic Species of Pteridophytes to the East African Coastal Forest.

Sr. No.	Family	Species name and authors	Distribution
1	Aspleniaceae	*Asplenium albersii* Hieron.	Kenya and Tanzania
2	Aspleniaceae	*Asplenium praegracile* Rosenst.	Tanzania, Madagasar
3	Cyatheaceae	*Gymnosphaera schliebenii* (Reimers) S. Y. Dong	Tanzania (Uluguru Mountains, Ukaguru Mountains)
4	Cyatheaceae	*Alsophila mossambicensis* (Baker) R. M. Tryon	Kenya, Mozambique, Tanzania, Zimbabwe
5	Cyatheaceae	*Alsophila tanzaniana* R. M. Tryon	Tanzania
6	Dennstaedtiaceae	*Blotiella stipitata* (Alston) Faden	Kenya and Tanzania
7	Dennstaedtiaceae	*Microlepia fadenii* Pic.Serm.	Tanzania
8	Polypodiaceae	*Ceradenia sechellarum* (Baker) Parris	Liberia, Seychelles, Tanzania
9	Polypodiaceae	*Stenogrammitis strangeana* (Pic. Serm.) Labiak	Kenya and Tanzania
10	Hymenophyllaceae	*Didymoglossum lenormandii* (Bosch) Ebihara & Dubuisson	Comoros, Madagascar, Tanzania
11	Dryopteridaceae	*Elaphoglossum phanerophlebium* C.Chr.	Madagascar, Tanzania
12	Marsileaceae	*Marsilea megalomanica* Launert	Tanzania
13	Ophioglossaceae	*Ophioderma pendulum* (L.) C. Presl	Tanzania
14	Polypodiaceae	*Hovenkampia liebuschii* (Hieron.) Li Bing Zhang & X. M. Zhou	Tanzania
15	Pteridaceae	*Pteris albersii* Hieron.	Tanzania
16	Pteridaceae	*Pteris woodwardioides* Bory	Madagascar, Mauritius, Réunion, Tanzania
17	Thelypteridaceae	*Pneumatopteris usambarensis* Holttum	Kenya and Tanzania
18	Athyriaceae	*Diplazium pseudoporrectum* Hieron.	Tanzania

Furthermore, the coastal forests of Eastern Africa, including their pteridophyte populations, face a complex web of present and future threats. Coastal forests are under constant pressure due to land use change, climate change, fires, hurricanes, and human activities such as fire, grazing, agriculture, logging, infrastructure development, and urbanization (Hall et al., 2009; Tabor et al., 2016). These activities result in habitat destruction and fragmentation, which directly affect pteridophytes by reducing their available living. Moreover, the reduction in suitable habitats for pteridophytes may be linked to the impact of climate change. Pteridophytes are such important plant species both for the ecosystem and for provisioning services and are least considered by the policy makers and other stakeholders of biodiversity conservation.

4.3.1 LAND USE CHANGE

As land is cleared for agriculture, logging, and infrastructure development, the natural habitats of pteridophytes are destroyed. Coastal forests are particularly vulnerable because they often contain valuable timber species, and the removal of these trees can result in the loss of the forest's understory flora, including pteridophytes. The conversion of coastal forest land into agricultural fields can directly impact pteridophytes. Pesticides, herbicides, and fertilizers used in agriculture can negatively affect these plants. Furthermore, agricultural practices can alter soil composition and hydrology, disrupting the habitats that pteridophytes rely on. Furthermore, the expansion of urban areas can lead to the fragmentation of coastal forests. Roads, buildings, and other urban infrastructure can isolate populations of pteridophytes, making it difficult for them to disperse and maintain genetic diversity. Additionally, urban pollution and runoff can degrade the quality of water sources, which is essential for many pteridophyte species.

4.3.2 CLIMATE CHANGE (HURRICANES, RAINFALL, TEMPERATURE)

Climate change can have significant effects on pteridophytes (ferns and fern allies) in coastal forests of Eastern Africa. Pteridophytes are particularly sensitive to environmental changes, including shifts in temperature, precipitation patterns, and sea-level rise. Increasing temperatures can affect the distribution and growth of pteridophytes. Some species may struggle to adapt to higher temperatures, especially those adapted to cooler and more stable climates. Warmer temperatures can also lead to an increase in the number and intensity of wildfires, which can destroy pteridophyte habitats. Climate change can result in altered rainfall patterns, including extended dry seasons and more intense rainfall events. Pteridophytes are highly dependent on consistent moisture levels. Extended dry seasons can lead to water stress, reduced growth, and even mortality in pteridophytes that require a consistent supply of moisture. Coastal forests are particularly vulnerable to sea-level rise, which can lead to saltwater intrusion into freshwater ecosystems. Many pteridophytes are adapted to freshwater conditions and cannot tolerate high salinity. As sea levels rise, coastal habitats that support pteridophyte populations may be submerged or converted into brackish or saltwater environments, threatening their survival.

4.4 CONSERVATION STATUS

Of 269 Pteridophyte species of the East African coastal forest for this chapter, conservation assessment was done based on the locality information obtained from GBIF (https://

www.gbif.org/), Plant of the world online (Powo) (https://powo.science.kew.org), inaturalist (https://www.inaturalist.org) and Flora of the British Isles (Flickr) (https://www.flickr.com).

We used Geocat tool to conduct an assessment of pteridophytes species found in coastal forest areas. The IUCN assessments use of area of occupancy for all the species, we found two species (*Asplenium albersii*) to be critically endangered (CR) this species is an endemic to Kenya and Tanzania coastal forests (Table 4.2) and *Adiantum stolzii* (CR) is distributed in Central African Republic, Mali, Nigeria, Sudan, Tanzania, and Togo; however, its habitat is highly fragmented due to land-use change. One hundred and thirty (130) are endangered. Thirty-six (36) are near threatened, while seventy-nine (79) are vulnerable. Twenty-two species had no data (data deficiency, DD) for calculating their area of occupancy. Some of these pteridophytes are preserved in herbarium as dried plant specimens in the East African Herbarium in Kenya and the National herbarium of Tanzania. More efforts are required for live collections of these plants into gardens.

4.5 TASK AHEAD

Further field studies are needed to elucidate a comprehensive checklist of Coastal East African pteridophytes for prioritization of conservation planning. Field studies will reveal the true picture of the habitats for these species especially those that had no field data and thus recorded as DD.

KEYWORDS

- **coastal forest**
- **Eastern Africa**
- **richness**
- **pteridophytes**
- **distribution**

REFERENCES

Aldasoro, J. J.; Cabezas, F.; Aedo, C. Diversity and distribution of ferns in Sub-Saharan Africa, Madagascar and some Islands of the South Atlantic. *J. Biogeogr.* **2004**, *31*(10), 1579–1604. https://doi.org/10.1111/j.1365-2699.2004.01106.x.

Anthony, N. C.; Schelpe, E. A. C. L. E. New species and new combinations in African Pteridophyta and Orchidaceae. *Contr. Bolus Herb.* **1982,** *10*, 143–161.

Burgess, N. D.; Clarke, G. P.; Rodgers, W. A. Coastal forests of eastern Africa: Status, endemism patterns and their potential causes. *Biol. J. Linn. Soc.* **1998,** *64*(3), 337–367. https://doi.org/10.1111/j.1095-8312.1998.tb00337.x

Du, S.; Wang, M.; Wei, N.; Mwachala, G.; Hu, G.; Wu, L.; Wang, S.; Wang, Q. Contributions to the flora of tropical East Africa. *Plants* **2023,** *12*(6), 1336. https://doi.org/10.3390/plants12061336

Dubuisson, J. Y.; Hennequin, S.; Droissart, V. *Hymenophyllum senterreanum* Dubuisson & Deblauwe, sp. nov. (Hymenophyllaceae) and its relatives in western Central Africa. *Phytotaxa* **2016,** *257*(3), 287–294. https://doi.org/10.11646/phytotaxa.257.3.7

Dubuisson, J. Y.; Rouhan, G.; Grall, A.; Hennequin, S.; Senterre, B.; Pynee, K.; Ebihara, A. New insights into the systematics and evolution of the filmy fern genus *Crepidomanes* (Hymenophyllaceae) in the Mascarene Archipelago with a focus on dwarf species. *Acta Botanica Gallica* **2013**, *160*(2), 173–194. https://doi.org/10.1080/12538078.2013.819294

Fischer, E.; Rembold, K.; Althof, A.; Obholzer, J.; Malombe, I.; Mwachala, G.; Onyano, J. C.; Dumbo, B.; Theisen, I. Annotated checklist of the vascular plants of Kakamega forest, western province, Kenya. *J. East African Nat. Hist.* **2010,** *99*, 129–226.

Hall, J.; Burgess, N. D.; Lovett, J.; Mbilinyi, B.; Gereau, R. E. Conservation implications of deforestation across an elevational gradient in the Eastern Arc Mountains, Tanzania. *Biological Conserv.* **2009,** *142*(11), 2510–2521. https://doi.org/10.1016/j.biocon.2009.05.028

Hemp, A. *Christella parasitica*-A new fern for Kenya. *J. East African Nat. Hist.* **2012,** *101*, 251–252.

Hemp, A.; Crouch, N. R. *Asplenium arcumontanum* (Aspleniaceae), a new species from the Eastern Arc Mountains of Tanzania. *Kew Bull.* **2018,** *73*, 1–9.

Jacobsen, W.; Jacobsen, N. Comparison of the pteridophyte floras of southern and eastern Africa, with special reference to high-altitude species. *Bulletin du Jardin Botanique National de Belgique/Bulletin van de NationalePlantentuin van Belgie* **1989,** *59*(3&4), 261–317.

Jeremy, A. C.; Schelpe, E. A. C. L. E. New species and new combinations in African Pteridophyta and Orchidaceae. *Contributions from the Bolus Herbarium* **1982,** *10*, 143–161.

Kornas, J. Adaptive strategies of African pteridophytes to extreme environments. *Proc. Roy. Soc. Edinburgh, Section B: Biol. Sci.* **1985,** *86*, 391–396.

Kornas, J. The significance of historical factors and ecological preference in the distribution of African pteridophytes. *J. Biogeogr.* **1993,** *20*(3), 281–286. https://doi.org/10.2307/2845636

Liu, H. M.; Zhang, S. Z.; Wan, T.; Kamau, P. W.; Wang, Z. W.; Grall, A.; Hemp, A.; Schneider, H. Exploring the pteridophyte flora of the Eastern Afromontane biodiversity hotspot. *J. Syst. Evol.* **2016,** *54*(6), 691–705. https://doi.org/10.1111/jse.12232

Lovett, J. C.; Thomas, D. The ecology of pteridophytes in the Mwanihana Forest Reserve, Tanzania. *Fern Gazette* **1986,** *13*, 103–107.

Lovett, J. C.; Wasser, S. K. (Eds). *Biogeography and Ecologv of the Rain Forests of Eastern Arc.* Cambridge University Press: Cambridge, **1993**.

Luke, Q. Annotated checklist of the plants of the Shimba Hills, Kwale District, Kenya. *J. East Afr. Nat. Hist.* **2005,** *94*(1), 5–120.

Mehltreter, K. Diversity of ferns and lycophytes at different spatial scales, along environmental gradients, and in the anthropogenic landscape. In: Marimuthu, J.; Fernández, H.; Kumar, A.; Thangaiah, S. (Eds.), *Ferns: Biotechnology, Propagation, Medicinal Uses and Environmental Regulation*. Springer Nature Singapore: Singapore, **2022**. pp. 603–625.

Moran, R. C. *The importance of mountains to Pteridophytes, with emphasis on Neotropical montane forests.* Paper presented at the Biodiversity and conservation of Neotropical montane forests. In: *Proceedings of a Symposium*, New York Botanical Garden, 21–26 June 1993, Bronx, New York, **1995.**

Myers, N.; Mittermeier, R. A.; Mittermeier, C. G.; Da Fonseca, G. A.; Kent, J. Biodiversity hotspots for conservation priorities. *Nature* **2000,** *403*(6772), 853–858.

Oloyede, F. Survey of ornamental ferns, their morphology and uses for environmental protection, improvement and management. *Ife J. Sci.* **2012,** *14*(2), 245–252.

Roux, J. P. A revision of the fern genus *Dryopteris* (Dryopteridaceae) in sub-Saharan Africa. *Phytotaxa* **2012,** *70*, 1–118.

Roux, J. P. *Conspectus of Southern African Pteridophyta*. Project Coordinator, Southern African Botanical Diversity Network, c/o National Botanical Institute, **2001**.

Roux, J. P. Synopsis of the Lycopodiophyta and Pteridophyte of Africa, Madagascar, and neighbouring islands. *Strelitzia* **2009,** *23*, 1–296.

Roux, J. P. The fern genera *Dryopteris* and *Nothoperanmea* (Dryopteridaceae) in Madagascar and neighbouring Indian Ocean Islands. *Adansonia.* **2011,** *33*, 7–67.

Roux, J. Pteridophyta: *Dryopteris filipaleata* (Pteropsida: Dryopteridaceae). A new species from tropical East Africa. *Bothalia* **2004,** *34*(1), 27–30.

Schwartsburd, P. B.; Prado, J. Subspecies of *Hypolepis rugosula* (Dennstaedtiaceae; Pteridophyta) around the world: Morphological and biogeographic perspectives. *Acta Botanica Brasilica* **2014,** *28*, 206–226. https://doi.org/10.1590/S0102-33062014000200008

Tabor, K.; Kashaigili, J. J.; Mbilinyi, B.; Wright, M. Forest cover and change for the Eastern Arc Mountains and Coastal Forests of Tanzania and Kenya circa 2000 to circa 2010. Cambridge: Final Report. Birdlife International **2016**.

Watuma, B.; Kipkoech, S.; Melly, D.; Ngumbau, V.; Rono, P.; Mutie, F.; Mkala, E.; Nzei, J.; Mwachala, G.; Gituru, R. An annotated checklist of the vascular plants of Taita Hills, Eastern Arc Mountain. *PhytoKeys* **2022,** 191, 1–158.

White, F. *The Vegetation of Africa*. UNESCO: Paris, **1983.**

CHAPTER 5

ANGIOSPERMS DIVERSITY IN EAST AFRICA: AN INSIGHT FROM THE TANZANIA COAST

JOHN K. BUKOMBE[1*], FRANK MBAGO[2], ALLY J. K. NKWABI[1], JOSEPHAT KALUGASHA[2], PIUS KAVANA[1], RAYMOND OKICK[1], SHADIA I. KILWANILA[3], JOHN SANARE[2], and WILLIAM KINDEKETA[4]

[1]*Tanzania Wildlife Research Institute (TAWIRI), Arusha, Tanzania*

[2]*Department of Botany, University of Dar es Salaam, Dar es Salaam, Tanzania*

[3]*Department of Zoology, University of Dar es Salaam, Dar es Salaam, Tanzania*

[4]*Tanzania Forest Research Institute, Morogoro, Tanzania*

*Corresponding author

ABSTRACT

Tanzania is an East African country with major higher plant diversity. The country has a corresponding diversity of plant associations, within wide geographic distribution, diverse climatic conditions, and soil types. In this chapter, all angiosperm species occurring along the Tanzania coast were assessed based on the completed "Flora of Tropical East Africa (FTEA)," documented in 1995. Species of conservation status for this area were assessed for being within the IUCN Red List or being endemic to the Tanzania coast. In total, there are 154 families, 884 genera, and 2332 angiosperm species, dominated by the dicotyledons with 1818 (78%) species, and 714 genera. The Fabaceae with 80 genera and 244 species is the largest family, followed by the Poaceae (Gramineae) with 75 genera and 216 species, and Euphorbiaceae with 46 genera and 164 species. Twenty one percent (494 species) of angiosperm of the Tanzania coast are of conservation concern, and are widely spread along the coast of Tanzania. Of these, 59% (292 species) were in the IUCN red list whereas 352 (71%) were endemic to Tanzania. Moreover, 32% (160 species) were both threatened and endemic.

5.1 INTRODUCTION

The vegetation of Tanzania is a complex mosaic comprised of extensive savannas, ranging from almost pure grassland to wooded savannas, bushes and scrubs, thickets, and montane

Biodiversity Hotspot of the Coastal Forests of Eastern Africa. T. Pullaiah and Ally J. K. Nkwabi (Eds.)

DOI: 10.1201/9781998511099-5

grassland and forest all of which are exposed to various types of soil and vegetation degradation. The diverse vegetation types of Tanzania include numerous native plant species, which include numerous endemic species (Temu and Andrew, 2008). Many vegetation works on Tanzania plants have been focused on specific groups and purposes such as medicinal plants (Chhabra et al., 1989, 1990, 1991, 1993). Books with broader coverage of the whole East Africa have however included descriptions of plant biodiversity in Tanzania, such as "A Classification of the vegetation of East Africa" (Greenway, 1973), "Collins Guide to the wildflowers of East Africa" (Blundell, 1987), "Field Guide to Acacias of East Africa" (Dharani, 2019), "Handbook of African medicinal plants" (Zhou et al., 2017), "Medicinal Plants of East Africa—An Illustrated Guide" (Dharani and Yenesew, 2022), "Field Guide to Common Trees and Shrubs of East Africa, 2nd Edition," "Field guide to trees of southern Africa" (Van Wyk, 2013). "Flora of Tropical East Africa (FTEA)" covers plants from Tanzania, Uganda, and Kenya, which has been edited several times since 1952 and completed in 2012. To date, FTEA records ca.12,154 plant species from the three countries, in which at least 10,000 are from Tanzania. Shortcomings of FTEA on Tanzania plants are that taxa are still under-represented, and with incomplete naming, just a few to mention. Of more concern in relation to this chapter, though plant systematics has been one of the hottest research areas elsewhere, including Kenya in East Africa (Zhou et al., 2017), reviews and documentation flora have been highly slow or not implemented in Tanzania, limiting accessibility for Up-to-date information and data for conservation planning and research.

Despite the existing documentations for plants of Tanzania, coastal forest of Tanzania coast is a rich area in angiosperm biodiversity and has a significant undocumented record. Documentations of angiosperm biodiversity focusing on coastal areas will contribute to the accurate understanding of the species composition in that important area in the region, which is currently facing habitat fragmentation as well as degradation from human activities and will foster research as well as strategic conservation. We anticipate that these contributions will help conservation managers, botanists, and researchers to stay up-to-date and use the information for management purposes. Furthermore, it will provide baseline information that will help to enhance research on the unknown and undocumented species in the area. In addition, the information will provide an assemblage of solid databases for future use and studies on angiosperm diversity across Tanzania and in the East African region. Therefore, this chapter summarizes families, genera, and species of Tanzania coast plants based on FTEA, focusing on the following questions: (1) What is the total number of plant species found in Tanzania coast? (2) Which families and genera are the largest?, and (3) What plants of conservation concern is found in the Tanzania coast?

5.2 THE TANZANIA COAST

Tanzania is located within Eastern Africa, the area now comprising Kenya, Tanzania, and Uganda. The coastal area of mainland Tanzania forms an area of 143,361 km^2 within five regions in mainland Tanzania namely Tanga, Coast, Dar es Salaam, Lindi, and Mtwara (Muhando and Rumisha, 2008; Nsemwa, 2013) (Figure 5.2). We chose an altitude range between 0 and 500 m.a.s.l (Figure 5.2), based on definition of coast area in Tanzania as were provided by previous authors (Volařík, 2010; Wang et al., 2005; Yang et al., 2015) (Figure 5.3). Therefore, the Angiosperm plants we report in this chapter are those found within elevation range between 0 and 500 m along the coastal areas of Tanzania (Figure 5.1).

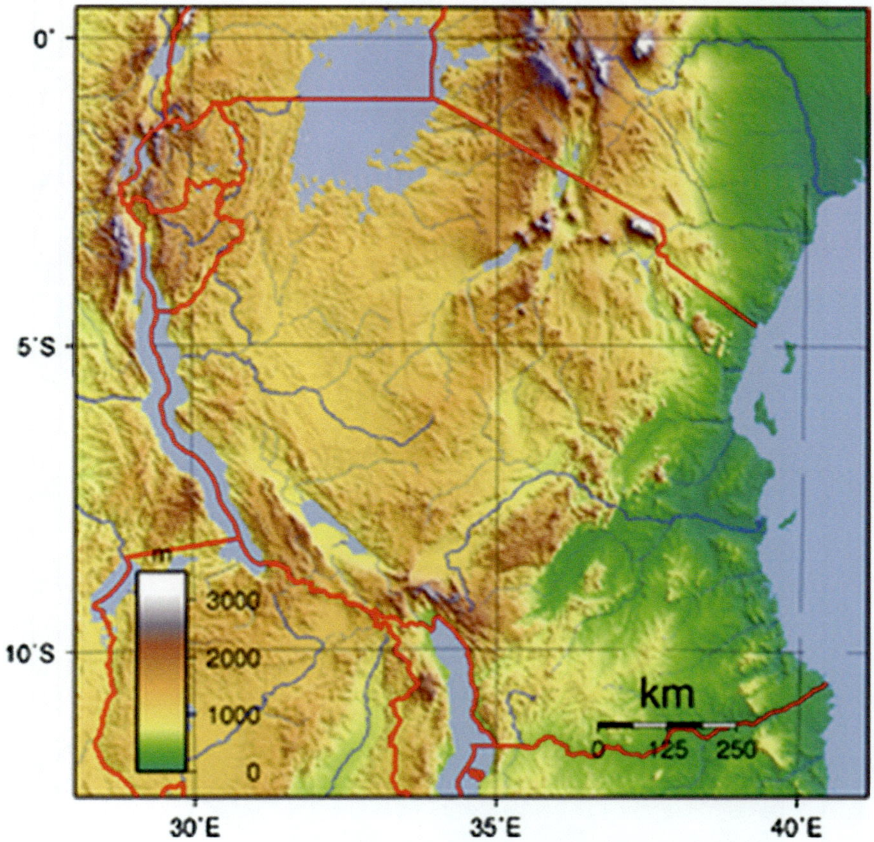

FIGURE 5.1 Topographic map of Tanzania highlighting in "green" color the coastal areas with altitude ranging between 0 and around 500 m, Map taken from other sources.

Source: Public domain

5.3 DATA SOURCES

Angiosperm species data from the LEAP of Tanzania were compiled. Species compilation considered all collection/sighting locations previously obtained and thus formed a wide angiosperm distribution along the coast area (Figure 5.2). Plant names were verified using online sources such as the Catalogue of Life, the African Plant Database, and the GBIF website (https://www.gbif.org/tools/data-validator) to identify missing or incorrect information in our dataset. We used "species matching" tool available under GBIF website (https://www.gbif.org/tools/species-lookup), the Plant List (http://www.theplantlist.org/), and World Flora Online (http://www.worldfloraonline.org/).

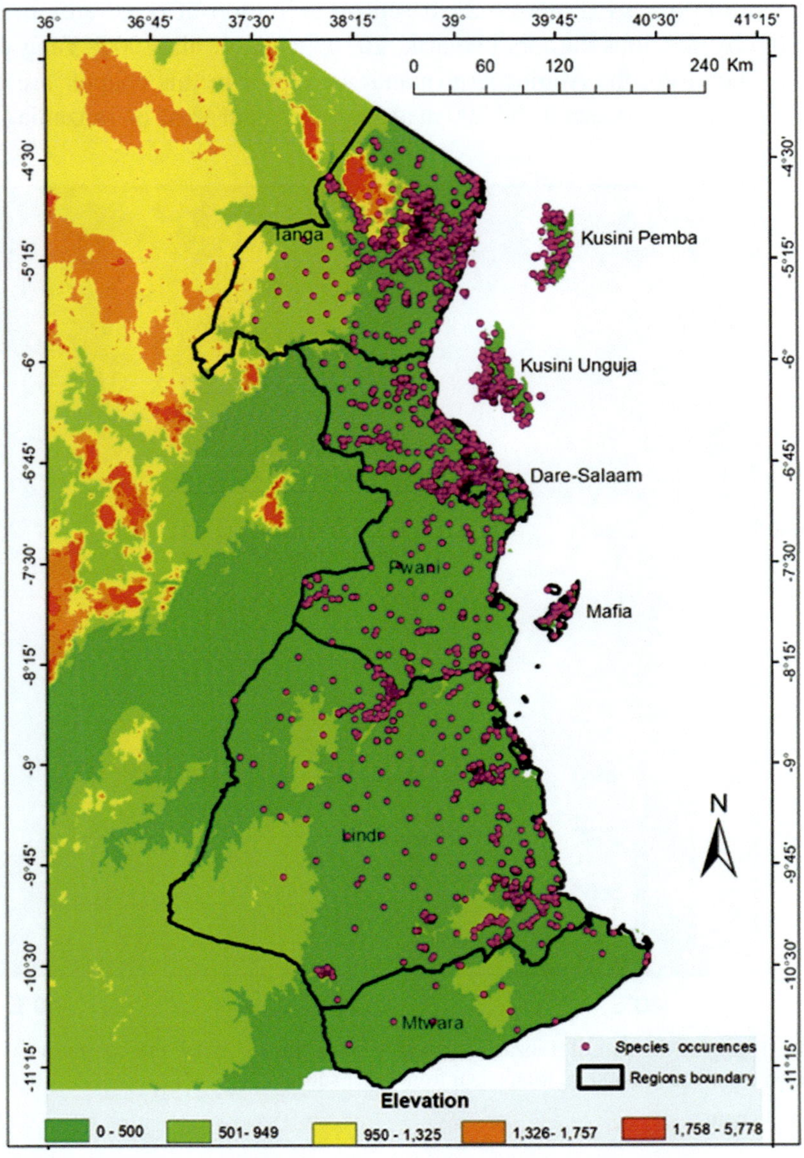

FIGURE 5.2 Map of collection effort for the Angiosperm plants of Tanzania coast.

5.4 ANGIOSPERMS COMPOSITION OF TANZANIA COAST

In total, we compiled 11,629 plant species that are known for Tanzania (list not attached). Of all these, the coastal area angiosperms comprise a total of approximately 2332 species, which are distributed across 884 Genera and 154 families. The topmost families considered had at least 50 species each. They were dominated by the family Fabaceae (224 species and 80 Genera), followed by the family Poaceae (216 species, 75 genera), Euphorbiaceae (216 species, 75 Genera), and Rubiaceae (153 species, 57 Genera) (Table 5.1). Of all the 2332 species, 1818 species were Dicotyledons, distributed across 714 genera and 120 families, and the monocotyledons were 514 species in total spread across 166 genera and 31 families.

TABLE 5.1 The Nine Top Families of Angiosperms of the Tanzania Coast

Sr. No.	Row Labels	Number of Genera	Number of species
1	Fabaceae	80	244
2	Poaceae	75	216
3	Euphorbiaceae	46	162
4	Rubiaceae	57	153
5	Cyperaceae	11	111
6	Acanthaceae	28	81
7	Asteraceae	40	76
8	Orchidaceae	21	53
9	Convolvulaceae	13	52

5.5 SPECIES OF CONSERVATION CONCERN ALONG COASTAL AREA OF TANZANIA

Species of conservation concern are those that either face a decline in their habitats as determined by the IUCN, or are endemic to a specific area. According to the IUCN, nearly 3% (approximately 3000) out of the more than 10,000 known higher plants in Tanzania are threatened. A total of 494 (about 21%) angiosperm species of Tanzania coast are of conservation concern (IUCN threatened and/or Endemic to Tanzania) (see Table 5.2). The 494 angiosperms are widely spread along the coast of Tanzania (Figure 5.3). Moreover, of the 494 angiosperm species, 292 (59%) were IUCN threatened species whereas 352 (71%) were endemic to Tanzania. A total of 160 (32%) angiosperm species were both threatened and endemic. The threatened species comprise of 32 critically endangered (CR), 85 are endangered (EN), 53 near threatened (NT) and 122 are Vulnerable (VU). Some endemic plants include *Tricalysia bridsoniana* var. *pandensis* (VU), *Ochna kirkii* ssp. *multisetosa* (VU), *Uvariodendron kirkii* (VU), *Uvaria pandensis* (EN), *Uvariodendron gorgonis* (EN), *Xylopia mwasumbii* (EN), and *Uvaria punguensis* (CR) (Figures 5.4–5.6).

5.6 ENDEMIC SPECIES OF COAST TANZANIA

A total of 352 species endemic to Tanzania are distributed along the coast Tanzania (Table 5.2), which are within 72 families and 194 genera. The 352 species are dominated by Rubiaceae (63 species), Fabaceae (48 species), Annonaceae (22 species), Euphorbiaceae (20 species), Malvaceae (13 species), and Lamiaceae (10 species). The genera were dominated by *Psychotria* (10 species) and *Pavetta* (10 species). Other dominant Genera include *Crotalaria, Cynometra, Dorstenia, Grewia, and Ochna*, each with six species, all these followed by *Aloe, Coffea, Millettia*, and *Tricalysia.*

TABLE 5.2 Angiosperm Species of Conservation Concern (Endemic and IUCN) Found in the Tanzania Coast Area (E = Endemic, NT = Near Threatened, VU = Vulnerable, EN = Endangered, CR = Critically Endangered)

Sr. No.	Family	Scientific Name	IUCN Group	Endemic
1	Acanthaceae	*Justicia fittonioides* S.Moore		E
2	Anacardiaceae	*Lannea schweinfurthii* (Engl.) Engl. var. *acutifoliolata* (Engl.) Kokwaro		E
3	Annonaceae	*Artabotrys modestus* Diels		E
4	Annonaceae	*Greenwayodendron suaveolens* (Engl. & Diels) Verdc. ssp. *usambaricum* Verdc.	EN	
5	Annonaceae	*Isolona cauliflora* Verdc.		E
6	Annonaceae	*Isolona heinsenii* Engl. & Diels	VU	E
7	Annonaceae	*Mkilua fragrans* Verdc.	VU	E
8	Annonaceae	*Monanthotaxis faulknerae* Verdc.	EN	E
9	Annonaceae	*Monanthotaxis fornicata (Baill.)* Verdc.		E
10	Annonaceae	*Monanthotaxis trichantha* (Diels) Verdc.	VU	E
11	Annonaceae	*Monodora carolinae* Couvreur	EN	
12	Annonaceae	*Monodora hastipetala* Couvreur	EN	E
13	Annonaceae	*Monodora minor* Engl.&Diels	NT	
14	Annonaceae	*Ophrypetalum odoratum* Diels	NT	
15	Annonaceae	*Sanrafaelia ruffonammari* Verdc.	EN	E
16	Annonaceae	*Toussaintia orientalis* Verdc.	EN	E
17	Annonaceae	*Uvaria decidua* Diels		E
18	Annonaceae	*Uvaria faulknerae* Verdc.	EN	E
19	Annonaceae	*Uvaria kirkii* Oliv. ex Hook. f.	NT	
20	Annonaceae	*Uvaria leptoclados* Oliv. subsp. *leptoclados*	NT	
21	Annonaceae	*Uvaria lungonyana* Vollesen	VU	E
22	Annonaceae	*Uvaria pandensis* Verdc.	EN	E
23	Annonaceae	*Uvaria puguensis* D.M. Johnson	CR	E
24	Annonaceae	*Uvaria tanzaniae* Verdc.	VU	
25	Annonaceae	*Uvariodendron gorgonis* Verdc.	EN	E
26	Annonaceae	*Uvariodendron kirkii* Verdc.		E
27	Annonaceae	*Uvariodendron oligocarpum* Verdc.	EN	E
28	Annonaceae	*Uvariodendron pycnophyllum* (Diels) R.E. Fr.	EN	E
29	Annonaceae	*Uvariodendron usambarense* R.E. Fr.		E
30	Annonaceae	*Xylopia arenaria* Engl.	VU	E
31	Annonaceae	*Xylopia mwasumbii* D.M. Johnson	EN	E
32	Apocynaceae	*Cryptolepis hypoglauca* K. Schum.		E
33	Apocynaceae	*Landolphia watsoniana* Romburgh	VU	
34	Apocynaceae	*Mondia ecornuta* (N.E. Br.) Bullock		E
35	Apocynaceae	*Sacleuxia newii* (Benth.) Bullock		E
36	Apocynaceae	*Secamone gracilis* N.E. Br.		E
37	Apocynaceae	*Stephanostema stenocarpum* K. Schum.	CR	E
38	Apocynaceae	*Strophanthus zimmermannianus* Monach.		E
39	Apocynaceae	*Strophanthus zimmermannii* Monach.	EN	E

TABLE 5.2 *(Continued)*

Sr. No.	Family	Scientific Name	IUCN Group	Endemic
40	Araceae	*Amorphophallus stuhlmannii* (Engl.) Engl. & Gehrm.		
41	Araceae	*Anchomanes abbreviatus* Engl.		E
42	Araceae	*Callopsis volkensii* Engl.	NT	E
43	Araceae	*Gonatopus marattioides* (Peter) Bogner	EN	E
44	Araceae	*Gonatopus petiolulatus* (Peter) Bogner	VU	
45	Araceae	*Stylochaeton bogneri* Mayo	EN	E
46	Araceae	*Stylochaeton crassispathus* Bogner	VU	E
47	Araceae	*Stylochaeton milneanus* Mayo	VU	E
48	Araceae	*Stylochaeton salaamicus* N.E. Br.		E
49	Arecaceae	*Dypsis pembana* (H.E. Moore) Beentje & J. Dransf.		E
50	Asparagaceae	*Asparagus humilis* Engl.	EN	
51	Asparagaceae	*Chlorophytum filipendulum* subsp. *amaniense* (Engl.) Nordal & A.D. Poulsen		E
52	Asteraceae	*Ethulia angustifolia* Bojer ex DC.		E
53	Asteraceae	*Ethulia faulknerae* C. Jeffrey	EN	E
54	Asteraceae	*Ethulia paucifructa* M.G. Gilbert	EN	
55	Asteraceae	*Grauanthus linearifolius* (O. Hoffm.) Fayed	EN	
56	Asteraceae	*Gutenbergia pembensis* S. Moore	VU	E
57	Asteraceae	*Microglossa hildebrandtii* O. Hoffm.		E
58	Asteraceae	*Sphaeranthus greenwayi* Ross-Craig	VU	
59	Asteraceae	*Sphaeranthus spathulatus* Peter		E
60	Asteraceae	*Vernonia muelleri* Wild. Subsp. *integra* C Jeffrey		E
61	Asteraceae	*Vernonia vollesenii* C. Jeffrey		E
62	Asteraceae	*Vernonia zanzibarensis* Less.		E
63	Begoniaceae	*Begonia engleri* Gilg	EN	E
64	Bignoniaceae	*Fernandoa lutea* (Verdc.) Bidgood	CR	
65	Boraginaceae	*Cordia faulknerae* Verdc.	NT	
66	Boraginaceae	*Cordia fissistyla* Vollesen	VU	E
67	Boraginaceae	*Cordia guineensis* subsp. *mutica* Verdc.		E
68	Boraginaceae	*Cordia peteri* Verdc.	EN	E
69	Boraginaceae	*Cordia trichocladophylla* Verdc.	CR	E
70	Burseraceae	*Commiphora fulvotomentosa* Engl.	VU	
71	Burseraceae	*Commiphora madagascariensis* Jacq	NT	
72	Burseraceae	*Commiphora pteleifolia* Engl.	NT	
73	Buxaceae	*Buxus cordata* (Radcl.-Sm.) Friis	CR	
74	Buxaceae	*Buxus obtusifolia* (Mildbr.) Hutch.	VU	E
75	Canellaceae	*Warburgia elongata* Verdc.	CR	E
76	Canellaceae	*Warburgia stuhlmannii* Engl.	VU	E
77	Capparaceae	*Capparis viminea* Hook. f. & Thomsen ex Oliv. var. *orthacantha* (Gilg-Ben.) DeWolf	CR	
78	Capparaceae	*Maerua holstii* Pax		E
79	Capparaceae	*Rinorea ilicifolia* (Welw. ex Oliv.) Kuntze var. *amplexicaulis* Grey-Wilson	NT	

TABLE 5.2 *(Continued)*

Sr. No.	Family	Scientific Name	IUCN Group	Endemic
80	Capparaceae	*Thilachium densiflorum* Gilg & Gilg-Ben.	VU	E
81	Capparaceae	*Thilachium paradoxum* Gilg		E
82	Celastraceae	*Gymnosporia livingstonei* Jordaan	VU	
83	Celastraceae	*Maytenus masindei* Gereau		E
84	Celastraceae	*Pristimera graciliflora* (Welw ex Oliv) N Hallé subsp. *newalensis* (Blakelock) N Hallé		E
85	Chrysobalanaceae	*Hirtella cf. zanzibarica* Oliv.		E
86	Clusiaceae	*Allanblackia stuhlmannii* (Engl.) Engl.		E
87	Combretaceae	*Combretum chionanthoides* Engl. & Diels	NT	E
88	Combretaceae	*Combretum goetzei* Engl. & Diels	VU	
89	Combretaceae	*Combretum tenuipetiolatum* Wickens	EN	E
90	Commelinaceae	*Aneilema nyasense var. brevisepalum* Brenan		E
91	Commelinaceae	*Aneilema taylorii* C.B. Clarke		E
92	Commelinaceae	*Pollia bracteata* K. Schum.		E
93	Connaraceae	*Cnestis confertiflora* Gilg		E
94	Connaraceae	*Vismianthus punctatus* Mildbr.	VU	
95	Convolvulaceae	*Convolvulus jefferyi* Verdc.	VU	E
96	Convolvulaceae	*Ipomoea irwiniae* Verdc.	NT	E
97	Convolvulaceae	*Ipomoea kilwaensis* Pilg.	CR	E
98	Convolvulaceae	*Ipomoea lapathifolia* Hallier f. var. *bussei* (Pilg) Verdc		E
99	Convolvulaceae	*Ipomoea zanzibarica* Verdc.		E
100	Convolvulaceae	*Stictocardia lutambensis* (Schulze-Menz) Verdc.	EN	E
101	Crassulaceae	*Kalanchoe lateritia* Engl. var. *prostrata* Raadts		E
102	Crassulaceae	*Kalanchoe nyikae* subsp. *nyikae* Engl.	NT	
103	Crassulaceae	*Kalanchoe obtusa* Engl.	VU	
104	Cucurbitaceae	*Coccinia ulugurensis* Harms		E
105	Cucurbitaceae	*Gerrardanthus grandiflorus* Gilg ex Cogn.	NT	
106	Cucurbitaceae	*Kedrostis heterophylla* A. Zimm.		E
107	Cucurbitaceae	*Momordica glabra* A. Zimm.	EN	E
108	Cucurbitaceae	*Momordica henriquesii* Cogn.	EN	
109	Cucurbitaceae	*Momordica peteri* A. Zimm.	EN	E
110	Cucurbitaceae	*Peponium leucanthum* (Gilg) Cogn.	VU	
111	Cucurbitaceae	*Thladiantha africana* C. Jeffrey	EN	
112	Cyperaceae	*Bulbostylis hispidula* (Vahl) R.W. Haines		E
113	Cyperaceae	*Bulbostylis microcarpa* (Lye) R.W. Haines	CR	E
114	Cyperaceae	*Cyperus holstii* Kük		E
115	Cyperaceae	*Cyperus micromariscus* Lye		E
116	Cyperaceae	*Cyperus obsoletenervosus* Peter & Kük		E
117	Cyperaceae	*Fuirena claviseta* Peter		E
118	Cyperaceae	*Fuirena microcarpa* Lye	NT	
119	Cyperaceae	*Kyllinga cartilaginea* K. Schum.		E
120	Dichapetalaceae	*Dichapetalum arenarium* Breteler		E

TABLE 5.2 *(Continued)*

Sr. No.	Family	Scientific Name	IUCN Group	Endemic
121	Dichapetalaceae	*Dichapetalum braunii* Engl. & Krause	VU	
122	Dichapetalaceae	*Dichapetalum edule* Engl.	NT	
123	Dichapetalaceae	*Dichapetalum lindicum* Breteler	EN	
124	Dichapetalaceae	*Dichapetalum macrocarpum* Engl. ex K. Krause	VU	
125	Dilleniaceae	*Tetracera litoralis* Gilg		E
126	Dipterocarpaceae	*Monotes lutambensis* Verdc.	EN	E
127	Ebenaceae	*Diospyros amaniensis* Gürke		E
128	Ebenaceae	*Diospyros bussei* Gürke	NT	
129	Ebenaceae	*Diospyros capricornuta* F. White	EN	
130	Ebenaceae	*Diospyros greenwayi* F. White	NT	E
131	Ebenaceae	*Diospyros mafiensis* F. White	NT	
132	Ebenaceae	*Diospyros shimbaensis* F. White	VU	
133	Ericaceae	*Erica mafiensis* (Engl.) Dorr	VU	
134	Eriocaulaceae	*Eriocaulon strictum* Milne-Redh.	EN	
135	Erythroxylaceae	*Nectaropetalum acuminatum* Verdc.	CR	
136	Erythroxylaceae	*Nectaropetalum kaessneri* Engl. var. *kaessneri*		E
137	Euphorbiaceae	*Acalypha bussei* Hutch.	EN	E
138	Euphorbiaceae	*Acalypha echinus* Pax & K. Hoffm.		E
139	Euphorbiaceae	*Acalypha engleri* Pax		E
140	Euphorbiaceae	*Acalypha gillmanii* Radcl.-Sm.	EN	E
141	Euphorbiaceae	*Argomuellera basicordata* Peter ex Radcl.-Sm.	EN	E
142	Euphorbiaceae	*Argomuellera mjikendae* Q. Luke		E
143	Euphorbiaceae	*Croton jatrophoides* Pax	VU	E
144	Euphorbiaceae	*Croton kilwae* Radcl.-Sm.	EN	
145	Euphorbiaceae	*Croton longipedicellatus* J Léonard subsp. *austrotanzanicus* Radcl.-Sm		
146	Euphorbiaceae	*Croton megalocarpoides* Friis & M.G. Gilbert	VU	
147	Euphorbiaceae	*Cyathogyne usambarensis* (Verdc) J Léonard		
148	Euphorbiaceae	*Dalechampia scandens* var. *hildebrandtii* (Pax) Pax		E
149	Euphorbiaceae	*Euphorbia nyikae* Pax		E
150	Euphorbiaceae	*Euphorbia quadrialata* Pax	VU	E
151	Euphorbiaceae	*Euphorbia selousiana* S. Carter		E
152	Euphorbiaceae	*Euphorbia wakefieldii* N.E. Br.	EN	E
153	Euphorbiaceae	*Micrococca holstii* (Pax) Prain	NT	
154	Euphorbiaceae	*Micrococca scariosa* Prain	VU	E
155	Euphorbiaceae	*Mildbraedia carpinifolia* var. *carpinifolia* (Pax) Hutch.		E
156	Euphorbiaceae	*Monadenium crispum* N.E. Br.		E
157	Euphorbiaceae	*Omphalea mansfeldiana* Mildbr.		E
158	Euphorbiaceae	*Paranecepsia alchorneifolia* Radcl.-Sm.	VU	
159	Euphorbiaceae	*Ricinodendron heudelotii* (Baill) Heckel var. *tomentellum* (Hutch & E A Bruce) Radcl Sm		E
160	Euphorbiaceae	*Shirakiopsis trilocularis* (Pax & K. Hoffm.) Esser	EN	E

TABLE 5.2 *(Continued)*

Sr. No.	Family	Scientific Name	IUCN Group	Endemic
161	Euphorbiaceae	*Suregada lithoxyla* (Pax & K. Hoffm.) Croizat		E
162	Euphorbiaceae	*Synadenium pereskiifolium* (Baill.) Guill.		E
163	Euphorbiaceae	*Tragia pogostemonoides* Radcl.-Sm.	CR	
164	Fabaceae	*Acacia etbaica* Schweinf subsp. *australis* Brenan		
165	Fabaceae	*Acacia latistipulata* Harms	VU	
166	Fabaceae	*Acacia taylorii* Brenan & Exell		E
167	Fabaceae	*Acacia zanzibarica* (S.Moore) Taub. var. *zanzibarica* Brenan		E
168	Fabaceae	*Adenopodia rotundifolia* (Harms) Brenan	EN	E
169	Fabaceae	*Aeschynomene nematopoda* Harms	NT	
170	Fabaceae	*Aeschynomene sansibarica* Taub.	CR	
171	Fabaceae	*Aeschynomene uniflora* E. Mey. var. *grandiflora* Verdc.	EN	
172	Fabaceae	*Alysicarpus vaginalis* (L.) DC.		E
173	Fabaceae	*Angylocalyx braunii* Harms	VU	E
174	Fabaceae	*Baikiaea ghesquiereana* J. Leonard		E
175	Fabaceae	*Baphia kirkii* Baker	VU	E
176	Fabaceae	*Baphia macrocalyx* Harms	VU	
177	Fabaceae	*Baphia puguensis* Brummitt	EN	E
178	Fabaceae	*Baphia punctulata* Harms subsp. *punctulata*	VU	E
179	Fabaceae	*Bauhinia loeseneriana* Harms		E
180	Fabaceae	*Berlinia orientalis* Brenan	VU	
181	Fabaceae	*Crotalaria deserticola* Taub ex Baker f. subsp. *orientalis* Polhill		E
182	Fabaceae	*Crotalaria emarginata* Bojer ex Benth.		E
183	Fabaceae	*Crotalaria laburnoides* var. *nudicarpa* Polhill		E
184	Fabaceae	*Crotalaria lukwangulensis* Harms		E
185	Fabaceae	*Crotalaria patula* Polhill	EN	E
186	Fabaceae	*Crotalaria pterocalyx* Harms		E
187	Fabaceae	*Cynometra brachyrrhachis* Harms	VU	E
188	Fabaceae	*Cynometra engleri* Harms	VU	E
189	Fabaceae	*Cynometra filifera* Harms	EN	E
190	Fabaceae	*Cynometra gillmanii* J Léonard		
191	Fabaceae	*Cynometra greenwayi* Brenan	VU	E
192	Fabaceae	*Cynometra suaheliensis* (Taub.) Baker f.	NT	E
193	Fabaceae	*Cynometra webberi* Baker f.		E
194	Fabaceae	*Dalbergia acariiantha* Harms		E
195	Fabaceae	*Dalbergia melanoxylon* Guill. & Perr.	NT	
196	Fabaceae	*Dalbergia vacciniifolia* Vatke	VU	E
197	Fabaceae	*Dialium orientale* Baker f.		E
198	Fabaceae	*Eriosema youngii* Baker f. var. *rufum* (Hepper) Verdc.	CR	
199	Fabaceae	*Erythrina sacleuxii* Hua	NT	E
200	Fabaceae	*Erythrina schliebenii* Harms	CR	
201	Fabaceae	*Galactia argentifolia* S. Moore	NT	E

TABLE 5.2 *(Continued)*

Sr. No.	Family	Scientific Name	IUCN Group	Endemic
202	Fabaceae	*Gigasiphon macrosiphon* (Harms) Brenan	EN	
203	Fabaceae	*Guibourtia schliebenii* (Harms) J. Léonard	VU	
204	Fabaceae	*Indigofera concinna* Baker	EN	
205	Fabaceae	*Indigofera fulgens* Baker subsp. *fulgens*	VU	
206	Fabaceae	*Indigofera longimucronata* Baker f.	VU	
207	Fabaceae	*Indigofera rhynchocarpa* Welw. ex Baker var. *rhynchocarpa*		E
208	Fabaceae	*Indigofera wituensis* Baker f. var. *wituensis*	VU	E
209	Fabaceae	*Indigofera zanzibarica* J.B. Gillett	VU	E
210	Fabaceae	*Isoberlinia scheffleri* (Harms ex Engl.) Greenway		E
211	Fabaceae	*Julbernardia magnistipulata* (Harms) G. Troupin	VU	E
212	Fabaceae	*Millettia impressa* Harms ssp. *goetzeana* (Harms) J.B. Gillett	VU	
213	Fabaceae	*Millettia makondensis* Harms	VU	
214	Fabaceae	*Millettia micans* Taub.	EN	E
215	Fabaceae	*Millettia puguensis* J.B. Gillett	CR	E
216	Fabaceae	*Millettia sacleuxii* Dunn	EN	E
217	Fabaceae	*Millettia schliebenii* Harms	EN	E
218	Fabaceae	*Millettia usaramensis* Taub. ssp. *usaramensis* var. *parvifolia* Dunn	VU	
219	Fabaceae	*Newtonia erlangeri* (Harms) Brenan	EN	E
220	Fabaceae	*Newtonia paucijuga* (Harms) Brenan		
221	Fabaceae	*Ormocarpum schliebenii* Harms	NT	
222	Fabaceae	*Platysepalum inopinatum* Harms	VU	
223	Fabaceae	*Pseudoeriosema borianii* (Schweinf) Hauman subsp. *longipedunculatum* Verdc		E
224	Fabaceae	*Pseudoprosopis euryphylla* Harms subsp. *puguensis* Brenan		E
225	Fabaceae	*Pterocarpus mildbraedii* Harms subsp. *usambarensis* (Verdc) Polhill		E
226	Fabaceae	*Rhynchosia braunii* Harms	CR	E
227	Fabaceae	*Rhynchosia calobotrya* Harms	CR	E
228	Fabaceae	*Rhynchosia congensis* Baker ssp. *orientalis* Verdc.	NT	
229	Fabaceae	*Rhynchosia holstii* Harms	CR	
230	Fabaceae	*Scorodophloeus fischeri* (Taub.) J Léonard		E
231	Fabaceae	*Sesbania hirtistyla* var. *hirtistyla* J.B. Gillett	VU	E
232	Fabaceae	*Sesbania speciosa* Taub. ex Engl.	VU	
233	Fabaceae	*Stuhlmannia moavi* Taub.		E
234	Fabaceae	*Tessmannia martiniana* var. *martiniana* Harms	EN	E
235	Fabaceae	*Xylia africana* Harms	EN	E
236	Fabaceae	*Xylia schliebenii* Harms	EN	E
237	Fabaceae	*Zenkerella egregia* J. Léonard		
238	Gelsemiaceae	*Mostuea microphylla* Gilg	NT	
239	Gesneriaceae	*Saintpaulia ionantha* H. Wendl.	NT	E
240	Gesneriaceae	*Saintpaulia tongwensis* Vel sp. *aff.* B.L. Burtt		E

TABLE 5.2 *(Continued)*

Sr. No.	Family	Scientific Name	IUCN Group	Endemic
241	Gesneriaceae	*Saintpaulia velutina* B.L. Burtt		E
242	Gesneriaceae	*Streptocarpus holstii* Engl.		E
243	Hydnoraceae	*Hydnora sinandevu* Beentje & Q. Luke		E
244	Hydroleaceae	*Hydrolea sansibarica* Gilg	VU	E
245	Icacinaceae	*Iodes usambarensis* Sleumer	EN	E
246	Lamiaceae	*Clerodendrum cephalanthum* Oliv. subsp. *impensum* (B Thomas) Verdc		E
247	Lamiaceae	*Clerodendrum lutambense* Verdc.	VU	E
248	Lamiaceae	*Clerodendrum tricholobum* Gürke	NT	
249	Lamiaceae	*Leucas densiflora* Vatke		E
250	Lamiaceae	*Leucas tsavoensis* Sebald var. *tsavoensis* Sebald		E
251	Lamiaceae	*Ocimum fischeri* Gürke	VU	
252	Lamiaceae	*Orthosiphon hanningtonii* (Baker) A.J. Paton	EN	E
253	Lamiaceae	*Orthosiphon scedastophyllus* A.J. Paton	CR	E
254	Lamiaceae	*Premna gracillima* Verdc.	VU	E
255	Lamiaceae	*Premna hans-joachimii* Verdc.		E
256	Lamiaceae	*Premna schliebenii* Werderm.	VU	
257	Lamiaceae	*Rotheca kissakensis* (Gürke) Verdc.	VU	
258	Lamiaceae	*Rotheca myricoides* (Hochst.) (Hochst.) Steane & Mabb. ssp. *mafiensis* (Verdc.) Verdc.	CR	
259	Lamiaceae	*Vitex amaniensis* W. Piep.	VU	E
260	Lamiaceae	*Vitex keniensis* Turrill	EN	
261	Lamiaceae	*Vitex mossambicensis* Gürke	VU	
262	Lamiaceae	*Vitex zanzibarensis* Vatke	VU	E
263	Leguminosae	*Indigofera bussei* Gillett		E
264	Leguminosae	*Millettia eriocarpa* Dunn	EN	E
265	Linaceae	*Hugonia castaneifolia* Engl.		E
266	Loganiaceae	*Strychnos xylophylla* Gilg	EN	E
267	Loranthaceae	*Agelanthus validus* Polhill & Wiens		E
268	Loranthaceae	*Erianthemum alveatum (Sprague) Danser*	VU	E
269	Loranthaceae	*Erianthemum lindense* (Sprague) Danser	VU	
270	Loranthaceae	*Erianthemum occultum* (Sprague) Danser	VU	
271	Loranthaceae	*Erianthemum rotundifolium* Wiens & Polhill		E
272	Loranthaceae	*Oncella curviramea* (Engl.) Danser	VU	
273	Loranthaceae	*Oncella schliebeniana* Balle ex Polhill & Wiens		E
274	Lythraceae	*Nesaea aurita* Koehne		E
275	Lythraceae	*Nesaea maxima* Koehne		E
276	Lythraceae	*Nesaea pedicellata* Hiern		E
277	Lythraceae	*Nesaea stuhlmannii* Koehne		E
278	Malpighiaceae	*Triaspis schliebenii* A. Ernst		E
279	Malvaceae	*Abutilon zanzibaricum* Bojer ex Mast.		E
280	Malvaceae	*Byttneria fruticosa* K. Schum. ex Engl.	VU	

TABLE 5.2 *(Continued)*

Sr. No.	Family	Scientific Name	IUCN Group	Endemic
281	Malvaceae	*Cola lukei* Cheek	CR	E
282	Malvaceae	*Cola pseudoclavata* Cheek	NT	
283	Malvaceae	*Cola scheffleri* K. Schum.		
284	Malvaceae	*Dombeya taylorii* Baker f.		E
285	Malvaceae	*Grewia calymmatosepala* K. Schum.		E
286	Malvaceae	*Grewia capitellata* Bojer		E
287	Malvaceae	*Grewia filipes* Burret	EN	E
288	Malvaceae	*Grewia goetzeana* K. Schum.		E
289	Malvaceae	*Grewia holstii* Burret		E
290	Malvaceae	*Grewia leptopus* Ulbr.	NT	
291	Malvaceae	*Grewia meizophylla* Burret	EN	E
292	Malvaceae	*Grewia pedunculata* K. Schum.	NT	
293	Malvaceae	*Hibiscus faulknerae* Vollesen		E
294	Malvaceae	*Hildegardia migeodii* (Exell) Kosterm.		E
295	Malvaceae	*Mansonia diatomanthera* Brenan	EN	E
296	Malvaceae	*Melhania angustifolia* K. Schum.	EN	E
297	Malvaceae	*Nesogordonia holtzii* (Engl.) Capuron ex L.C. Barnett & Dorr	NT	
298	Malvaceae	*Sterculia schliebenii* Mildbr.	VU	
299	Melastomataceae	*Dissotis aprica* Engl.		E
300	Melastomataceae	*Memecylon cogniauxii* Gilg	VU	
301	Melastomataceae	*Memecylon semseii* A. Fern. & R. Fern.		E
302	Melastomataceae	*Tristemma schliebenii* Markgr.	EN	E
303	Melastomataceae	*Warneckea melindensis* (A. Fern. & R. Fern.) R.D. Stone & Q. Luke	EN	
304	Meliaceae	*Pseudobersama mossambicensis* (Sim) Verdc.	NT	
305	Meliaceae	*Trichilia aff. lovettii* Cheek		E
306	Meliaceae	*Turraea mombassana* C. DC. subsp. *schliebenii* (Harms) Styles & F White		E
307	Menispermaceae	*Albertisia exelliana* (G. Troupin) Forman	EN	
308	Menispermaceae	*Anisocycla blepharosepala* subsp. *tanzaniensis* Vollesen	VU	E
309	Menispermaceae	*Cissampelos nigrescens* var. *nigrescens* Diels	VU	E
310	Menispermaceae	*Tinospora oblongifolia* (Engl.) G. Troupin		E
311	Moraceae	*Dorstenia bicaudata* Peter		E
312	Moraceae	*Dorstenia goetzei* Engl.	NT	E
313	Moraceae	*Dorstenia hildebrandtii* var. *hildebrandtii* Engl.	NT	
314	Moraceae	*Dorstenia holstii* Engl. var. *longistipulata* Hijman		E
315	Moraceae	*Dorstenia tayloriana* Rendle var. *tayloriana*	VU	E
316	Moraceae	*Dorstenia tenuiradiata* Mildbr.		E
317	Moraceae	*Dorstenia warneckei* Engl.	NT	E
318	Moraceae	*Ficus faulkneriana* C.C. Berg	VU	E
319	Moraceae	*Mesogyne insignis* Engl.	NT	
320	Myristicaceae	*Cephalosphaera usambarensis* (Warb.) Warb.	NT	E

TABLE 5.2 *(Continued)*

Sr. No.	Family	Scientific Name	IUCN Group	Endemic
321	Myrtaceae	*Syzygium cordatum* Hochst. ssp. *shimbaense* Verdc.	EN	
322	Ochnaceae	*Gomphia lutambensis* (Sleumer) Verdc.		E
323	Ochnaceae	*Gomphia sacleuxii* (Tiegh.) Verdc.		E
324	Ochnaceae	*Ochna holstii* Engl.		E
325	Ochnaceae	*Ochna holtzii* Gilg		E
326	Ochnaceae	*Ochna kirkii* Oliv. ssp. *multisetosa* Verdc.		E
327	Ochnaceae	*Ochna polyarthra* Verdc.	CR	E
328	Ochnaceae	*Ochna pseudoprocera* Sleumer	VU	E
329	Ochnaceae	*Ochna schliebenii* Sleumer	CR	E
330	Olacaceae	*Olax pentandra* Sleumer	NT	
331	Oleaceae	*Jasminum ellipticum* Knobl.		E
332	Oleaceae	*Olea woodiana* Knobl. subsp. *disjuncta* P. S. Green		E
333	Orchidaceae	*Polystachya albescens* Ridl. ssp. *kraenzlinii* (Rolfe) Summerh.		E
334	Orchidaceae	*Polystachya caudata* Summerh.		E
335	Orobanchaceae	*Harveya buchwaldii* Engl.	NT	E
336	Pandanaceae	*Pandanus kirkii* Rendle		E
337	Pandanaceae	*Pandanus rabaiensis* Rendle	NT	E
338	Passifloraceae	*Adenia schliebenii* Harms	EN	
339	Passifloraceae	*Basananthe zanzibarica* (Mast.) W.J. de Wilde	VU	E
340	Passifloraceae	*Paropsia braunii* Gilg	NT	
341	Phyllanthaceae	*Cleistanthus beentjei* Luke		E
342	Phyllanthaceae	*Meineckia fruticans* var. *engleri* (Pax) G.L. Webster		E
343	Phyllanthaceae	*Meineckia paxii Brunel ex* Radcl.-Sm.		E
344	Phyllanthaceae	*Phyllanthus frazieri* Radcl.-Sm.	VU	E
345	Phyllanthaceae	*Phyllanthus kaessneri* Hutch. var. *kaessneri*	VU	
346	Phyllanthaceae	*Phyllanthus maderaspatensis* L. var *exasperatus* Radcl Sm		E
347	Phyllanthaceae	*Phyllanthus schliebenii* Mansf. ex Radcl.-Sm.	CR	E
348	Phyllanthaceae	*Phyllanthus wingfieldii* Radcl.-Sm.	VU	E
349	Picrodendraceae	*Aristogeitonia magnistipula* Radcl.-Sm.	CR	E
350	Picrodendraceae	*Aristogeitonia monophylla* Airy Shaw	NT	E
351	Poaceae	*Eragrostis perbella* K. Schum.	VU	
352	Poaceae	*Humbertochloa greenwayi* C.E. Hubb.	CR	E
353	Poaceae	*Hylebates chlorochloe* (K. Schum.) Napper		E
354	Poaceae	*Panicum nudiflorum* Renvoize		E
355	Poaceae	*Panicum pinifolium* Chiov.	VU	
356	Poaceae	*Panicum pleianthum* Peter	NT	
357	Poaceae	*Panicum vollesenii* Renvoize	VU	
358	Polygonaceae	*Oxygonum schliebenii* Mildbr.		E
359	Polypodiaceae	*Pyrrosia liebuschii* (Hieron.) Schelpe	EN	
360	Putranjivaceae	*Drypetes natalensis* (Harv.) Hutch. var. *natalensis*	VU	
361	Putranjivaceae	*Drypetes sclerophylla* Mildbr.	EN	E
362	Putranjivaceae	*Drypetes usambarica* (Pax) Hutch. var. *usambarica*	VU	E

TABLE 5.2 *(Continued)*

Sr. No.	Family	Scientific Name	IUCN Group	Endemic
363	Putranjivaceae	*Sibangea pleioneura* Radcl.-Sm.	EN	E
364	Rubiaceae	*Afrocanthium kilifiense (Bridson)* Lantz	NT	E
365	Rubiaceae	*Afrocanthium peteri (Bridson)* Lantz	VU	E
366	Rubiaceae	*Afrocanthium rondoense* (Bridson) Lantz	EN	E
367	Rubiaceae	*Afrocanthium siebenlistii* (K. Krause) Lantz	VU	
368	Rubiaceae	*Afrocanthium vollesenii* (Bridson) Lantz	VU	E
369	Rubiaceae	*Aidia abeidii* S.E.Dawson & Gereau	VU	
370	Rubiaceae	*Aoranthe penduliflora* (K. Schum.) Somers		E
371	Rubiaceae	*Burttdavya nyasica* Hoyle	NT	
372	Rubiaceae	*Canthium oligocarpum* subsp. captum (Bullock) Bridson	VU	
373	Rubiaceae	*Chamaepentas hindsioides* (K. Schum.) Kårehed & B. Bremer var. *glabrescens* (Verdc.) Kårehed & B. Bremer	EN	
374	Rubiaceae	*Chassalia zimmermannii* Verdc.		E
375	Rubiaceae	*Cladoceras subcapitatum* (K. Schum. & K. Krause) Bremek.	EN	E
376	Rubiaceae	*Coffea costatifructa* Bridson		E
377	Rubiaceae	*Coffea pocsii* Bridson		E
378	Rubiaceae	*Coffea pseudozanguebariae* Bridson	NT	E
379	Rubiaceae	*Coffea schliebenii* Bridson	VU	E
380	Rubiaceae	*Coffea sessiliflora* Bridson		E
381	Rubiaceae	*Cuviera migeodii* Verdc.	VU	E
382	Rubiaceae	*Cuviera schliebenii* Verdc.	EN	
383	Rubiaceae	*Cuviera semseii* Verdc.	VU	
384	Rubiaceae	*Cuviera tomentosa* Verdc.	EN	
385	Rubiaceae	*Didymosalpinx norae* (Swynn.) Keay	NT	
386	Rubiaceae	*Heinsia bussei* Verdc.		E
387	Rubiaceae	*Keetia koritschoneri* Bridson		E
388	Rubiaceae	*Keetia purpurascens* (Bullock) Bridson		E
389	Rubiaceae	*Kraussia kirkii* (Hook. f.) Bullock		E
390	Rubiaceae	*Leptactina delagoensis* K Schum subsp. *bussei* (K Schum & K Krause) Verdc		E
391	Rubiaceae	*Leptactina oxyloba* K. Schum.		E
392	Rubiaceae	*Leptactina papyrophloea* Verdc.		E
393	Rubiaceae	*Morinda asteroscepa* K. Schum.	VU	
394	Rubiaceae	*Multidentia castaneae* (Robyns) Bridson & Verdc.	Endemic	
395	Rubiaceae	*Multidentia kingupirensis* Bridson	Endemic	
396	Rubiaceae	*Multidentia sclerocarpa* (K. Schum.) Bridson	EN	
397	Rubiaceae	*Mussaenda microdonta* Wernham subsp. *microdonta*		E
398	Rubiaceae	*Mussaenda monticola* K. Krause var. *monticola*	VU	E
399	Rubiaceae	*Oldenlandia aegialodes* Bremek.	VU	E
400	Rubiaceae	*Oldenlandia flosculosa* Hiern	EN	E
401	Rubiaceae	*Oldenlandia patula* Bremek.		E
402	Rubiaceae	*Oldenlandia wiedemannii* K. Schum.		E

TABLE 5.2 *(Continued)*

Sr. No.	Family	Scientific Name	IUCN Group	Endemic
403	Rubiaceae	*Oxyanthus haerdii* Bridson	VU	E
404	Rubiaceae	*Oxyanthus pyriformis* (Hochst.) Skeels ssp. *longitubus* Bridson	EN	
405	Rubiaceae	*Pavetta amaniensis* Bremek var. *trichocephala* (Bremek) Bridson		E
406	Rubiaceae	*Pavetta crebrifolia* var. *kimbozensis* (Bremek.) Bridson	VU	
407	Rubiaceae	*Pavetta decumbens* K. Schum. & K. Krause	VU	
408	Rubiaceae	*Pavetta diversipunctata* Bridson	EN	E
409	Rubiaceae	*Pavetta holstii* K. Schum.		E
410	Rubiaceae	*Pavetta kyimbilensis* Bremek. var. *kyimbilensis*	VU	
411	Rubiaceae	*Pavetta linearifolia* Bremek.		E
412	Rubiaceae	*Pavetta lutambensis* Mildbr. ex Bremek.	EN	E
413	Rubiaceae	*Pavetta macrosepala* Hiern var. *macrosepala*	EN	E
414	Rubiaceae	*Pavetta matumbiensis* Bridson	CR	
415	Rubiaceae	*Pavetta sansibarica* K. Schum. subsp. *sansibarica*	EN	E
416	Rubiaceae	*Pavetta schliebenii* Mildbr. ex Bremek.		E
417	Rubiaceae	*Pavetta sphaerobotrys* K Schum subsp. *lanceisepala* (Bremek) Bridson		E
418	Rubiaceae	*Pavetta stenosepala* K Schum subsp. *kisarawensis* (Bremek) Bridson		
419	Rubiaceae	*Pavetta tendagurensis* Bremek. var. *tendagurensis*		E
420	Rubiaceae	*Phellocalyx vollesenii* Bridson	VU	
421	Rubiaceae	*Polysphaeria braunii* K. Krause		E
422	Rubiaceae	*Polysphaeria ntemii* S.E. Dawson & Gereau	EN	
423	Rubiaceae	*Psychotria amboniana* K Schum var *velutina* (EMA Petit) Verdc		E
424	Rubiaceae	*Psychotria faucicola* K. Schum.		E
425	Rubiaceae	*Psychotria holtzii* (K. Schum.) E.M.A. Petit var. *holtzii*		E
426	Rubiaceae	*Psychotria griseola* K. Schum.	NT	E
427	Rubiaceae	*Psychotria leucopoda* E.M.A. Petit	VU	
428	Rubiaceae	*Psychotria pandurata* Verdc.	VU	E
429	Rubiaceae	*Psychotria pocsii* Borhidi & Verdc. subsp. *pocsii*		E
430	Rubiaceae	*Psychotria porphyroclada* K. Schum.		E
431	Rubiaceae	*Psychotria schliebenii* E.M.A. Petit var. *schliebenii*		E
432	Rubiaceae	*Psychotria tanganyicensis* Verdc. subsp. *tanganyicensis*	CR	E
433	Rubiaceae	*Psychotria triclada* E.M.A. Petit	VU	E
434	Rubiaceae	*Psydrax faulknerae* Bridson	VU	E
435	Rubiaceae	*Psydrax kibuwae* Bridson	VU	E
436	Rubiaceae	*Rothmannia macrosiphon* (Engl.) Bridson	VU	
437	Rubiaceae	*Rytigynia binata* (K. Schum.) Robyns	VU	E
438	Rubiaceae	*Rytigynia celastroides* (Baill) Verdc. var. *nuda* Verdc.		E
439	Rubiaceae	*Sericanthe odoratissima* (K. Schum.) Robbr.	NT	
440	Rubiaceae	*Tapiphyllum schliebenii* Verdc.		E

TABLE 5.2 *(Continued)*

Sr. No.	Family	Scientific Name	IUCN Group	Endemic
441	Rubiaceae	*Tarenna drummondii* Bridson		E
442	Rubiaceae	*Tricalysia allocalyx* Robbr.		E
443	Rubiaceae	*Tricalysia bridsoniana* Robbr. var. *pandensis* Robbr		E
444	Rubiaceae	*Tricalysia elegans* Robbr.	CR	E
445	Rubiaceae	*Tricalysia microphylla* Hiern		E
446	Rubiaceae	*Tricalysia ovalifolia* Hiern		E
447	Rubiaceae	*Tricalysia schliebenii* Robbr.	VU	
448	Rubiaceae	*Vangueria bicolor* K. Schum.	VU	
449	Rubiaceae	*Vangueria pallidiflora* (Bullock) Lantz	VU	E
450	Rubiaceae	*Vangueria randii* S. Moore subsp. *vollesenii* Verdc		E
451	Rutaceae	*Vepris amaniensis* (Engl.) Mziray	VU	E
452	Rutaceae	*Vepris morogorensis* (Kokwaro) Mziray	NT	
453	Rutaceae	*Vepris sansibarensis* (Engl.) Mziray		
454	Rutaceae	*Vepris schliebeni* Mildbr.		E
455	Rutaceae	*Zanthoxylum gilletii* (De Wild.) P.G.Waterman	EN	
456	Rutaceae	*Zanthoxylum holtzianum* (Engl.) P.G. Waterman	VU	
457	Rutaceae	*Zanthoxylum lindense* (Engl.) Kokwaro	VU	E
458	Rutaceae	*Zanthoxylum paracanthum* (Mildbr.) Kokwaro	VU	
459	Santalaceae	*Thesium panganense* Polhill		E
460	Sapindaceae	*Allophylus chirindensis* Baker f.	NT	
461	Sapindaceae	*Allophylus vestitus* F.G. Davies	EN	E
462	Sapindaceae	*Chytranthus obliquinervis* Engl.		E
463	Sapotaceae	*Mimusops acutifolia* Mildbr.	VU	E
464	Sapotaceae	*Mimusops riparia* Engl.	VU	E
465	Sapotaceae	*Pouteria alnifolia* (Baker) Roberty var. *sacleuxii* (Lecomte) L. Gaut.		E
466	Sapotaceae	*Pouteria pseudoracemosa* (J.H. Hemsl.) L. Gaut.	VU	E
467	Sapotaceae	*Synsepalum kaessneri* (Engl.) T.D. Penn.	VU	E
468	Sapotaceae	*Vitellariopsis cuneata* (Engl.) Aubrév.	VU	E
469	Sapotaceae	*Vitellariopsis kirkii* (Baker) Dubard	VU	E
470	Solanaceae	*Solanum kitivuense* vel sp. aff. Dammer		E
471	Solanaceae	*Solanum zanzibarense* Vatke		E
472	Thymelaeaceae	*Dicranolepis usambarica* Gilg		E
473	Urticaceae	*Urera cf. fischeri* Engl.		E
474	Verbenaceae	*Clerodendrum robustum* Klotzsch var *latilobum* Verdc		E
475	Violaceae	*Rinorea angustifolia* (Thouars) Baill. subsp. *albersii* (Engl) Grey-Wilson		E
476	Violaceae	*Rinorea scheffleri* Engl.		E
477	Violaceae	*Rinorea welwitschii* (Oliv) Kuntze subsp. *tanzanica* Grey-Wilson		E
478	Vitaceae	*Ampelocissus africana* (Lour.) Merr. var. *migeodii* Verdc.	CR	
479	Vitaceae	*Cissus egestosa* Werderm.		E

TABLE 5.2 *(Continued)*

Sr. No.	Family	Scientific Name	IUCN Group	Endemic
480	Vitaceae	*Cissus migeodii* Verdc.	EN	E
481	Vitaceae	*Cissus rotundifolia* (Forssk.) Vahl		E
482	Vitaceae	*Cissus sylvicola* Masinde & L.E. Newton		E
483	Vitaceae	*Cyphostemma bidgoodiae* Verdc.		E
484	Vitaceae	*Cyphostemma feddeanum* (Gilg & M. Brandt) Desc.		E
485	Vitaceae	*Cyphostemma gracillimum* (Werderm.) Desc.		E
486	Vitaceae	*Cyphostemma zanzibaricum* Verdc.		E
487	Xanthorrhoeaceae	*Aloe boscawenii Christian*		E
488	Xanthorrhoeaceae	*Aloe flexilifolia* Christian	CR	E
489	Xanthorrhoeaceae	*Aloe leachii* Reynolds	VU	E
490	Xanthorrhoeaceae	*Aloe massawana* Reynolds	VU	E
491	Xanthorrhoeaceae	*Aloe pembana* L.E. Newton	CR	E
492	Zamiaceae	*Encephalartos hildebrandtii* A Braun & Bouché		E
493	Zingiberaceae	*Aframomum amaniense* Loes.		E
494	Zingiberaceae	*Aframomum orientale* Lock	EN	E

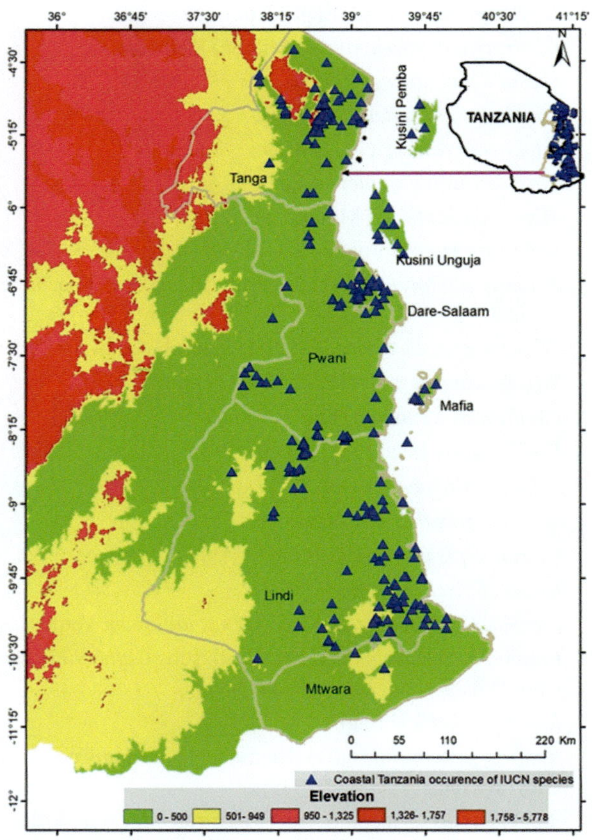

FIGURE 5.3 Distribution of threatened plant species long the Tanzania coast.

FIGURE 5.4 *Uvariodendron gorgonis*, Endangered (EN) family Annonaceae.

Source: Authors

FIGURE 5.5 *Ochna kirkii* ssp. *multisetosa*, Vulnerable (VU), Ochnaceae family.

Source: Authors

FIGURE 5.6 *Xylopia mwasumbii*, Endangered (EN) family Annonaceae.

Source: Authors

There are five species of *Uvaria* and five species of *Uvariodendron*, with each genus being represented by five unique species. These endemic plants can be found in various regions along the coast (Figure 5.7).

5.7 CONCLUSIONS AND RECOMMENDATIONS

The initial focus of this chapter was to gather fundamental species data along the Tanzania coast, providing essential information for research and conservation efforts. The chapter demonstrates that the Tanzania coast is of greater importance in conservation due to its exceptionality in the diversity of angiosperms, which include those that need deliberate conservation efforts because they are already threatened and/or endemic. The area's abundant biodiversity, varied microhabitats, and absence of systematic monitoring programs suggest the presence of unidentified plant species. Therefore, additional survey efforts are necessary to catalogue plants, especially in unsurveyed areas. Furthermore, due to clear evidences of anthropogenic fragmentation which are commonly observed through our visual experience in the Tanzania coastal area, species loss can be inevitable. Therefore, to avoid precedented biodiversity loss, we urge that (1) There is high need for monitoring and reassessments of species status based on local availability to ensure conservation decisions are always anchored on reliable data and information. (2) There is high need for deliberate conservation and awareness efforts for communities, including both the adults and the youth in the areas. (3) Habitat restoration programs are necessary to ensure sustainability in the areas.

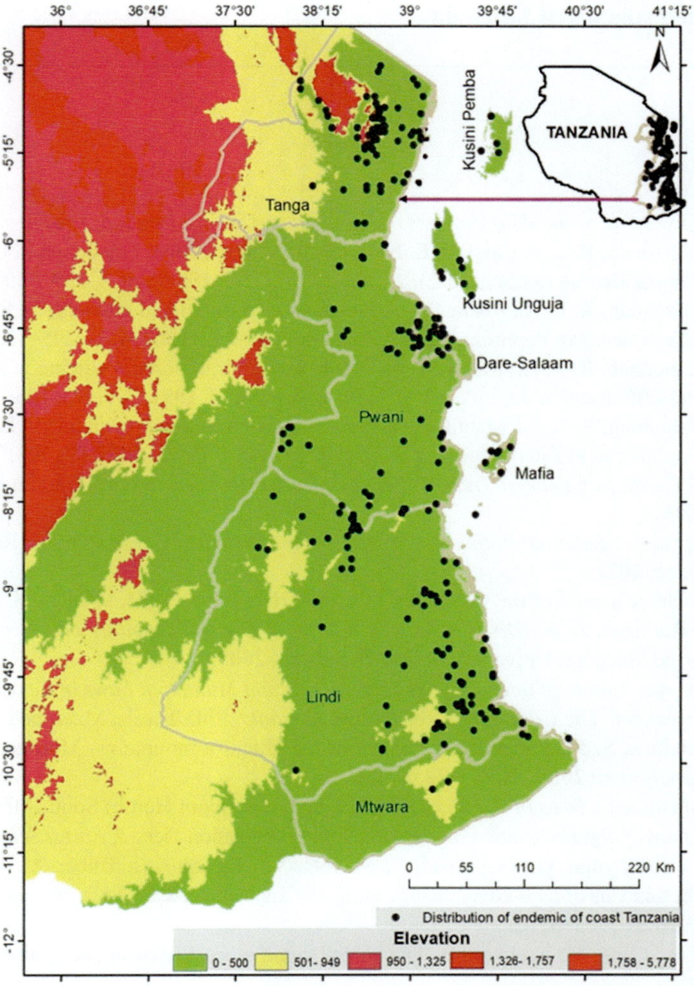

FIGURE 5.7 Distribution of angiosperms endemic to Tanzania along the Tanzania coast.

Source: Authors

ACKNOWLEDGMENTS

We express our gratitude to Mr. Nelson Mungoya for cross-referencing the plant lists with the IUCN red list. Appreciation is extended to all the researchers who contributed to the development of this chapter. The University of Dar es Salaam in Tanzania has been instrumental in the long-term research efforts aimed at comprehending the significance of these forests.

KEYWORDS

- **East Africa flora**
- **angiosperms of Tanzania**
- **threatened species of Tanzania**

REFERENCES

Blundell, M. *Collins Guide to the Wild Flowers of East Africa*. Collins: London, **1987**.

Chhabra, S. C.; Mahunnah, R. L. A.; Mshiu, E. N. Plants used in traditional medicine in Eastern Tanzania. II. Angiosperms (Capparidaceae to Ebenaceae). *J. Ethnopharmacol.* **1989,** *25*(3), 339–359.

Chhabra, S. C.; Mahunnah, R. L. A.; Mshiu, E. N. Plants used in traditional medicine in Eastern Tanzania. IV. Angiosperms (Mimosaceae to Papilionaceae). *J. Ethnopharmacol.* **1990,** *29*(3), 295–323.

Chhabra, S. C.; Mahunnah, R. L. A.; Mshiu, E. N. Plants used in traditional medicine in Eastern Tanzania. V. Angiosperms (Passifloraceae to Sapindaceae). *J. Ethnopharmacol.* **1991,** *33*(1–2), 143–157.

Chhabra, S. C.; Mahunnah, R. L. A.; Mshiu, E. N. Plants used in traditional medicine in Eastern Tanzania. VI. Angiosperms (Sapotaceae to Zingiberaceae). *J. Ethnopharmacol.* **1993,** *39*(2), 83–103.

Dharani, N. *Field Guide to Common Trees & Shrubs of East Africa*. Penguin Random House: Cape Town, South Africa, **2019**.

Dharani, N.; Yenesew, A. *Medicinal Plants of East Africa: An Illustrated Guide*. Penguin Random House: Cape Town, South Africa, **2022**.

Greenway, P. J. A Classification of the vegetation of East Africa. *Kirkia* **1973,** *9*(1), 1–68.

Muhando, C. A.; Rumisha, C. K. *Distribution and Status of Coastal Habitats and Resources in Tanzania.* Institute for Marine Sciences, University of Dar Es Salaam, **2008**.

Nsemwa, C. E. *The Assessment of International Conventions and Municipal Laws in Relation to the Protection of Coastal Environment: The Case of Dar Es Salaam Tanzania.* PhD Thesis, Mzumbe University, **2013**.

Temu, R. P. C.; Andrew, S. M. Endemism of plants in the Uluguru mountains, Morogoro, Tanzania. *Forest Ecology and Management* **2008,** *255*(7), 2858–2869.

Van Wyk, B. *Field Guide to Trees of Southern Africa*. Penguin Random House: South Africa. **2013**.

Volařík, D. Application of digital elevation model for mapping vegetation Tiers. *J. Forest Sci.* **2010,** *56*(3), 112–20.

Wang, Y.; Tobey, J.; Bonynge, G.; Nugranad, J.; Makota, V.; Ngusaru, A.; Traber, M. Involving geospatial information in the analysis of land-cover change along the Tanzania coast. *Coast. Manag.* **2005,** *33*(1), 87–99. doi: 10.1080/08920750590883132.

Yang, W.; Seager, R.; Cane, M. A.; Lyon, B. The annual cycle of East African precipitation. *J. Climate* **2015,** *28*(6), 2385–2404.

Zhou, Y.; Liu, B.; Mbuni, Y.; Yan, X.; Mwachala, G.; Hu, G.; Wang, Q. Vascular flora of Kenya, based on the flora of Tropical East Africa. *PhytoKeys* **2017,** *90*, 113.

CHAPTER 6

DIVERSITY OF INVERTEBRATES IN THE COASTAL FORESTS OF EASTERN AFRICA

KELVIN NGONGOLO

Department of Biology, College of Natural and Mathematical Sciences, The University of Dodoma, Dodoma, Tanzania

ABSTRACT

The chapter delves into the intricate and often underappreciated world of invertebrate life thriving within the diverse ecosystems of Eastern African coastal forests. Based on an extensive literature review, this chapter delves into the extensive variety of invertebrates found in these essential habitats. Spanning from mangroves to tropical rainforests, these coastal forests harbor a plethora of invertebrate taxa, including insects, arachnids, mollusks, and crustaceans. It was noted that the species of freshwater snail in coastal forests in East Africa are about 290, of which 6.2% (n = 18) were identified in Kenya, including *Gonaxis quadrilateralis, Gulella vicina vicina, Trochonanina mozambicensis, Thapsia curvatula, Trochonanina shimbiensis, Pseudoglessula boivini, Achatina fulica rodatzi, Rhachidina braunsi*, and *Maizania hildebrandtii.* In Tanzania, a total of 161 species of invertebrates, comprising 86 species of Mollusca and 75 species of butterflies, are considered endemic in the coastal forest. However, the chapter also underscores the imminent threats facing invertebrates and their habitats, including agriculture, charcoal making, and firewood harvesting. Synthesizing a wealth of information, the chapter emphasizes the urgent need for holistic conservation strategies that prioritize the preservation of both invertebrate species and their unique habitats.

6.1 INTRODUCTION

Invertebrates are the largest group of living organisms in the world, making up 97% of animal species, with the majority being insects (>1,000,000 species) and nematodes (>1,000,000 species) (May, 1988). The most familiar invertebrates include the Protozoa, Porifera, Coelenterata, Platyhelminthes, Nematoda, Annelida, Echinodermata, Mollusca, and Arthropoda. Arthropoda include insects, crustaceans, and arachnids. There is no

Biodiversity Hotspot of the Coastal Forests of Eastern Africa. T. Pullaiah and Ally J. K. Nkwabi (Eds.)

DOI: 10.1201/9781998511099-6

denying that invertebrates make up a significant portion of biodiversity, accounting for the majority of species and biomass (1,250,000–1,359,367, or 72% of all documented species) (Chapman, 2009). Invertebrates make up 75% of the 18,000 new species described in 2007, and an estimated 5,396,465 invertebrate species still need to be documented globally, making them a neglected component in comparison to the far less diversified (Chapman, 2009). By far, the largest number of described invertebrate species is insects. Along the coastal areas of Eastern Africa, an invertebrate group has been described. For instance, in Tanzania, 20 orders of anthropoids have been described in the coastal forest (Burgess et al., 1999).

Among the coastal forests in Kenya are Shimba Hills, Lower Tana River forests, Witu Forest Reserve, Arabuko, Sokoke Forest, Diani Forest, and Kaya Ribe (Matiku, 2023). In Tanzania, the coastal forests are Genda Genda North Forest Reserve, Genda South Forest Reserve, Handeni Hill Forest Reserve, Kusumba Forest Reserve, Mbuzini Forest, Mgambo Forest, Mkongo Forest Reserve, Mkuli Forest Reserve, Mtunguru Forest Reserve, Katundu Forest Reserve, Kichi Hills Forest Reserve, Kikale Forest Reserve, Jozani Forest, Muyuni Forest, Ngaramia Riverine Forest, and Pande Game Reserve (TFCG, 2006; Lovett, 1989). In Mozambique, the coastal forest includes Cabo Delgado (Burgess et al., 1998).

Habitats and ecosystems along the coastal forests of Eastern Africa support diverse species of invertebrates. The coastal forests of Eastern Africa constitute unique and intricate ecosystems that provide critical ecological services and support a remarkable diversity of life forms (Burgess et al., 1998). The Eastern African coastal forests exhibit a concentration of biodiversity hotspots owing to their rich array of habitats and ecological niches. From the vibrant coral reefs along the Indian Ocean to the lush canopies of tropical rainforests, these hotspots encompass a range of environments that provide the ideal conditions for a multitude of invertebrate species to flourish (Burgess et al., 1998; Ragavan et al., 2019; Hemp, 2005; Kiwia, 2006). In these lush landscapes, invertebrates play an essential, yet often understated, role in shaping the fabric of these habitats. For instance, the Taita Hill ecosystem, which is among the coastal forest ecosystems in Kenya, revealed a high diversity of butterflies. In total, 1820 genera and 19,000 species of butterflies have been identified worldwide. According to Kioko et al. (2021), there are 318 genera and 4325 species (or nearly 23% of all species worldwide) in the Afrotropical region. In Kenya, there are 903 species of butterflies, accounting for 4.8% of all species globally and 20.9% of all Afrotropical butterflies. About 63.3% of all butterfly species in Kenya, including endemic and endangered species, can be found in the Taita Hills coastal forest (Kioko et al., 2021). They participate in intricate relationships, influencing pollination, nutrient cycling, and decomposition processes that underpin the overall health and resilience of the forests (Garbach et al., 2014). Studies in pollination revealed the importance of invertebrates in pollination services in the coastal forest of Tanzania. A study in Wazo Hill revealed an increase in seed setting in *Leucaena leucocephala* when it is pollinated. The potential pollinators were *Apis mellifera, Trichopoda* sp., and *Megachile* sp., which were the dominant pollinators. Also, pollination advantages were noted for local communities that were cultivating sesame crops adjacent to the Kichi forest, which is among the coastal forests in Tanzania (Ngongolo et al., 2014a, b, 2015a, b).

6.1.1 METHODS USED IN THIS CHAPTER

The book chapter offers a comprehensive exploration of the often-overlooked invertebrate world within these vital ecosystems. Through a meticulous literature review, the chapter uncovers the invertebrate diversity, endemism, and threats in East African coastal forests. This chapter provides an invaluable resource for researchers, conservationists, and ecologists interested in Eastern African coastal forests, shedding light on the intricate relationships between invertebrates and their environment. Its well-rounded approach offers a holistic view of biodiversity, fostering a deeper understanding of these unique ecosystems' dynamics and challenges.

6.2 GEOGRAPHICAL CONTEXT

Geographically, these coastal forests span the eastern seaboard of the African continent, stretching along the Indian Ocean coastline. Encompassing countries such as Kenya, Tanzania, Mozambique, and parts of Somalia, this region boasts a diverse array of ecosystems, landscapes, and habitats that support a rich variety of invertebrate species (Figure 6.1). From the lush coastal forests of Kenya's Shimba Hills to the mangrove-fringed estuaries of Tanzania's Rufiji Delta, the geographical context of this chapter covers a spectrum of ecosystems. These coastal forests are characterized by their proximity to the ocean, influenced by the ocean's climate-regulating effects and the interplay of land and sea. They have unique adaptations for invertebrates in these coastal habitats, which often experience fluctuations in salinity, moisture, and temperature due to tidal movements. Mangroves, sand dunes, and tropical rainforests are all integral components of these coastal forests, each contributing to the varied habitats and niches that support an impressive diversity of invertebrate life (Lovett, 1989).

6.3 INVERTEBRATE DIVERSITY AND DISTRIBUTION

East Coast forests are among the tropical forests found in East Africa that are known to harbor diverse species, as it has been indicated that tropical ecosystems contain half of the world's species (Newmark, 2002). The diversity of invertebrates within these coastal forests is immense and includes a wide range of taxonomic groups such as arachnids (spiders, scorpions), insects (butterflies, ants, beetles), mollusks (snails, slugs), crustaceans (crabs, shrimps), and more (Gereau et al., 2016). Each of these groups occupies specific ecological niches within the forest, contributing to various ecological processes and interactions.

6.3.1 PHYLUM ARTHROPODA

The members phylum Arthropoda consists of different classes such as Chelicerata (arachnids), Crustacea (crustaceans), Hexapoda (insects and springtails), and Myriapoda (millipedes and centipedes). Costal forest, being the hotspot, is anticipated to host the majority of species (Huber and Warui, 2012). A study in Kenya, for instance, has shown that most East African species of spiders (90%) belong to one of only six genera: *Buitinga* (21 species); *Smeringopus* (18 species); *Pholcus* (17 species); *Spermophora* (12 species); *Leptopholcus* (5 species); and *Quamtana* (4 species). In Tanzania, a study in lowland coastal forests of Eastern Tanzania revealed diverse species of anthropoids, with ants (Hymenoptera: Formicidae, 31%) scoring higher, followed by grasshoppers and crickets (Orthoptera, 22%), beetles (Coleoptera, 17.39%), and spiders (Araneae, 13.48%) (Burgess et al., 1999). These arthropods in coastal forest were found within 20 groups, among which are: Acarina, Hymenoptera (ants), Collembola, Coleoptera, Diptera, Orthoptera, Pscoptera, Araneae, Heteroptera, Lepidoptera, Homoptera, Isopoda, Diplopoda, Diplura, Anoplura, Siphonoptera, Annelida, and Isoptera (Burgess et al., 1999).

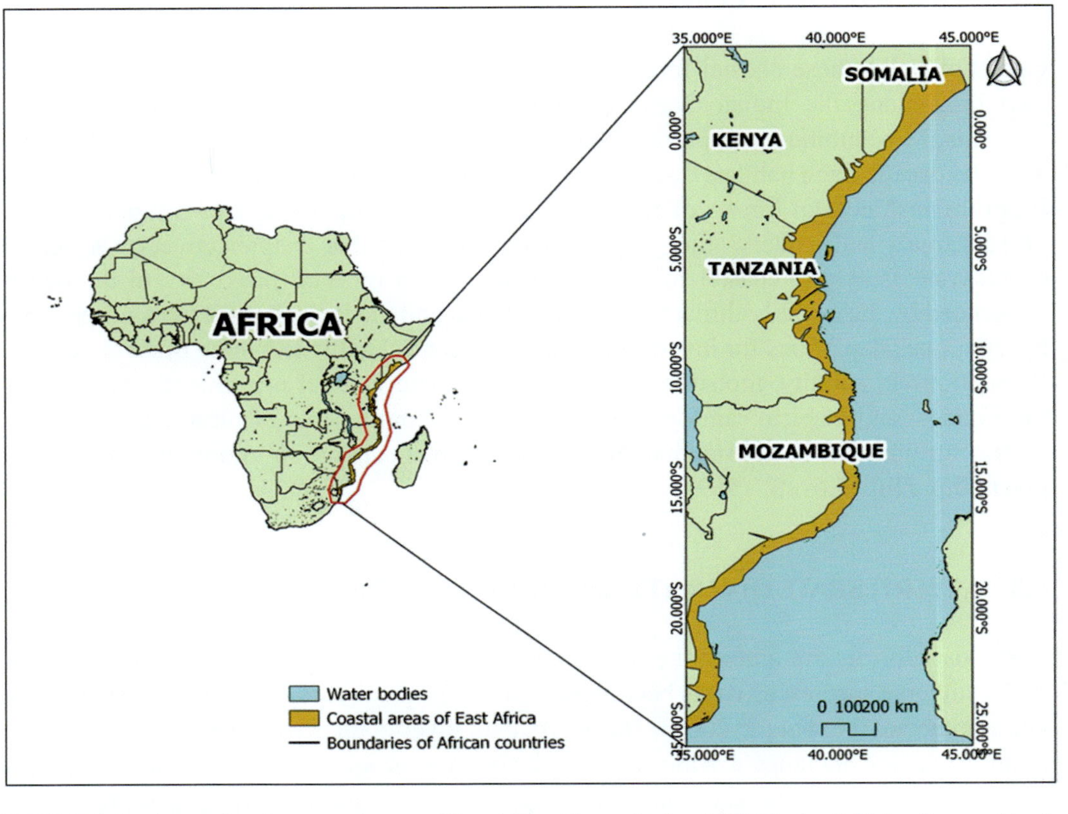

FIGURE 6.1 A map showing coastal areas of East Africa (Created using QGIS Version 3.26.3—Buenos Aires).

According to a study conducted in the Eastern African Tropical Forest (EACF), ten of the world's most endangered species of gastropods are found there. Eight locations, including the Taita Hills Forests accommodates four species of gastropods, Wami River (two species), Arabuko-Sokoke Forest (one species), Kingani (Kingani, lower Ruvu) River has only one species of gastropod, the Muheza District coastal forests, Pangani District coastal forests, Shimba Hills, and Zanzibar coastal areas all have only one species of gastropod. Generally coastal areas of Eastern African are considered home to these threatened gastropod species. Also, *Gulella amboniensis* and *Lanistes alexandri* are two of these vulnerable species which can be found along the coastal forest of East Africa (Gereau et al., 2016).

Insects are diverse species, which occur in different habitats, including the East African coastal forest. On the other side, the EACF has more than 25 insect species that are endangered. Seven EACF sites, including the Zanzibar coast, the East Usambara Mountains (17), the West Usambara Mountains (5), and the Nguru Mountains (3), contain these bug species (Gereau et al., 2016). The coastal forest harbors diverse insects. For instance, a study in Kichi Forest, which is among the coastal forests in Tanzania, revealed 84 species of insects, including *Streblognatus* spp., *Sisyphus* spp., *Cophogryllus* sp., *Abisares* sp., *Aporochin* sp., *Atradopolis* sp., *Apanteles* sp., *Colotis* spp., *Eurema brigitta,* and *Sarcophaga* spp. Millipeds, snails, spiders, and worms observed during the survey but unidentified to species level, were threatened by agricultural practices and fire occurrence in the area (Ngongolo et al., 2015a). From the observed species, other invertebrates were beneficial to local communities living adjacent to the Kichi Forest as they were providing pollination services to their crops, specifically the sesame plants, on which *A. mellifera* L., *Xylocopa caffra* L. while species like *Polistes africanus* are good floral visitors (Figure 6.2*)* and potential pollinators (Ngongolo et al., 2015a).

A study in the Shimba Hills, which is among the coastal forests in Kenya, revealed that the area has potential for conservation of butterflies as the area supports a high number of species in Kenya (Lehmann and Kioko, 2000). Also, Shimba Hill National Reserve is reported to have invertebrates, which are of economic importance to livestock keepers, as well as other vectors such as ticks and tsetse flies. For example, species of tsetse flies like *Glossina pallidipes, Glossina austeni,* and *Glossina brevipalpis* are reported in the area and have significant impacts on the ecosystem while transmitting trypanosome parasites to livestock (Ebhodaghe et al., 2022). Other species of vectors reported in East Africa coastal ecosystem includes mosquitoes like *Anopheles gambiae s.s*, *Anopheles arabiensis*, and *Anopheles funestus, Anopheles gambiae s.l*, *Anopheles pharoensis*, *Anopheles ardensis*, *Anopheles azamiae, Anopheles christyi*, *Anopheles maculpalpis*, *Anopheles hamoni*, and *Anopheles sergentii Culex quinquefasciatus*, and *Cx. pipiens* that are important vector of parasites for Malaria, lymphatic filariasis, chikungunya, and dengue *(*Dida, et al., 2018).

Class Arachnids is among of the class found in Arthropoda that can be identified in coastal forest of East Africa. For instance, the checklist of spiders that are found in Tanzania, a significant number were reported from coastal forests like Amani Nature Reserve, Kwangumi Forest Reserve. These species of spiders include *Cyphonisia affinitata, Cyphonisia rastellata, Gasteracantha falcicornis, Apochinomma formicaeforme, Copuetta naja* Haddad, and *Orchestina ampulla. Leptopholcus gurnahi, Pehrforsskalia shambaa* (Smith, 2020). A study in Shimbi Hill has reported tick species like *Amblyomma eburneum*, *Amblyomma tholloni, Rhipicephalus maculatus,* and a novel *Rhipicephalus* sp. that transmit parasites to livestock (Mwamuye et al., 2017).

FIGURE 6.2 *Polistes africanus* (Vespidae).
Source: Kelvin Ngonolo.

Crustaceans are another important class from phylum Arthropoda that can be observed in the coastal forest of East Africa. The previous study show that there are 209 species of freshwater crabs that are globally endangered found in EACF, 28 of which are found in the Afrotropics (16 are vulnerable, 10 are endangered, 2 are critically endangered), and 9 are found in East Africa (7 are vulnerable, and 2 are endangered). The species comprises *Potamonautes* species such as *Potamonautes raybouldi, Potamonautes xiphoidus, Potamonautes infravallatus,* and *Potamonautes unisulcatus,* as well as those belonging to the Class Malacostraca (Gereau et al., 2016). For instance, two species including *P. raybouldi* are found in the coastal forest of Shimba Hills in Kenya and the East Usambara Mountains in Tanzania; and two species, *P. infravallatus* and *P. xiphoidus* are threatened species along the Coastal forest of East Africa (Gereau et al., 2016).

In Kenya, the report indicated that among the species that can be found in coastal forests are invertebrates. For instance, the invertebrates for Mangrove forests found in the Tana River near Kipini as well as at the Ramisi River include crabs, prawns, and mollusks (Matiku, 2023). These species include a number of prawns (*Penaeus indicus, P. monodon,*

P. semisulcatus, and *Matapenaeus monoceros*); crabs (*Scylla serrata, Uca* spp., *Sesarma* spp., and *Birgus latro);* and mollusca (oysters such as *Brachydontes* spp. and *Crassostrea cucullata* and cockles, *Donax* spp.) (Matiku, 2023). The endangerment of these species is attributed to factors such as poverty, ineffective governance and management, insufficient technological resources, lack of knowledge, and incentives for adopting alternative livelihoods and promoting environmental consciousness. Other contributing factors include population growth, limited coordination, and lack of focus on the landscape, as well as the overexploitation of forests on privately owned land (Matiku, 2023).

6.3.2 PHYLUM MOLLUSCA

Phylum Mollusca is among the invertebrates found in coastal of East Africa. The group comprises both snails and slugs. About 28 species of snail that belong to 8 families have been reported in Shimba Forest Reserve. The following species from the family Streptaxidae are included: *Gonaxis quadrilateralis*, *Gulella vicina vicina*, *Gonaxis kibweziensis*, *Gonaxis gibbonsi*, *Gonaxisen neoides*, *Gonaxis* sp., and *Gulella radius*. Additionally, species from the family Urocyclidae are present, including *Trochonanina mozambicensis*, *Thapsia curvatula*, *Trochonanina shimbiensis*, and *Trochozonites* sp. For the family Subulinidae, the species include *Pseudoglessula boivini*, and *Opeas lamoense*. For the family Achatinidae, it includes *A. fulica rodatzi*, and *Achatina albopicta*, while for the family Enidae, it includes *Rhachidina braunsi*, and *Edouardia alycaeoides*. Other species include *Maizania hildebrandtii* (Maizaniidae), *Tropidophora letourneuxii* (Pomatiasidae), and *Trachycystis arid* (Dalila, 2009). These snails inhabited diverse vegetation habitats such as forest, grass, scrub, and plantation grassland, with forest, particularly in wet seasons, exhibiting the highest diversity and abundance of snails in the ecosystem (Dalila, 2009).

6.3.3 PHYLUM PORIFERA

Sponges are a type of simple invertebrate animal categorized in the Porifera phylum. They predominantly inhabit marine environments, although some can also be found in freshwater habitats. These organisms are commonly found in shallow ocean environments and have been known to thrive at depths of up to 5 km. Within the database, there are a total of 9589 recognized species, encompassing both marine and nonmarine varieties, out of nearly 20,000 taxon names (de Voogd et al., 2023). A study on Porifera from Ponta do Ouro in Mozambique from coastal ecosystems revealed 26 species, including *Hyattella sulfurea, Hyattella pedunculata, Amphimedon palmata, Phoriospongia mozambiquensis, Chondrosia corticate, Callyspongia aerizusa, Clathria hirsute, Ciocalypta heterostyla, Callyspongia abnormis, Callyspongia pulitzeri, and Amphimedon brevispiculifera* (Calcinai et al., 2020). Ponta do Ouro is characterized by high endemism, which suggests that there is a possibility of finding endemic species of sponge in the area. Also, other taxa reported in Quirimba Archipelago were *Aplysina* sp., *Axinyssa* sp., *Biemma* sp., *Callyspongia* sp., *Carteriospongia foliascens., Cinachyrella voeltzkowi, Ciocalypta* sp., *Dysidea* sp., *Gelliodes* sp., *Geodia crustosa, Haliclona* sp., *Iotrochota* sp., *Jaspis* sp., *Lendenfeldia dendyi., Lissodendoryx* sp. (Barnes, 1999).

6.3.4 PHYLUM COELENTERATA

The sea animals that make up the phylum Coelenterata have radially symmetrical bodies and sensory tentacles on their mouths that make it easier to catch prey. The Phylum Coelenterata includes the hydra, comb jellies, genuine jellies, marine pens, coral animals, sea anemones, and other creatures (Moore, 2001). In the coastal area of East Africa, different species of this phylum have been described. For instance, in Kenya, *Hydra mariana,* a species of hydra, was reported (Cox and Young, 1973). A study in Taa Reef along the coast of the Tanga region, Tanzania, showed that *Scleractinia* coral were among the animals that were found in the area, yet their recruitment was affected by fishing activities, sedimentation rates, and competition for space with other organisms (Nzali et al., 1998).

6.3.5 PHYLUM PLATYHELMINTHES

Phylum platyhelminthes, acoelomates in the kingdom Animalia, comprises 13,000 species of flatworms, ranging in size from single-celled to 2–3 feet long. Members of this phylum are found in three classes: Turbellaria (e.g., *Planaria, Otoplana)*, Trematoda (*Fasciola hepatica, Diplozoon)*, and Cestoda (*Taenia* spp., *Convoluta)* (Collins, 2017). The members of the group play a great role in the ecosystem, while others act as parasites for vertebrates and nonvertebrate. For instance, in the coastal area of Kenya, the important species were *Fasciola gigantica* and *Stilesia hepatica*, which were found in slaughtered animals (Mungube et al., 2006).

6.3.6 PHYLUM NEMATODA

Nematoda, a diverse animal phylum with more than 28,000 species identified, includes roundworms, plant-parasitic nematodes, or eelworms, which inhabit various environments (Justine and Jamieson, 2000). Classes for this phylum include the primarily parasitic Phasmidia or Secernentea, characterized by the absence of caudal glands, the presence of unicellular, pouch-like sensory organs called plasmids, and the presence of paired lateral canals in the excretory system. *Ascaris* and *Enterobius* are two stellar examples from the group (Justine and Jamieson, 2000). A study done in the Tanga region of the coastal ecosystem of Tanzania revealed the availability of *Cryptosporidium* spp., which affects cattle from smallholder farmers (Ogden et al., 2007). Species of nematodes have also been observed in the coastal area of Kenya. For instance, species such as *Chromaspirina okemwai, Pseudochromadora interdigitatum*, *Eubostrichus africanus,* and *Eubostrichus longosetosus* have been observed from intertidal sediments of the *Ceriops* mangroves in Gazi Bay, Kenya (Muthumbi et al., 1995). In addition, another study to assess the diversity of nematodes in mangroves in Kenya revealed the presence of different genera of nematodes whose diversity was affected by seasons and species of mangrove, organic litter, sediment depth, oxygen levels, sediment composition, macro- and meiobenthic interactions, mangrove species, and human activities (Waweru et al., 2022). The genera of nematodes observed in this study were as follows: *Paralongicyatholaimus, Anoplostoma, Paralinhomoeus, Metalinhomoeus, Viscosia, Neochromadora, Desmolaimus, Spilophorella, Bathylaimus, Sphaerolaimus, Tubolaimodes, Spirinia, Paracomesoma, Terschellingia, Microlaimus, Desmodora, Dichromadora, Daptonema, Theristus Chromaspirina, Linhomoeus, Polysigma, Metacyatholaimus,* and *Molgolaimu* (Waweru et al., 2022).

Furthermore, a study in coastal forest of Kenya to understand the community structure of nematodes revealed high abundance of *Daptonema, Dorylaimopsis, Terschellingia,* in the shelf with a silty sediment-dwelling community while *Microlaimus* and *Halalaimus* dominated a sandy sediment-dwelling community (Muthumbi et al., 2004). The slope communities were dominated by the genera *Monhystera, Acantholaimus, Sabatieria, Molgolaimus,* and *Halalaimus* (Muthumbi et al., 2004).

A study on nematodes conducted in the Incomati River Estuary, Mozambique revealed a total of 35 nematode genera that included *Haliplectus, Axonolaimus*, *Terschellingia, Theristus, Sabatiera, Daptonema,* and *Parodontophora*. The species were noticed as good indicators of water body pollution in the area (Soko and Gyedu-Ababio, 2019). Their diversity and abundance in the study area were observed to be affected by the presence of very coarse sand, coarse sand, and fine sand, salinity, and grain size of sediments. Other factors that were recommended to affect the diversity of nematodes were organic content and Chl-a (Soko and Gyedu-Ababio, 2019).

6.3.7 PHYLUM ANNELIDA

Animalia's phylum Annelida is one of the largest in the animal kingdom with more than 15,000 species, including earthworms, sandworms, bristle worms, clam worms, fan worms, and leeches (Holland, 2011). Both aquatic and terrestrial settings support annelids (Holland, 2011). These invertebrate species have bilateral symmetry. They can be distinguished from other organisms by their segmented bodies (Holland, 2011). The various classes of annelids are as follows: Polychaeta, Oligochaeta, Hirudinea, and Archiannelida. Different species of Annelida have been reported in the coastal forests of East Africa. For instance, in Tanzania, earthworms (*Haplotaxida*) were observed to be among the invertebrates that were affected by forest clearance, season, and soil moisture in Dar es Salaam Forest (Middleton, 1987). Also, a study on the impacts of heavy metal pollution on benthic communities in the coastal forest of Zanzibar, Tanzania, revealed 25 species of benthic invertebrates, including annelida (Pankaj et al., 2013). Out of the 60 invertebrate taxa harvested for bait and food in Kenya and the South African coastal area, annelida species, particularly *Marphysa mossambica*, were among the dominant species (Simon et al., 2021). Furthermore, in a study on marine invertebrates that are found in Mozambique and So Tomé and Principe, the results showed that Annelia was among the group that was found in both countries (Bento et al., 2023).

6.3.8 PHYLUM ECHINODERMATA

The members of the phylum Echinodermata, such as *Sterias* (starfish), *Echinus* (sea urchin), *Antedon* (sea lily), *Cucumaria* (sea cucumber), and *Ophiura* (brittle star), are exclusively marine animals. The larval forms show bilateral symmetry and the adult forms show radial symmetry. They are triploblastic, exhibit organ system grade of organization, have a true coelom, are uniquely shaped and can be star like, elongated or spherical, are unsegmented without a head, have a surface covered with calcareous spicules, and have a distinguishing water vascular system (Toppr, 2023). Also, they have tube feet that help in locomotion. The brain is absent, but a nervous system is present with a nerve ring and radial nerve cords. Respiration occurs through the tube feet and gills. Sense organs are poorly developed and include tactile organs, chemoreceptors, terminal tentacles, sexes are separate, fertilization is external, and the lost parts can be regenerated (Toppr, 2023). Phylum Echinodermata has the following classes: Asteroidea, Ophiuroidea, Echinoidea, Holothuroidea, and Crinoidea (Toppr, 2023). Members of this phylum have been reported in the coastal areas of Eastern Africa. The species reported in Tanzania include Stalked crinoids (sea lilies), Hyocrinid, Bathycrinid, Ophiuroidea (brittle stars), and *Thoracaster cylindratus* (Gates, 2016). A checklist of Echinodermata in Kiunga Marine National Reserve, Kenya, revealed 37 species, which are found in 13 families and 19 genera, that make the park potential as a sanctuary for the conservation of marine invertebrate fauna. Among these species are *Clypeaster rarispinus, Phyllacanthus imperialis, Microcyphus rousseaui* (Echinoidea), *Holothuria (Cystipus) rigida, Holothuria (Platyperona) difficilis, Labidodemas pertinax, Stichopus chloronotus, Stichopus cf. monotuberculatus* (Holothuroidea), and *Holothuria (Theelothuria) turriscelsa* (Samyn and Berghe, 2000). Furthermore, invertebrate species under the phylum Echinodermata have been reported in Mozambique. For instance, sixteen species of shallow-water starfish have been reported in Inhaca Island and Maputo Bay of Mozambiques (Walenkamp, 1990). Among the species recorded are *Astropecten monacanthus, Patiriella exigua, Fromia milleporella, Monachaster sanderi, Ophidiaster hemprichii,* and *Stellaster equestri* (Walenkamp, 1990). Also, the fissiparous asterinid seastar *Aquilonastra chantalae* was described from Europa Island in Mozambique (O'loughlin and Mackenzie, 2013).

6.4 BIODIVERSITY HOTSPOTS

These forests emerge as thriving hubs of biological richness, hosting a plethora of unique and often endemic invertebrate species that contribute to the global tapestry of life. Biodiversity hotspots are specific geographic regions that host an exceptionally high concentration of species, many of which are found nowhere else in the world (Myers, 1988). Remarkable levels of biodiversity characterize these areas and often face significant threats, making them conservation priorities. Biodiversity hotspots are not only rich in species diversity but also contain unique ecosystems and contribute to global ecological resilience. The concept of biodiversity hotspots was developed by Norman Myers in the 1980s and later refined by Conservation International (Myers et al., 2000). In order to be designated as a hotspot, a location must satisfy two primary criteria: species richness and endemism. Species richness refers to the abundance of plant species unique to the area, while endemism signifies the presence of species that have evolved in isolation and are at heightened risk of habitat loss.

6.4.1 ENDEMIC INVERTEBRATES IN COASTAL FOREST OF EAST AFRICA

The East African coastal forests are diverse in endemism for invertebrates. For instance, there are 66 coastal forest patches spread across 700 square kilometers in Tanzania, which are home to a variety of endemic plant and animal species. As an illustration, there are 86 mollusks and 75 butterflies (TFCG, 2006). This brings the total number of endemic invertebrate species to roughly 161. The endemic species in Coastal forests include butterflies such as *Anthene rubrimaculata* and *Amanipodagrion gilliesi,* found in Amani nature reserve Muheza; wandering spiders such as *Afroneutria tanga* (Araneae, Ctenidae); *Babycurus gigas* (Scopion; and a jumping spider, *Tomomingi nywele,* found in Tanga (LNT, 2023). In addition, endemic species have included freshwater snails that are found in these coastal forests of Tanzania, such as *Potadomoides pelseneeri,* which is also considered endangered under the IUCN red list, and *Dendrolimax vangoethemi,* which is endemic in Pemba (Warren and Ngereza, 2010). The invertebrates that are endemic to the coastal forest of Kenya include butterflies, including *Aslauga marshalli, Aslauga latifurca*, and *Aslauga orientalis* (Libert and Collins, 2013). *Papilio desmondi teita* is endemic and endangered in the Taita hills of Kenya, while *Acraea matuapa, Lepidochrysops lukenia,* and *Spialia kituina* are endemic in the hills, but their conservation status has not been evaluated (Kioko et al., 2021). Also, land snails, *Gulella taitensis* (Gastropoda: Streptaxidae), are endemic and endangered; for instance, it has been reported to be endemic in Taita Hill forest (Lange, 2006). In addition, two species have become endemic and critically endangered in the area, namely, The Streptaxidae land snails *Gulella ndiwenyiensis* and *Gulella nictitans* (Rowson and Lange, 2007).

Apparently, there are 20,000 species of Mollusca, 7000 species of millipedes, and 20,000 species of butterflies. Of these, 125 species of Mollusca and 400 species of butterflies are found in East Africa's coastal forest (Burgess et al., 1998). According to estimates by Burgess et al. (1998), the coastal forest of East Africa is home to 550 endemic species, of which (n = 185) 33.63% are invertebrates, such as butterflies (79 species), snails (>86 species), and millipedes (>20 species). Lower Tana River, Arabuko-Sokoke, and Shimba Hills in Kenya; lowland East Usambara, Pugu Hills, Matumbi Hills, Rondo and Litipo; and other plateaus near Lindi in Tanzania; Pemba, a small island off the coast of Tanzania; and the Bazaruto archipelago in Mozambique are the forests with the highest concentrations of endemic species (Burgess et al., 1998). Examples of endemic butterfly species found in East Africa's coastal forest (Arabuko-Sokoke and "Usambara-Kwale") include *Iolaus bhda hyae, Ioluc mantimus mantimus, Aphnaeus comnae littoralis, Acraea mahapa, Coeliades h'thloa,* and *Gotyra diu,* and those found in Lindi are *Ticlna himboza* and *Celaenotrhinu kimboza* (Burgess et al., 1998).

6.5 THREATS TO INVERTEBRATES IN THE EAST AFRICA COASTAL FOREST

The second factor to consider is threats, whereby biodiversity hotspots are also areas facing significant habitat loss and degradation. These threats often result from human activities such as deforestation, urbanization, and agriculture, mining, and fire. The high level of species endemism combined with severe threats makes conservation in these areas a priority (Pascual et al., 2011). This has made several species of invertebrates vulnerable and endangered in the ecosystem. For instance, The East African coastal forests of Taita Hills Forests (four species), Muheza District Coastal Forests (one species), Pangani District Coastal Forests (one species), Shimba Hills (one species), and Zanzibar Coast (one species) are home to eight globally vulnerable species of gastropods, according to the 2015 Red List (Gereau et al., 2016). According to Gereau et al. (2016), there are about 26 insect species in East Africa's coastal forests that are globally vulnerable. These species are found on the Zanzibar coast, the East Usambara Mountains, and the West Usambara Mountains. Potamonautes, a group of freshwater crabs, and Malacostraca, a group of vulnerable freshwater crabs East Usambara Mountains (three species), West Usambara Mountains (two species), and Shimba Hills (one species) are the locations of the six vulnerable species in the East African coastal forest (Gereau et al., 2016). One species (*P. raybouldi*) is found in the Shimba Hills in Kenya and the East Usambara Mountains in Tanzania, while two of the vulnerable species (*P. infravallatus* and *P. xiphoidus*) are endemic to the East and West Usambara Mountains (Bayliss, 2002; Cumberlidge and Vannini, 2004).

In Wazo Hill, anthropogenic activities, particularly quarrying for cement production, Dar-es-Salam urbanization, and settlements, were revealed to have negative impacts on insects, with an improvement in diversity when restoration practices are carried out (Ngongolo and Mtoka, 2013). The 36 species of butterflies were observed to be affected by quarrying while benefiting from restoration programs, including *Eurema brigitta, Colotis euppe omphale,* and *Colotis euppe complexivus,* which occurred in both revegetated and un-mined sites; *Colotis ione, Anthene ligures, Eurema regularis*, *Desmolycaena rogersi, Gegenes nisobrevicornis,* and *Bicyclus safitze* (Ngongolo and Mtoka, 2013).

On the other hand, the assessments indicated that despite their importance, coastal forests in Kenya and Tanzania are still in danger (TFCG, 2006). More than 25% of Tanzania's coastal forest has been lost as a result of heavy pressure from fires, logging, and the production of charcoal (TFCG, 2006). Charcoal burning, fuel wood collection, destructive mining practices, expanding agriculture, settlement, urbanization, lack of legal protection, uncontrolled fires, mining (Quarrying), and wildlife–human conflicts (elephants) are the main threats to the coastal forests in the hotspot. Poverty is a root cause behind many of these issues, and with more than 20 million people depending on the natural resources of the coastal forest, the pressures are high (Ngongolo and Mtoka, 2013; TFCG, 2006). For instance, Pugu forest reserve and Kazimzumbwi forest reserve in Tanzania are facing anthropogenic activities like agriculture, logging, pole cutting, firewood, charcoal making, and firewood collection (Clarke and Dickinson, 1995; TFCG, 2006).

In the last decade, coastal forests near Dar es Salaam (Pande Game Reserve, Pugu, Kazimzumbwi, and Ruvu South Forest Reserves) have lost significant areas of forest due to charcoal burning, while unsustainable, illegal logging using pit sawing techniques is an increasing problem, particularly in Southern Tanzania and Northern Mozambique (Milledge et al., 2007). Also, Mining is another challenge to the habitats for invertebrates in the East African coastal forest. The coastal forests have a wealth of mineral resources (gas, gemstones, iron, manganese, titanium, limestone, and kaolin) that have been exploited using destructive mining practices (Mwamfupe, 1998). For example, limestone, kaolin, rubies, and other precious and semiprecious stones are mined from a few Tanzanian coastal forests (Mwamfupe, 1998), while iron and manganese have been extracted from the Kwale Kaya forests in Kenya (Abuodha, 2002). In Mozambique, it has been observed that uncontrolled fires, the mining industry, an increase in the human population, increased demand for natural resources, overexploitation of resources, involvement in illegal activities (poaching, corruption), and infrastructure development on coastal forests increase (Albano, 2004).

To ensure the survival of invertebrates, different approaches have been taken in this coastal forest in East Africa. For instance, Inhamitanga Forest Reserve in Mozambique was gazetted in 1957 and managed by the National Directorate of Land and Forestry (TFCG, 2007). Saadani National Park in Tanzania was designated a National Park in 2005, which incorporated the Saadani Game Reserve, Zaraninge Forest Reserve, and Mwkaja Ranch, managed by Tanzania National Parks, while Pugu Forest is annexed as Part of Kisarawe District Coastal Forests IBA 47 (TFCG, 2006). On the other hand, in the Kaya coastal forests in Kenya, a conservation initiative is being implemented by the National Museums of Kenya's Coastal Forest Conservation Unit with support from World Wildlife Fund for nature, which has implemented conservation initiatives in the Kaya Forests from 1992 to 2001 (TFCG, 2006).

6.6 CONCLUSION

In conclusion, this chapter provides a deep insight into the complex and diverse world of invertebrate life thriving in the coastal forests of this region. By delving into this subject, it has shed light on the crucial ecological functions performed by these frequently unnoticed creatures and their significant influence on the well-being and sustainability of these ecosystems. The invertebrates within these forests, spanning arachnids, insects, mollusks, and crustaceans, form an interconnected web of life that contributes to pollination, nutrient cycling, decomposition, and overall ecosystem resilience. However, their existence is underlined by the looming threats of habitat degradation, climate change, and human activities. The journey underscores the urgent need for conservation efforts that prioritize the protection of these invertebrate species and their habitats. Engaging in interdisciplinary research, promoting community involvement, and advocating for sustainable practices are essential for protecting the delicate equilibrium that supports the diverse biodiversity of coastal forests and the welfare of local communities. In the tapestry of life, invertebrates are intricate threads that weave the story of these ecosystems. As conclusive in this chapter, we are reminded of the profound interconnectedness of all life forms and the responsibility we share to ensure their survival for generations to come.

KEYWORDS

- **invertebrates**
- **forest conservation**
- **species diversity**
- **benefits and threat**

REFERENCES

Abuodha, J. O. Z. Environmental impact assessment of the proposed titanium mining project in Kwale District, Kenya. *Mar. Georesources Geotechnol.* **2002,** *20*(3), 199–207.

Albano, G. *Coastal Forests of Mozambique. Socio-economic Aspects—Review*. WWF Eastern Africa Regional Programme Office, **2004,** p. 22.

Barnes, D. K. High diversity of tropical intertidal zone sponges in temperature, salinity and current extremes. *African J. Ecol.* **1999,** *37*(4), 424–434.

Bayliss, J. The East Usambara tree-hole crab (Brachyura: Potamoidea: Potamonautidae)—a striking example of crustacean adaptation in closed canopy forest, Tanzania. *African J. Ecol.* **2002,** *40*, 26–34.

Bento, M.; Niza, H.; Cartaxana, A.; Bandeira, S.; Paula, J.; Correia, A. M. Mind the gaps: Taxonomic, geographic and temporal data of marine invertebrate databases from mozambique and são tomé and príncipe. *Diversity* **2023,** *15*(1), 70.

Burgess, N. D.; Clarke, G. P.; Rodgers, W. A. Coastal forests of eastern Africa: Status, endemism patterns and their potential causes. *Biol. J. Linn. Soc.* **1998,** *64*(3), 367, https://doi.org/10.1111/j.1095-8312.1998.tb00337.x

Burgess, N. D.; Ponder, K. L.; Goddard, J. Surface and leaf-litter arthropods in the coastal forests of Tanzania. *African J. Ecol.* **1999,** *37*(3), 355–365.

Calcinai, B.; Belfiore, G.; Pica, D.; Torsani, F.; Palma, M.; Cerrano, C. Porifera from Ponta do Ouro (Mozambique). *European J. Taxon.* **2020,** (698).

Chapman, A. D. *Number of Living Species in Australia and in the World.* 2nd ed. Department of Sustainability, Environment, Water, Population and Communities: Canberra, Australia, **2009**.

Clarke, G. P.; Dickinson, A. *Status Reports for 11 Coastal Forests in Coast Region, Tanzania*. Frontier-Tanzania Technical Report No. 17. The Society for Environmental Exploration: U.K. and The University of Dar es Salaam: Tanzania, **1995**.

Collins, J. J. Platyhelminthes. *Curr. Biol.* **2017,** *27*(7), R252–R256.

Cox, N.; Young, J. O. A new species of *Hydra* (Coelenterata) from Kenya, East Africa. *J. Zool.* **1973,** *170*(4), 441–449.

Cumberlidge, N.; Vannini, M. Ecology and taxonomy of a tree-living freshwater crab (Brachyura: Potamoidea: Potamonautidae) from Kenya and Tanzania, East Africa. *J. Nat. Hist.* **2004,** *38*(6), 681–693.

Dalila, M. N. *The Distribution and Diversity of Land Snails in Shimba Hills National Reserve, Kenya.* Doctoral dissertation, University of Nairobi, **2009**.

de Voogd, N. J.; Alvarez, B.; Boury-Esnault, N.; Carballo, J. L.; Cárdenas, P.; Díaz, M.-C.; Dohrmann, M.; Downey, R.; Goodwin, C.; Hajdu, E.; Hooper, J. N. A.; Kelly, M.; Klautau, M.; Lim, S. C.; Manconi, R.; Morrow, C.; Pinheiro, U.; Pisera, A. B.; Ríos, P.; Rützler, K.; Schönberg, C.; Vacelet, J.; van Soest, R. W. M.; Xavier, J. World Porifera Database, **2023**. https://www.marinespecies.org/porifera Accessed August 19, 2023. https://doi.org/10.14284/359.

Dida, G. O.; Anyona, D. N.; Abuom, P. O.; Akoko, D.; Adoka, S. O.; Matano, A. S.; Owuor, P. O.; Ouma, C. Spatial distribution and habitat characterization of mosquito species during the dry season along the Mara River and its tributaries, in Kenya and Tanzania. *Infect. Dis. Poverty* **2018,** *7*(1), 1–16.

Ebhodaghe, F. I.; Bastos, A. D.; Okal, M. N.; Masiga, D. K. Entomological assessment of tsetse-borne trypanosome risk in the Shimba Hills human-wildlife-livestock interface, Kenya. *Front. Vet. Sci.* **2022,** *9*, 931078.

Garbach, K.; Milder, J. C.; Montenegro, M.; Karp, D. S.; DeClerck, F. A. J. Biodiversity and ecosystem services in agroecosystems. In: *Encyclopedia of Agriculture and Food Systems*, vol. 2. **2014,** pp. 21–40.

Gates, A. *Deep-sea Life of Tanzania*. National Oceanography Centre: Southampton, **2016**. https://nora.nerc.ac.uk/id/eprint/513680/1/Deep%20sea%20life%20of%20Tanzania_2016_reduced.pdf. Accessed 25 August 2023.

Gereau, R. E.; Cumberlidge, N.; Hemp, C.; Hochkirch, A.; Jones, T.; Kariuki, M.; Lange, C. N.; Loader, S. P.; Malonza, P. K.; Menegon, M.; P. Kariuki Ndang'ang'a, P.; Rovero, F.; Shirk, P. Globally threatened biodiversity of the Eastern Arc Mountains and coastal forests of Kenya and Tanzania. *J. East African Nat. Hist.* **2016,** *105*(1), 115–201.

Hemp, C. The influence of fire on Saltatoria diversity in coastal habitats near Pangani, Tanzania (East Africa). *Ecotropica* **2005,** *11*(1–2), 53–61.

Holland, P. *The Animal Kingdom: A Very Short Introduction.* Oxford University Press: Oxford, **2011**.

Huber, B. A.; Warui, C. M. East African pholcid spiders: An overview, with descriptions of eight new species (Araneae, Pholcidae), **2012,** http://repository.mut.ac.ke:8080/xmlui/handle/123456789/2959. Accessed 14 August 2023.

Justine, J. L.; Jamieson, B. G. *Nematoda. Reproductive Biology of Invertebrates. Voume 9, Part B: Progress in Male Gamete Ultrastructure and Phylogeny,* **2000,** pp. 183–266.

Kioko, E. N.; Musyoki, A. M.; Luanga, A. E.; Genga, O. C.; Mwinzi, D. K. *Fluttering Beauty with Benefits: The Butterflies of Taita Hills. A Field Guide.* National Museums of Kenya: Nairobi, Kenya, **2021**.

Kiwia, H. Y. D. Species richness and abundance estimates of small mammals in Zaraninge coastal forest in Tanzania. *Tanzania J. Sci.* **2006,** *32*(2), 51–60.

Lange, C. N. The endemic land snail *Gulella taitensis* of the Taita Hills forests, Kenya: On the brink of extinction. *Oryx* **2006,** *40*(3), 362–364

Lehmann, I.; Kioko, E. Preliminary survey on butterflies and moths and their habitats in two Kaya forests of the Kenya coast. *Metamorphosis* **2000,** *4*, 1–52.

Libert, M.; Collins, S. C. Description of two new species of African Lycaenidae in the collection of the African Butterfly Research Institute. *Metamorphosis* **2013,** *24*, 3–6.

LNT (Living nation treasure). Tanzania, **2023,** https://lntreasures.com/tanzania.html. Accessed 15 2023.

Lovett, J. C. The botanical importance of the East Usambara Forests. In: *Forest Conservation in the East Usambara Mountains, Tanzania* vol. *15*, **1989,** p. 207.

Matiku, P. *The Coastal Forests of Kenya Forests Data, Threats, Socio-economic Issues, Values, Stakeholders, Challenges, Strategies, Investment and Enabling Environment.* A national synthesis report for the development of the WWF-EARPO Eastern Africa Coastal Forests Ecoregion Programme, **2023,** https://coastalforests.tfcg.org/pubs/National-Synthesis-Ken.pdf. Accessed 13 August 2023.

May, R. M. How many species are there on earth? *Science* **1988,** *241*(4872), 1441–1449. https://doi.org/10.1126/science.241.4872.1441.

Middleton, J. Effect of forest clearance, season and soil moisture on the activity of the larger soil-dwelling invertebrates in Dar es Salaam, Tanzania. *Afr. J. Ecol.* **1987,** *25*(4), 247–254.

Milledge, S. A. H.; Gelvas, I. K.; Ahrends, A. *Forestry, Governance and National Development: Lessons Learned from a Logging Boom in Southern Tanzania*. TRAFFIC East/Southern Africa/Tanzania Development Partners Group/Ministry of Natural Resources of Tourism: Dar es Salaam, Tanzania, **2007,** p. 252.

Moore, J. *An Introduction to the Invertebrates*. Cambridge University Press: Cambridge, **2001**.

Mungube, E. O.; Bauni, S. M.; Tenhagen, B. A.; Wamae, L. W.; Nginyi, J. M.; Mugambi, J. M. The prevalence and economic significance of *Fasciola gigantica* and *Stilesia hepatica* in slaughtered animals in the semi-arid coastal Kenya. *Trop. Anim. Health Prod.* **2006,** *38*, 475–483.

Muthumbi, A. W.; Vanreusel, A.; Duineveld, G.; Soetaert, K.; Vincx, M. Nematode community structure along the continental slope off the Kenyan Coast, Western Indian Ocean. *Intern. Rev. Hydrobiol.* **2004,** *89*(2), 188–205.

Muthumbi, A.; Verschelde, D.; Vincx, M. New Desmodoridae (Nematoda: Desmodoroidea): Three new species from *Ceriops* mangrove sediments (Kenya) and one related species from the North Sea. *Cahier de Biol. Marin* **1995,** *36*, 181–195

Mwamfupe, D. Demographic impacts on protected areas in Tanzania and options for action. *Parks* **1998,** *8*(1), 3–14.
Mwamuye, M. M. Kariuki, E.; Omondi, D.; Kabii, J.; Odongo, D.; Masiga, D.; Villinger, J. Novel Rickettsia and emergent tick-borne pathogens: A molecular survey of ticks and tick-borne pathogens in Shimba Hills National Reserve, Kenya. *Ticks Tick-borne Dis.* **2017,** *8*(2), 208–218.
Myers, N. Threatened biotas: "hot spots" in tropical forests. *Environmentalist* **1988,** *8*, 187–208.
Myers, N.; Mittermeier, R. A.; Mittermeier, C. G.; Da Fonseca, G. A.; Kent, J. Biodiversity hotspots for conservation priorities. *Nature* **2000,** *403*(6772), 853–858.
Newmark, W. D. *Conserving Biodiversity in East African Forests: A Study of the Eastern Arc Mountains*. Springer Science & Business Media, **2002,** p. 155
Ngongolo, K.; Mtoka, S. Using butterflies to measure biodiversity health in Wazo hill restored quarry. *J. Entomol. Zool. Stud.* **2013,** *1*(4), 81–86.
Ngongolo, K.; Mtoka, S.; Mahulu, A. The Abundance and pollinators' impact on seed setting of *Leucaena leucocephala* in Wazo Hill restored Quarry, Tanzania. *J. Zool. Biosci. Res.* **2014a**, *1*(2), 6–10.
Ngongolo, K.; Mtoka, S.; Mahulu, A.; Sigala, A. The *Leucaena leucocephala* floral visitors, pollinators and their predators in the restored Wazo Hill Quarry, Tanzania. *Entomol. Appl. Sci. Lett.* **2014b**, *1*, 36–42.
Ngongolo, K.; Mtoka, S.; Rubanza, C. D. Diversity and abundance of insects in Kichi Forest and adjacent local communities' farms. *J. Zool. Biosci. Res.* **2015a**, *2*(2), 47–53.
Ngongolo, K.; Mtoka, S.; Rubanza, C. D. Floral visitors and pollinators of sesame (*Sesamum indicum* L.) from Kichi Forest to the adjacent local communities' farms. *Entomol. Appl. Sci. Lett.* **2015b**, *2*(2), 32–39.
Nzali, L. M.; Johnstone, R. W.; Mgaya, Y. D. Factors affecting scleractinian coral recruitment on a nearshore reef in Tanzania. *Ambio* **1998,** *27*, 717–722.
O'loughlin, P. M.; Mackenzie, M. Asterinid seastars from the Mozambique channel (Echinodermata: Asteroidea: Asterinidae). *Zootaxa* **2013,** *3613*(2), 176–180.
Ogden, N. H.; Kambarage, D. M.; Bryant, M. J.; Fitzpatrick, J. L.; Karimuribo, E. D.; French, N. P.; Swai, E. S. Prevalence and determinants of *Cryptosporidium* spp. infection in smallholder dairy cattle in Iringa and Tanga Regions of Tanzania. *Onderstepoort J. Vet. Res.* **2007,** *74*(1), 23–29.
Pankaj, P.; Elskens, M.; Leermakers, M.; Sheikh, M.; Kochzius, M. Impacts of heavy metal pollution on benthic communities on the coast of Zanzibar, Tanzania. In: Mees, J.; Seys, J. (Eds.), *Book of Abstracts—VLIZ Young Scientists' Day. Brugge, Belgium, 15 February 2013*. VLIZ Special Publication 63. Vlaams Instituut voor de Zee – Flanders Marine Institute (VLIZ): Oostende, Belgium. **2013,** pp. 168 + xii.
Pascual, L. L.; Luigi, M.; Alessandra, F.; Emilio, B.; Luigi, B. Hotspots of species richness, threat and endemism for terrestrial vertebrates in SW Europe. *Acta Oecologica* **2011,** *37*(5), 399–412.
Ragavan, P.; Dubey, S. K.; Dagar, J. C.; Mohan, P. M.; Ravichandran, K.; Jayaraj, R. S. C.; Rana, T. S. Current understanding of the mangrove forests of India. *Res. Dev. Saline Agric.* **2019,** 257–304.
Rowson, B.; Lange, C. N. Two new species of *Gulella* (Mollusca: Pulmonata: Treptaxidae) from the Taita Hills, Kenya. *Afr. Invertebr.* **2007,** *48*(2), 21–32.
Samyn, Y.; Berghe, E. V. Annotated checklist of the echinoderms from the Kiunga Marine National Reserve, Kenya. Part I: Echinoidea and Holothuroidea. *J. East African Nat. Hist.* **2000,** *89*(1), 1–36.
Simon, C. A.; Muthumbi, A. W.; Kihia, C. M.; Smith, K. M.; Cedras, R. B.; Mahatante, P. T.; Wang'ondu, V. W.; Katikiro, R. A review of marine invertebrates used as fishing baits and the implications for national and regional management in the Western Indian Ocean. *Afr. Zool.* **2021,** *56*(4), 237–263.
Smith, R. A checklist of the spiders of Tanzania. *Environ. Sci.* **2020,** *109*(1), 1–41
Soko, M. I.; Gyedu-Ababio, T. K. Free-living nematodes as pollution indicator in Incomati River Estuary, Mozambique. *Open J. Ecol.* **2019,** *9*(5), 117–133.
TFCG (Tanzania Forest Conservation group for the future). *Tanzania - Coast (Pwani) Region*. Districts in Coast (Pwani) Region, **2006**. http://coastalforests.tfcg.org/tz-coast.html#Pugu. Accessed 15 August 2023.
TFCG. Mozambique. Coastal forest of Kenya and Tanzania, **2007,** http://coastalforests.tfcg.org/mozambique.html. Accessed 16 August 2023.
Toppr. Animal kingdom. *Phylum Echinodermata,* **2023**. https://www.toppr.com/guides/biology/animal-kingdom/phylum-echinodermata/#Examples. Accessed 21 August 2023.

Walenkamp, J. H. C. Systematics and zoogeography of Asteroidea (Echinodermata) from Inhaca Island, Mozambique. *Zoologische Verhandelingen* **1990,** *261*(1), 1–86.

Warren, H. B. B. R.; Ngereza, F. C. Terrestrial molluscs of Pemba Island, Zanzibar, Tanzania, and its status as an "oceanic" island. *ZooKeys* **2010,** *70*, 1–39.

Waweru, B. W.; Muthumbi, A. W.; Vanreusel, A.; Wangondu, V.; Mutua, A. Free-living marine nematode communities in *Rhizophora mucronata* Lam. (Rhizophoraceae) forest at Mida Creek, Kenya. *Western Indian Ocean J. Mar. Sci.* **2022,** *21*(2), 33–43.

CHAPTER 7

BIODIVERSITY, CHALLENGES, AND PRESERVATION EFFORTS FOR INSECTS IN THE COASTAL REGIONS OF EAST AFRICA

KELVIN NGONGOLO, FRANCIS CHEBBY, and NAZA MMBAGA

Department of Biology, College of Natural and Mathematical Sciences, The University of Dodoma, Dodoma, Tanzania

ABSTRACT

The Coastal Area ecosystem is of higher concern for biodiversity conservation as it is endowed with high species diversity and endemism, both of which are highly threatened. Limited data is available regarding species diversity in this region, with existing information being widely scattered. The following chapter provides an overview of the challenges and conservation measures for insect species found in the coastal forests of East African nations such as Tanzania, Kenya, and Mozambique. The coastal areas of East Africa are well known as biodiversity hotspots. Insects of these areas have been underrepresented although some literature has tried to disseminate information on their distribution and diversity. For instance, it was reported that about 79 butterfly species are endemic to coastal forests of East Africa. The major threats to the insects in the coastal forests of east Africa are associated with anthropogenic factors such as logging, industrialization, settlement, urbanization, agriculture, and unprescribed burning (fire). Conservation strategies are being developed to ensure that the biodiversity integrity of this coastal forest is preserved. Among the conservation strategies are the establishment of protected areas such as game reserves (Pande game reserve) and forest reserves with the support of governmental and nongovernmental organizations. Because local communities are the custodians of biodiversity destruction in this east African coastal forest, it is recommended that they participate in all conservation strategies and raise awareness through conservation education.

7.1 INTRODUCTION

East African coastal areas are among the mega biodiversity hotspots well known for its richness in biodiversity and beautiful lands (Critical Ecosystem Partnership Fund, 2003, 2007).

Biodiversity Hotspot of the Coastal Forests of Eastern Africa. T. Pullaiah and Ally J. K. Nkwabi (Eds.)

DOI: 10.1201/9781998511099-7

It encompasses high mountains, grasslands, mangroves and healthy coastal forests which harbor a lot of plant and animal species which belong to different taxonomic ranks. Moreover, it has about 15% of endemic species whereby 60–70% of the wildlife is only found in the Indo-Pacific zone. The East Africa coastal zone extends to about 4600 km from Southern Africa to the Natal shores of South Africa (Figure 7.1) (Gereau et al., 2016). The coast of East Africa remains a potential biodiversity hotspot due to its supportive natural habitats of woodlands and forests including the healthy Eastern Arc Mountain forests which is well known for all its richness in endemic species (Burgess and Clarke, 2000; Salehe, 2011).

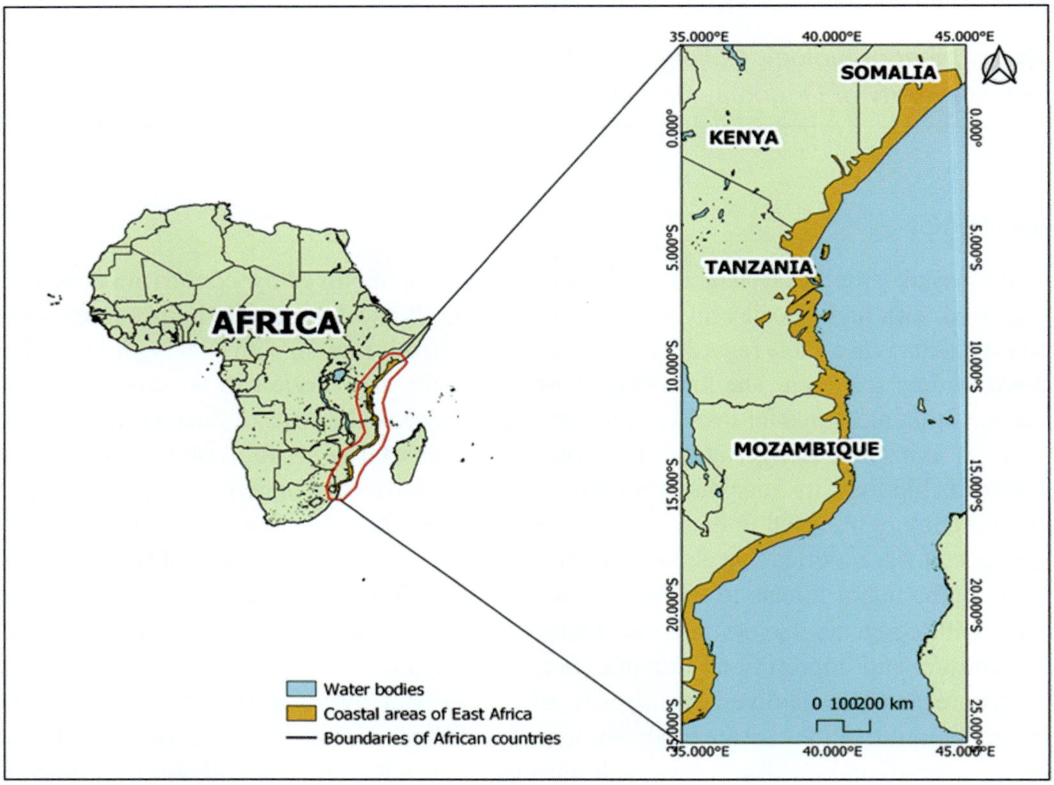

FIGURE 7.1 A map showing coastal areas of East Africa (Created using QGIS Version 3.26.3 Buenos Aires).

The richness of East Africa coastal areas involves the diversity ranging from high to lower flora and fauna species which are either native or invasive. Moreover, East Africa coastal forests are well known for their endemism. For instance, according to some studies, the number of endemic species found in selected zones of East African coastal areas which are Somalia, Kenya, Tanzania and Mozambique include plants (550), birds (9), reptiles (26), amphibians (2), mollusks (86), and butterflies (79) (Burgess et al., 1998). A vast literature has reported the information on habitats, diversity and distribution of mega flora and fauna including birds and fish found in the East African coastal areas.

Insects are very important in many aspects of life as they bring both joy and sorrow in human life and the natural environment (McKelvey, 1975). Their contribution has to be valued due to the diverse ecological roles they play. Scientists rely heavily on insects to gather information and reach conclusions on a wide range of issues and phenomena in various fields such as paleolimnology, developmental biology, genetics, ecology, biomechanics, and climate change (Scudder, 2017). There are no accurate estimates of the exact number of insects, although some studies have predicted it to be around 2–80 million globally (Hodkinson and Casson, 1991; Stork, 1993; Erwin, 2004). According to Scudder (2017), the current estimate on the number of insects globally is about 1,060,704, however only about 20% of all insects have been named so far (Grimaldi and Engel, 2005).

The ability of insects to facilitate the occurrence of some beneficial interactions in the environment through pollination, alteration of microhabit structures and seed dispersal to name a few, places them at the center of scientific studies especially during this century of massive climate change (Miller, 1993; Cock et al., 2012). Apart from having environmental and ecological contributions, insects are threatened in multiple ways. In the Red List of threatened species released by IUCN in 2015, about 43 insect species are threatened globally (IUCN, 2015). A total of 35 species previously unrecorded were added to the list during the years 2013–2015, with no documented alterations during that period (Gereau et al., 2016). It was also reported that the number of threatened species in East Africa coastal forests varies from one forest to another. For instance, East Usambara Mountains and Uluguru Mountains had a large number of threatened species of 17 and 15, respectively. The trend followed by West Usambara Mountains (five), Ukaguru Mountains (five), Zanzibar Coast (four), Udzungwa Mountains (three), and Nguru Mountains (three) (Gereau et al., 2016).

The orders of insects found in coastal East Africa is not much studied, however with the few available literature, Coleoptera, Lepidoptera, Orthoptera, Diptera and Hymenoptera are among the most prominent orders found in coastal East Africa. This review thoroughly examines the literature on the major insect orders found in the coastal zones of East Africa, encompassing both historical and contemporary studies. The data for this review was collected from a variety of sources, with a primary focus on documents detailing the diversity, threats, and conservation efforts surrounding coastal forest insects in East Africa. The target sources of information to be reviewed include posters, articles, reports, proceedings, and other unpublished documents. Most of the Internet-based information was retrieved from scholarly search databases such as Journal Storage (JSTOR) and Google Scholar. Keywords were used to search for relevant information for this chapter. Keywords include "diversity of insects in East African coastal areas," "distribution of insects in East African coastal zones," "threats to insects along East African coastal forests," and "conservation strategies for insects in East African coastal forests." Moreover, the search for specific species or specific coastal forests was employed whenever necessary to get detailed information about some of their required features, including morphology, distribution, and threats. The documents that had relevant information were included in the review, while those without the required information were excluded.

7.2 INSECTS DIVERSITY IN EAST AFRICAN COASTAL AREAS

The number of species of insect worldwide has been reported to be higher compared to other animals. For instance, more than 1,000,000 species of insects have been described, and 95% of them are known. The dominant order of insects is Coleoptera, Diptera, Hymenoptera, and Lepidoptera, with an estimated 670,000 species (Stork, 2018). In Africa, specifically East Africa, there are many species of insects. However, in this chapter, we highlighted a few species that are commonly found in eastern coastal ecosystems and the conservation threats they face.

7.2.1 ORDER DIPTERA: GRASS FLIES

Grass flies belong to Order Diptera in the family Chloropidae (Brachycera, Calyptrate). The species is commonly known as frit or grass flies. This class has nearly 3200 species which have been globally identified including in East African coastal areas (Nartshuk et al., 2020) (Figure 7.2). Chloropidae is among the successful families of Order Diptera because of its ability to live in a variety of habitats including steppe, wetlands, ponds, forests, and swamps (Nartshuk et al., 2020). Moreover, grass flies possess the capability to make use of different substrates, thereby contributing significantly to the ecological balance within the ecosystem.

FIGURE 7.2 Some East African Coastal Diptera: (A). Calyptrate flies (Zoosubsection Calyptratae), (B). Brown flies (*Drosophila melanogaster*).

Source: Field data.

7.2.3 ORDER HEMIPTERA: MEALYBUGS (PSEUDOCOCCIDAE)

Pseudococcidae is a family whose members are commonly known as mealybugs. One of the characteristics possessed by the members of this family is scale in their bodies. They are known as insect pests of herbaceous and woody plants. Pseudococcidae promotes sooty mold growth, transmits plant viruses, and affects the host plant directly (Daane et al., 2012). Pseudococcidae has been listed as a second large family after Diaspididae which has been documented to have not less than 2653 species and 418 genera (García et al., 2016). According to the most recent documentation, there are approximately 2032 species of Pseudococcidae spanning across 258 genera worldwide (García et al., 2016).

7.2.4 ORDER MANTODEA: PRAYING MANTIS (MANTIS RELIGIOSA)

The praying mantis includes a group of a large number of species which have successfully distributed all over the world in different habitats ranging from arid areas to rainforests. Out of 20 families of the order Mantodea, Mantidae is the largest family with 21 subfamilies and almost 50% of all species found in this order. Praying mantids have a variety of behavioral, habitat and morphological adaptations which enable them to live in diverse environmental conditions (Patel and Singh, 2016).

The praying mantids found in East African coastal areas, like other species of the same kind, have a specific predatory lifestyle associated with their habitats. They are capable of camouflaging for easy prey capturing and aggressive resemblance which help them to avoid their potential predators. Moreover, genus *Sphodromantis*, *Tenodera,* and *Miomantis* to name a few, were once reported to be found in these East African coastal areas especially in Kenya, Somalia, Zanzibar, Tanzania and Natal (Patel and Singh, 2016).

7.2.5 ORDER ORTHOPTERA: BROWN-SPOTTED LOCUST (CYRTACANTHACRIS TATARICA)

The species belong to the family Acrididae and are sometimes referred to as true grasshoppers. In East African coastal regions, approximately 42 genera of Acrididae family species have been identified (Popov et al., 2019). Most of the Acrididae members are potential pests of vegetables, fruits, forests, and other agricultural crops. The family Acrididae has been documented as the most diverse group compared to the rest in order orthoptera (Panhwar, 2018). In general terms, grasshoppers which are found in East African coastal areas can be categorized into two forms which are; long-horned (Ensifera) and short-horned (Caelifera) including members of family Acrididae and pyrgomorphidae (Memon and Panhwar, 2021) (Figure 7.3). Among the two, Caelifera (short horned grasshoppers) are the dominant and diverse group in the order orthoptera having an estimate of about 25,000 members globally.

FIGURE 7.3 One of the East African Coastal Orthoptera species—a member of Gaudy Grasshoppers (*Phymateus viridipes*) (Family pyromorphidae).

Source: Field data.

7.2.6 ORDER ODONATA: COMMON CITRIL (CERIAGRION GLABRUM)

Coenagrionidae is the family that includes common citrils which is a species of damselfly. Coenagrionidae are common and by far widespread in different areas of Africa including the East African coastal areas (Döring, 2020). It is reported to be found in a variety of habitats of freshwater including marshes, ponds, rivers, and streams. Moreover, common citril prefers the areas with reeds, sedges, or grasses. Most species of the family Coenagrionidae are usually found near water sources and sometimes away from the water sources.

7.2.7 ORDER HEMIPTERA: TRUE BUGS (PIEZOSTERNUM CALIDUM)

A large number of true bugs belong to the family Tessaratomidae. About 240 species of true bugs are contained in this family which has been further divided into 3 subfamilies and 56 genera (Magnien et al., 2008). Other species of order Hemiptera that have been observed in the coastal areas of East Africa include African Cotton Stainer (*Dysdercus fasciatus*), *Dysdercus nigrofasciatus* and African Giant Water Bug (*Lethocerus cordofanus*) (Figure 7.4).

FIGURE 7.4 Some East African Coastal Hemiptera: (A) *Acanthocephala alata* (Subfamily Coreinae), (B). *Acanthocephala terminalis* (Tribe Acanthocephalini), (C) *Cermatulus nasalis* (Family Pentatomidae), (D) *Murgantia histrionica* (Family Pentatomidae) (*Source: Field data*).

Tessaratomidae members are plant eaters found in coastal areas of East Africa that have some adaptive reproductive behaviors and mechanisms which enable them to catch their prey while avoiding their potential enemies (parasitoid wasps and some hemipterans) (Rider, 2008). They usually possess an ability of producing the chemicals (allomones) which once released can deter fellow arthropods or cause harm to human skin or even temporary blindness once sprayed onto the human eyes (Magnien et al., 2008).

7.2.8 ORDER LEPIDOPTERA: VIOLET-SPOTTED CHARAXES (CHARAXES VIOLETTA)

Charaxes violetta is among the members of family Nymphalidae which belongs to the subfamily Papilionoidea. Nymphalidae is the largest group of butterflies that is widely spread across the world dominating a variety of habitats including the East African coastal areas. In the coastal regions of Kenya in East Africa, a research conducted by Rogo and Odulaja (2001) indicated that butterfly species from the Nymphalidae family were prevalent, accounting for over 60% of all butterfly species documented in the study across various coastal zones.

Other butterfly species that are found in the coastal areas of Eastern Africa include *Gorgyra subflavidus* (Hesperiidae), *Charaxes varanes vologeses* (Nymphalidae), *Hypolimnas usambara* (Nymphalidae), *Pentila tropicalis* (Lycaenidae), *Acraea boopis ama* (Nymphalidae), *Amauris niavius* (Nymphalidae), *Bicyclus safitza safitza* (Nymphalidae), *Byblia anvatara* (Nymphalidae), *Charaxes candiope* (Nymphalidae), *Graphium Colonna* (Papilionidae) and *Appias epaphia orbona, Eurema brigitta brigitta, Colotiseuppe omphale, Colotis euppe complexivus, Colotis ione, Antheneligures, Eurema regularis regularis, Desmolycaena rogersi, Gegenes nisobrevicornis*, and *Bicyclus safitze safitze* (Rogo and Odulaja, 2001; Ngongolo and Mtoka, 2013) (Figure 7.5).

FIGURE 7.5 One of the East African Coastal Lepidoptera species—Guineafowl butterfly (*Hamanumida daedalus*).

Source: Field data.

7.2.9 ORDER COLEOPTERA: ZIG-ZAG FRUIT CHAFER (ANISORRHINA FLAVOMACULATA)

Anisorrhina flavomaculata which belong to family Scarabaeidae is among the species of order Coleoptera that are found in coastal areas of East Africa. Coleoptera is the most diverse and largest order in the animal kingdom globally (Having more than 37,000 species globally). The genus *Pachnoda* includes approximately 108 species found throughout Africa and specifically in the coastal regions of East Africa. Their well-developed chewing mouthparts and hard pair of wings covering the abdomen are among the body features that enable them to adapt in a variety of habitats world widely (Basson et al., 2019)

Apart from *A. flavomaculata*, the other species found in order Coleoptera that can be found in the East African coastal areas include lady beetle (*Coccinella septempunctata*), Dune Tortoise Beetle (*Aspidimorpha puncticosta*), Rhinoceros Beetles (subfamily Dynastinae) and Tortoise and Hispine Beetles (*Cassidinae*), June Beetles (subfamily *Melolonthinae*), and White-Spotted Fruit Chafer (*Mausoleopsis amabilis*) (Figure 7.6).

FIGURE 7.6 One of the East African Coastal Coleoptera species—White-Spotted Fruit Chafer (*Mausoleopsis amabilis*).

Source: Field data.

7.2.10 ORDER HYMENOPTERA: BEEWOLF (PHILANTHUS TRIANGULUM DIADEMA)

Beewolf which is sometimes known as bee-killer or bee-hunter *Philanthus* belongs to family Crabronidae and genus *Philanthus* and have more than 135 species (Kroiss et al., 2010). In the East African coastal areas, some members of the family Crabronidae including Beewolf can be easily located in lowlands and coastal dunes. Moreover, globally, species in this order have been reported to be in Afrotropical and Western Palearctic zoogeographical areas (Smit and Boer, 2008). In East African coastal regions, various species belonging to the Order Hymenoptera can be observed, such as bees (superfamily Apoidea), wasps (vespoides), and ants (formicidae) (Figure 7.7). A significant number of these species possess venom that they use for defensive purposes (Zoltani, 2018).

FIGURE 7.7 One of the East African Coastal Hymenoptera species: (A) *Camponotous* spp. (Subfamily Formicinae), (B) Western honey bee (*Apis mellifera*).

Source: Field data.

7.2.11 ORDER NEUROPTERA: COMMON GREEN LACEWING

Green Lacewings are found in the family Chrysopidae. This family is categorized into three subfamilies which are Nothochrysinae, Chrysopinae, and Apochrysinae. Green lacewings are large with delicately veined and have translucent wings (Winterton and Gupta, 2020). The group have 85 genera and an estimate of about 1300–2000 species which have widely spread all over the world and so does in East African coastal areas. The species feed mainly nectar, pollen and honey dew whereas others have been reported to be predatory (*Chrysopa*). The predators are mostly attacking the prey of their sizes including some species of aphids, insect larvae and caterpillars (Winterton and Brooks, 2002).

7.3 OVER VIEWS ON THREATS FACING THE INSECT DIVERSITY IN COASTAL AREAS OF EAST AFRICA

The existence of insects in the East African coastal areas is primarily threatened by different anthropogenic activities which are undertaken near the coastal zones. The climate change impact on the insects' diversity in East African coastal areas can also be triggered from unplanned land usage and illegal exploitation of natural resources. The rapid increase in the population in coastal regions increases the demand for shelter and expansion of different socioeconomic activities which in turn threaten biodiversity. The response of arthropods to the environmental disturbance differs. For instance, sun-loving butterflies may benefit from the disturbance of coastal forests as the canopies open while the typical forest insects may be harmed or disturbed (Hayes et al., 2009). This being a case, East African coastal areas have to be kept free from disturbance to ensure safety to all species around the coastal zones.

7.3.1 HABITAT LOSS AND ALTERATIONS

Habitat loss has been defined by Airoldi and Beck (2007) as the measurable reduction of coverage, abundance and distribution of the habitats. Habitat creates the structural complexity in the environment that ensures the proper interaction and flow of potential energy within the particular ecosystem. Most of the insects in East African coastal areas are threatened by anthropogenic stressors that lead to the destruction of natural habitats, making them unsuitable for the particular species (Halpern et al., 2008; Fraschetti et al., 2011). Triggering the point source of habitat loss may sometimes be tricky, wherever different anthropogenic activities are conducted at or near the coastal areas, the existence of some important insect species are jeopardized. Habitat loss can stand as a major contributor to biodiversity loss as reported by Brooks et al. (2002), although the impacts get bigger when all factors are acting together on the environment. According to the recent estimations, it has been reported that about 22% of natural land has been converted into other anthropogenic land uses causing disturbance to a variety of species including insects both in local and region level (Hoekstra et al., 2005; Stork et al., 2017). In case of coastal regions including East African coastal areas, dredging, eutrophication, and reduced water quality are among the major contributors of insect's disturbance, especially those which one or more of their life cycle stages occur in or near the water source. According to Airoldi et al. (2008), the habitat loss can bring about loss of native species, food resources, and loss of structural complexity of nature which in turn hamper the normal ecosystem functioning in the natural environment.

7.3.2 OVER-EXPLOITATION OF NATURAL RESOURCES

The coastal communities play a crucial role in preserving the health of coastal ecosystems. Unfortunately, rather than acting as stewards of these environments, they have become significant drivers of threats to biodiversity in these areas. Since ever, coastal ecosystems have played a vital role to mankind by supplying them with essential needs including medicines, foods, and recreational materials to name a few (Dugan, 1990; Martens, 1992; Saenger et al., 1993). As the population increases, mankind's desire to fulfill the essential demands for their lives becomes higher, creating an ever-growing exploitation of coastal natural resources which affect biodiversity including arthropods.

The existing relationship between species acts as a fundamental component of the healthy ecosystem. The over-harvest of any species may directly or indirectly affect the well-being of another species. Many coast and marine products including corals have been reported to be over-exploited to the point that threaten the existence of insects and other species that depend on them when they are either alive or dead (Wells, 1988). On top of that, in many coastal regions across the world including coast of East Africa, the over-harvesting of mangroves is a serious threat to biodiversity (Hamilton and Snedaker, 1984). A study by Mchenga and Ali (2013) revealed that a large number of insects inhabit mangroves. During their study, 10 orders (Hymenoptera, Diptera, Coleoptera, Orthoptera, Hemiptera, Odonata, Lepidoptera, Ephimeroptera, Plecoptera, and Dictyoptera) and 49 families of insects were identified to inhabit the mangroves of some coastal areas of East Africa. In this case, the destruction of mangroves may cause structural damage of the habitats and loss of important coastal insects.

7.3.3 POLLUTION

Many countries around the world have initiated various projects which aim at reducing pollution in the coastal areas, although the well-being of insects is still questionable (Majewski, 2019). According to some literature, insects are very sensitive to any environmental change. It was reported that car exhaust fumes and industrial pollution increase the level of NOx in the atmosphere, thus reducing the viability of insects (Whittaker, 2011; Campbell and Vallano, 2018). The industrial release of heavy metals and chemicals to the water streams and rivers, aquatic pollution as well as light pollution have been known to create disturbance to the insects in and near coastal zones (Owens and Lewis, 2018).

Moreover, agricultural runoff and sewage discharged by humans (rich in phosphorus and nitrogen) have been reported to affect coastal biodiversity. The accumulation of the disposed nutrients in or near the water sources initiate other biological processes which later become harmful to both marine and upstream biodiversity including insects. For instance, eutrophication may change the plant species composition and thereafter disrupt the ecological chain of the specific ecosystem (Pearson and Rosenberg, 1978).

7.3.4 GLOBAL CLIMATE CHANGE

The massive release of greenhouse gases (mainly carbon dioxide and chlorofluorocarbons) in the atmosphere lead to the global rise in temperature level (global warming). On top of that, these gases tend to alter the distribution of heat on the surface of the earth resulting in negative change in East African coastal areas circulation and precipitation level. These changes will thereafter affect most of biodiversity found in the coastal areas including insects (Robinet and Roques, 2010). The impacts of global climate change on biodiversity have been reported to be higher as far as phenological synchronicity adaptation capability are concerned (Forrest, 2016).

According to some literature, some insects including butterflies and moths have shown some phenological shifts due to climate change (Forrest, 2016; Bell et al., 2019). Other insects have experienced earlier flight periods and irregular change in their developmental stages due to climate change. Moreover, climate change may with no doubt force some insects which are not tolerant to high temperature to shift their range to other regions with moderate temperature conditions for their reproduction, distribution and survival (Hickling et al., 2005).

7.4 INSECTS' DIVERSITY, THREATS, AND CONSERVATION STRATEGIES IN THE COASTAL FOREST IN TANZANIA

The 11 coastal forests of Tanzania span across six regions: Dar es Salaam, the coastal region, Tanga, Lindi, Mtwara, and Zanzibar. Endemism thrives in these forests, with the majority of species facing extinction. For instance, Pande game reserves are among the coastal forests found in Dar es Salaam, with more than 14 species that are endemic and threatened (Tanzania Forest Conservation Group (TFCG), 2006). The list of species classified as threatened includes *Anthus sokokensis* (E), *Baphia kirkii* (V), *Circaetus fasciolatus* (NT), *Coffea pseudozanguebariae* (V), *Croton jatrophoides* (V), *Cynometra webberi* (V), *Galagoides rondoensis* (E), *Gardenia transvenulosa* (V), and *Rhynchocyon petersi* (E). Search efforts to understand the insects found in this coastal forest in Tanzania are few. Perhaps because insects have received less attention than amphibians, mammals, birds, and reptiles, despite their high abundance and diversity, in this report, we found that 48 species of insects (2 endemic, 2 unknown conservation status, and 44 widespread) were reported in four forests, namely, Pande Game Reserve, Wazo Hill Reforested Area, Kazimzumbwi Forest Reserve, and Kichi Forest Reserve (Doggart, 2003; Hemp, 2017; TFCG, 2006). Among the reported species, 39.58% were from Pande Game Reserve. Also, three out of four forest reserves were in Dar es Salaam. This suggests that the closeness to an academic and Research University like the University of Dar es Salaam provided the opportunity to study insects in these areas.

The species of insects reported in Pande game reserve were *Acraea zonata* (Endemic), *Catopsilia florella* (Widespread), *Aterica galena* (Widespread), *Bebearia chriemhilda* (Endemic), *B. safitza* (Widespread), *Byblia ilithya* (Widespread), *Danaus chrysippus* (Widespread), *Hypolimnas anthedon* (Widespread),, *Hypolimnas deceptor* (Widespread), *C. varanes* (Widespread), *C. violetta maritimes* (Widespread), *Euphaedra orientalis* (Widespread), *Charaxes zoolina* (Widespread), *Euphaedra neophron littoralis* (Widespread), *Euxanthe wakefieldi* (Widespread), *Eurytela dryope* (Widespread), *Charaxes etesipe* (Widespread), *Hamanumida daedalus* (Widespread) (Doggart, 2003). The Species reported in Wazo Hill reforestation included the following: *Freyeria trochylustrochylus* (Widespread), *Cacyreus lingeus* (Widespread), *Cacyeus* sp. (Widespread), *Gegenesnis obrevicornis* (Widespread), *C. florella* (Widespread), *Eurema senegalensis* (Widespread), *Colotis vestaliscastalis* (Widespread) (Ngongolo and Mtoka, 2013). In Kazimzumbwi Forest Reserve, two species of unknown conservation status were reported, namely, *Eurycorypha ligata* n. sp (Unknown) and *Eurycorypha annexata* n. sp (Unknown) (Hemp, 2017; TFCG, 2006). Kichi forest had the following species of insects being recorded: *Apanteles* sp. (Widespread), *Atradopolis* sp. (Widespread), *Aporochin* sp. (Widespread), *Abisares* sp. (Widespread), *Cophogryllus* sp., *Sisyphus* spp.,

Streblognatus aethiopicus (Widespread), *C. florella* (Widespread), *Pycnonotus barbatus,* Dragonflies (*ordonata*) (Widespread), *D. nigrofasciatus* (Widespread), *Eurema brigitta, Colotis* spp. (Widespread), *Sarcophaga* spp. (Unknown), *Xylocopa caffra* (Widespread), and *H. daedalus* (Widespread) (Ngongolo et al., 2014b, 2015; TFCG, 2006).

The insect species in the coastal forests are threatened by different anthropogenic activities such as agriculture, settlement, lumbering, and charcoal burning. For instance, insects in Pande Forest Reserve were reported to be threatened by urbanization, logging, agriculture encroachment, and fire, while in Kazimzumbi, they were threatened by logging, fire, wood collection, agriculture encroachment, and charcoal production (Doggart, 2003; Hemp, 2017; TFCG, 2006). Developments such as industrialization and human settlement threaten insect species in Wazo Hill reforested areas, while logging, roads, charcoal, shifting cultivation, and fire threaten insect species in Kichi Forest reserves.

Different conservation strategies have been adopted in Tanzania to conserve insects and their habitats in coastal forests. Among the conservation efforts made is the protection of forests to improve their status. As the forests are protected, different species such as insects are also well conserved. For instance, Kichi Forest, Pande, and Kazimzumbi are considered forest reserves. The conservation efforts for these coastal forests are being supported by governmental and nongovernmental institutions such as the Ministry of Natural Resources and Tourism, district local government authorities, the WWF, and the Wildlife Society of Tanzania.

For instance, the main threats facing the Pande game reserve include urbanization, logging, agriculture, encroachment, fire; with different conservation efforts being done in the reserve through Conservation nongovernmental organizations (NGOs) (e.g., Tanzania Forest Conservation Group, WWF) and government supports (Doggart, 2003). The main threats for insects in Wazo Hill afforestation are urbanization and settlement, industrialization while the effort for conservation is being done by Twiga Cement company (reforestation of the degraded areas) (Ngongolo and Mtoka, 2013). Logging, fire, wood collection, agriculture encroachment, charcoal production were reported to be the main threats for insects in Kazimzumbi Forest Reserve, the conservation efforts for the forest is being implemented by the support from District government authority and NGOs such as Wildlife Conservation Society, FAO, and GEF (Hemp, 2017; TFCG, 2006). Threats for insects in Kichi Forest included logging, roads, charcoal, shifting cultivation, fire while the conservation efforts are being done by support from District government authority and NGO such as Worldwide fund for nature (WWF) (Ngongolo et al., 2014a, 2015; TFCG, 2006).

7.5 INSECTS' DIVERSITY, THREATS, AND CONSERVATION STRATEGIES IN THE COASTAL FOREST IN KENYA

East Africa Coastal forests extend in the East Africa strip having a large number of different flora and fauna species. Kenyan coastal forests extend into four different districts which include Kwale (Southern part), Malindi and Kilifi (Middle), and Lamu (Northern part) (Matiku, 2005). The northern end of the Kenyan coastal forests is at the border of Kenya and Somalia. The Kenya coastal forests are estimated to cover an area of about 787 km^2 with

120000 ha of woodland/ bushland (Younge et al., 2002; Ngumbau et al., 2020). A total of 99 coastal forests in Kenya were reported in 1992. But the number increased in a report of 2002 where a total of 103 coastal forest sites have been documented (Younge et al., 2002). According to Matiku (2005), the current report shows that Kenya has a total of 107 coastal forests, of which 49 host globally threatened species of flora and fauna. Moreover, a total of 15 coastal forests have been named as priority forests (Table 7.1) although all other coastal forests need to be recognized and given some protection status (Matiku, 2005).

TABLE 7.1 Priority Kenyan Coastal Forests and Their Coverage Areas and Legal Status

Forest	Area Covered (km^2)
Arabuko Sokoke	370
Dodori/Boni	220
Shimba Hills	214
Madunguni	53
Kaya—47 Sites	28.4
Medium Kwale	51
Marafa Brachystegia	30
Tana River Delta	10
Witu Lamu	15
Boni/Lungi	95
Tana Gallery Forest	11
Ras Tenawi	20
Kilibasi	2
Mwangea	15
Mwangea Hill	5

Data source: Kenya National CF Task Force proceedings.

The Eastern coastal zone plays a vital role in conservation of biodiversity, as they harbor a lot of species, mostly endemic and some threatened species. For instance, in general, it has been reported that, a total of about 554 plants, 86 mollusks, 24 reptiles, 5 birds, and 3 mammals are endemic species found in the eastern strip of coastal forests. Of the estimates reported, a total of 333 species are globally threatened species whereby 105 species are from Kenya and the rest from Tanzania. Some of the threatened species include *Cercocebus galeritus galeritus*, *Colobus badius rufomitratus*, *Laephotis wintoni*, *Coffea sessiliflora*, *Populus ilicifolia*, *Pavetta sphaerobotrys*, *Sheppardia gunnifi*, and Malindi pipit.

Insects have been given a little concern by most of the conservationists despite their importance in the ecosystem. Many studies on the Kenyan coastal areas have focused on understanding the abundance and distribution of other species including reptiles, large mammals and plants, giving less concern to insects. Matiku (2005) reported that most Kenyan coastal forests have biodiversity value as they comprise different species, some being endemic to the particular areas. For instance, Arabuko Sokoke forest, which is ranked second as the most important forest for bird conservation has more than 250 butterfly species of which 4 are endemic. Some of the insect species recorded from Kenyan coastal forests include *Coryphagrion grandis*, *Aliciella subnuda*, *Peronura hildebrandtiana*, *Andronymus Caesar philander*, *Kedetes wallengreni wallengreni*, *Acraea neobule neobule*, *Acraea natalica natalica*, *Phalanta Phalantha aethiopica*, *Ypthima asterope asterope*, *Graphium angolanus*, *A. epaphia orbona, Hadrothemis scabrifrons*, and *Colotis euippe omole* (Rogo and Odulaja, 2001; Clausnitzer, 2003).

The distribution and abundance of insect species like any other species are threatened by several factors, mostly anthropogenic activities. In Kenya coastal forests, the reported threats include cultivation activities (e.g., Kaya Kinondo), encroachment (e.g., Mangea Hill), fires (e.g., Taita Hills), increased forest resources harvest (e.g., Arabuko-Sokoke forest reserve, Taita Hills, Dakatcha woodlands, and Shimba Hills National Reserve), and urbanization (e.g., Madunguni forest) (Burgess et al., 2003).

Different strategies and initiatives have been made so far to conserve the various species and insects in particular. Among the established strategies in Kenya coastal areas is establishment of special conservation units (Coastal Forest Conservation Unit) responsible for various issues relating to conservation in the defined protected zones (Matiku, 2005). On top of that, the National Forest Department has established the joint management of some forests with the local communities within the country. One of the forest reserves that have this initiative operating is Arabuko-Sokoke forest reserve. Moreover, the issues policy and legislation frameworks have been taken into consideration regularly to ensure they allow the participatory and cooperative conservation of forests from local communities, nongovernment organizations, and other stakeholders (Matiku, 2005).

In Kenya insects have been reported in following coastal forest: Shimba Hills, Arabuko Sokoke, Mwangea, Marafa Brachystegia. The species of insects reported in Shimba Hills were *Papilio dardanus, Amauris ochlea* ssp. *ochlea*, Short-horned Grasshoppers (Family Acrididae), *Dacus durbanensis, Agiades flesus,* Mantids (Family Mantidae*)* all being widespread status (Rogo and Odulaja, 2001; Younge et al., 2002; Matiku, 2005; Gereau et al., 2016). The species of insects reported in Arabuko Sokoke were *Phalanta phalantha,* Slantface Grasshoppers (Genus *Acrida*), *Papilio demodocus* ssp. *demodocus, Polyrhachis schistacea, L. cordofanus, Stethodesma strachani* ssp. *servillei* all with a status of widespread except *S. strachani* ssp. s*ervillei* with unknown conservation status (Rogo and Odulaja, 2001; Younge et al., 2002; Matiku, 2005; Gereau et al., 2016). Marafa Brachystegia reported single species of insect *Junonia hierta* which is threated by Grass fires, Inadequate management, cutting poles for timber, fuel and carving materials (Rogo and Odulaja, 2001; Younge et al., 2002; Matiku, 2005; Gereau et al., 2016). Also, single species of insects (*Epiphora lugardi)* has been reported in Mwangea with unknown conservation status and being threatened by inadequate management and encroachment (Rogo and Odulaja, 2001; Younge et al., 2002; Matiku, 2005; Gereau et al., 2016). Insects in Shimba Hills face significant threats including encroachment, fires, and the harvesting of poles for fuel, timber, and carving materials due to inadequate management. In contrast, Arabuko Sokoke is challenged by inadequate management, illegal hunting and tree harvest, encroaching human settlements, charcoal production, and conservation efforts. The conservation efforts are being implemented with the support form National Forest Department; Regional task force formed (KWS-Kenya); NGOs such WWF, IUCN, UNEP, and UNESCO for these coastal forests in Kenya (Rogo and Odulaja, 2001; Younge et al., 2002; Matiku, 2005; Gereau et al., 2016).

7.6 INSECTS' DIVERSITY, THREATS, AND CONSERVATION STRATEGIES IN THE COASTAL FOREST IN MOZAMBIQUE

A portion of East Africa coastal coastline extends to Mozambique ecoregion by approximately 2500 km long. Mozambican coastal forests are dominated by a diverse range flora and fauna species. The areas are primarily covered by several types of vegetation including mangroves, lowland forests, dune vegetation, thickets and grasslands. Albano (2004) pointed out that, about two-third of the coastal forests have been contributed by the Mozambican region whilst the rest of the East African coastal forests are contributed by the other East African countries. The data about the diversity of different species in Mozambique coastal regions are scarce as there were no current visits made to most of the Mozambican coastal areas especially in the northern part of coastal regions where a large portion of coastal forest is believed to be left (Younge et al., 2002).

In 1992, Mozambican coastline was reported to have a total of 18 forests covering an area of about 1790 km^2. As per the latest figures, Mozambican coastline has been reported to have a total of 55 forests, covering an area of approximately 4778 km^2 (Younge et al., 2002). Some of the closed forests have been reported to fall under the Gorongosa and Quirimbas National Parks while are categorized as forest reserves managed by National Directorate of Forest and Wildlife within the Ministry of Agriculture (Albano, 2004). The forest reserves in Mozambican coastal regions include Baixo Pinda Forest Reserve, Matibane Forest Reserve, Derre Forest Reserve, Inhamitanga Forest Reserve, and Nampacué Forest Reserve. Most of these reserves were gazetted in 1957.

Different flora and fauna species are found in coastal Mozambique. Plant species found in Mozambique coastline include *Crinum aurantiacum*, *Chlorophytum amplexicaule*, *C. polytrichus*, *Premna hans-joachimii*, *Sansevieria cf. metallica*, *Lepidagathis plantaginea*, *Pavetta lindina*, *Rothmannia macrosiphon*, and *Commiphora fulvotomentosa* (Timberlake et al., 2011). The fauna species found in the area include *Arthroleptis stenodactylus*, *Leptopelis flavomaculatus*, *Leptopelis mossambicus*, *Poyntonophrynus beiranus*, *Boulengerula uluguruensis*, *Boulengerula changamwensis*, *Chiromantis kelleri*, and *Scolecomorphus vittatus*. The abundance and distribution of insects have been barely studied and documented in Mozambican coastal areas. Some of the few reported insect species in this zone include *Scarabaeus bornemisszai*, *Neosisyphus mirabilis*, *Metacatharsius zuluanus*, *Scarabaeus (Scarabaeolus) clanceyi*, and *Anachalcos convexus*.

The biodiversity of Mozambican coastal forests has been reported to be threatened in current years compared to the past. The main driving factors for this situation include urbanization (e.g., Cheringoma complex, Bilene/Matutuíne, and Mecufi/Memba Complex), agriculture (e.g., Macomia/Palma complex, Bilene/Matutuíne, and Cheringoma Complex), uncontrolled fires (e.g., Mossuril, Cheringoma complex, and Mecufi/Memba Complex), and encroachment (e.g., Mossuril, Bilene/Matutuíne, and Macomia/Palma Complex).

Poverty has a primary source leading to unsustainable and illegal harvest of natural resources by the community adjacent to the Mozambican coastal forests. The Mozambican government is aware of the situation thus decided to initiate and implement poverty reduction strategies to the local communities. On top of that the Mozambican government has instituted several projects which aim at protecting, conserving, and utilizing forestry, agriculture and wildlife resources. Some of the implemented projects include Integrated Coastal Zone Management, Mecufi Coastal Zone Management, and Miombo Project (Kanji et al., 1998).

In Mozambique, various insect species have been documented within the coastal forests of Mossuril, Cheringoma, and Bilene/Matutuíne. The species in Mossuril were Kitty beetles (Genus *Lycus*), *Agrius convolvuli, Lampides boeticus, Nephele argentifera*, all being widespread. The challenges for conservation of insects in this forest were uncontrolled fires, encroachment, intensive agriculture, illegal harvest of trees for firewood and charcoal (Kanji et al., 1998; Younge et al., 2002; Albano, 2004; Timberlake et al., 2011). The species of insects for Cheringoma forest were *Leuconitocris nigricornis, Orthetrum stemmale, Ischnura senegalensis, H. daedalus, Earias insulana* with a widespread status. The threats for insects in Cheringoma were Illegal logging, timber harvesting, Intensive agriculture and uncontrolled fires (Kanji et al., 1998; Younge et al., 2002; Albano, 2004; Timberlake et al., 2011). In Bilene/Matutuíne, the observed species of insects were *Lamellothyrea descarpentriesi, Oecophylla longinoda, I. senegalensis* that are threatened by Intensive agriculture and urbanization, illegal harvest of trees for firewood and charcoal and encroachment (Kanji et al., 1998; Younge et al., 2002; Albano, 2004; Timberlake et al., 2011). The conservation efforts of these coastal forests in Mozambique are being implemented by Ministry of Agriculture and rural development (MADER); Ministry for Coordination of Environmental Affairs (MICOA); NGOs such as WWF, IUCN, EWT, UNESCO, and GECORENA (Kanji et al., 1998; Younge et al., 2002; Albano, 2004; Timberlake et al., 2011).

7.7 THE IMPORTANCE OF INSECTS IN EAST AFRICA

The importance of insects can be grouped into two major categories: ecological benefits and socioeconomic benefits. The ecological importance includes the ecosystem services they offer the ecosystem, such as pollination, and the fact that they act as a source of food for other vertebrates, such as birds and animals. The socioeconomic importance includes their use as food, such as termites, honey products from honey bees, and the increase in crop production through the pollination services they offer (Table 7.2).

7.8 THE WAY FORWARD AND CONCLUDING REMARKS

By referring to their role in the ecosystem and their distribution, insect species have to be at all cost protected especially in East African coastal areas. In the present era of heightened biodiversity crisis, it is crucial for protected area authorities, local communities, conservation stakeholders, and other relevant parties to actively participate in conservation efforts.

TABLE 7.2 Example of Insects and Their Importance in East African Coastal Forests

Sr. No.	Species of Insects	Orders, Family	Methods for Collection	Ecological Importance	Socioeconomic Importance	References
1	Frit flies/chloropid fly	(Diptera, Chloropidae)	Sweep net, Pitfall, beat sheet	Feed to vertebrates, scavenging	Agricultural pests	Nartshuk et al. (2020)
2	Mealy bugs (*Phenacoccus manihoti)*	Hemiptera, Pseudococcidae	Sweep net, Pitfall, beat sheet	Feed to vertebrates	Agricultural pests	Yonow et al. (2017) and Daane et al. (2012)
3	Target mantis (*Pseudocreobotra wahlbergii)*	Mantodea:	Pitfall, beat sheet	Feed to vertebrates,	Aesthetic value, Spiritual value	Patel and Singh (2016)
4	Brown-spotted Locust (*Cyrtacanthacris tatarica*)	Order Orthoptera, Acrididae	Sweep net, beat sheet	Feed to vertebrate	Agricultural pest, Sources of Protein	Panhwar (2018) and Memon and Panhwar (2021)
5	Honey bee (*Apis mellifera* L.)	Order Hymenoptera; Apidae	Direct observation, trap	Pollinators, vertebrate feed	Pollinator of crop, food (honey), medicine	Ngongolo et al. (2015)
6	Carpenter bee *(Xylocopa caffra L)*	Order Hymenoptera; Apidae	Direct observation, trap	Pollinators, vertebrate feed	Pollinator crop, food (honey), medicine	Ngongolo et al. (2015)
7	*Hypolimnas usambara*	Order Lepidoptera; Nymphalidae	Direct observation, pitfall and sweep net for larvae, sweep net, baited traps	Pollinators, vertebrate feed	Feed (larvae in some parts are eaten), pollination of crops, aesthetic value, income generation through tourism, butterfly farming, Bioindicator to environmental pollution	Ngongolo and Mtoka (2013) and Ngongolo et al. (2014a, 2015)

Various initiatives and strategies need to be implemented to mitigate the threats faced by insects in East African coastal regions and globally.

Up to date, there is deficiency in knowledge of the insects found in the coastal areas of East Africa. Moreover, there is no adequate documentation about the diversity, distribution, and reproduction patterns of the insects found in coastal areas. Lack of adequate information on insects hinders decision-makers and policymakers from formulating policies and strategies that encompass scientific, economic, and social considerations. These challenges increase public reluctance to accept established initiatives for the purpose of conserving biodiversity.

In order to prevent the loss of insect species, the responsible authorities and organs have to go far by designing feasible insect conservation programs that target the improvement of habitats for insect diversity, growth and distribution within the coastal zones. These strategies may include the initiation of the projects that will increase plant diversity in the coastal areas while keeping dead/dying wood for insects which are saproxylic in nature.

On top of that further scientific research on insect diversity, abundance, and distribution should be taken seriously. This will provide baseline information about different species inhabiting the coastal areas of East Africa. For better results and records, the researchers have to include DNA analysis using metabarcoding and barcoding techniques to gather relevant and distinctive information about the particular insect species.

Moreover, it is a time to cooperate and boost relations with local communities living in or near the coastal areas of East Africa (collaborative community management of biodiversity) so that they can fully participate in the conservation of insects and other species. Of great concern is comanagement of coastal biodiversity by the responsible institutions across the countries making up the coast of East Africa. This will help to design the projects, develop, and enforce the legislations that are feasible and applicable across the participating coastal regions.

KEYWORDS

- **coastal forest**
- **insect diversity**
- **East Africa**
- **threats**
- **conservation strategies**

REFERENCES

Airoldi, L.; Balata, D.; Beck, M. W. The gray zone: Relationships between habitat loss and marine diversity and their applications in conservation. *J. Exp. Mar. Biol. Ecol.* **2008,** *366*(1–2), 8–15.

Airoldi, L.; Beck, M. W. Loss, status and trends for coastal marine habitats of Europe. *Oceanogr. Mar. Biol.: Ann. Rev.* **2007,** *45*, 345–405.

Albano, G. Coastal Forests of Mozambique: Socio-economic aspects—review, **2004**. Accessed at coastalforests.tfcg.org/pubs/SocioEco-MZQ.pdf on 26th October 2022.

Basson, R. J.; Allison, J. D.; Roets, F. Recent records of fruit chafers (Scarabaeidae: Cetoniinae: Cetoniini) in the southwestern Cape region of South Africa suggest that range expansions were facilitated by human-mediated jump-dispersal and pre-adaptation to transformed landscapes. *Afr. Entomol.* **2019,** *27*(1), 135–145.

Bell, J. R.; Botham, M. S.; Henrys, P. A.; Leech, D. I.; Pearce-Higgins, J. W.; Shortall, C. R.; Brereton, T. M.; Pickup, J.; Thackeray, S. J. Spatial and habitat variation in aphid, butterfly, moth and bird phenologies over the last half century. *Glob. Chang. Biol.* **2019,** *25*(6), 1982–1994.

Brooks, T. M.; Mittermeier, R. A.; Mittermeier, C. G.; Da Fonseca, G. A.; Rylands, A. B.; Konstant, W. R.; Flick, P.; Pilgrim, J.; Oldfield, S.; Magin, G.; Hilton-Taylor, C. Habitat loss and extinction in the hotspots of biodiversity. *Conserv. Biol.* **2002,** *16*(4), 909–923.

Burgess, N. D.; Clarke, G. P. (Eds). *Coastal Forests of Eastern Africa*. IUCN Publications, **2000**.

Burgess, N.; Butynski, T.; Gordon, I.; Luke, Q.; Sumbi, P.; Watkin, J. Eastern Arc Mountains & coastal forests of Tanzania & Kenya. *Biodiversity Hotspots* **2003**.

Burgess, N. D.; Clarke, G. P.; Rodgers, W. A. Coastal forests of eastern Africa: Status, endemism patterns and their potential causes, *Biol. J. Linn. Soc.* **1998,** *64*, 337–367.

Campbell, S. A.; Vallano, D. M. Plant defences mediate interactions between herbivory and the direct foliar uptake of atmospheric reactive nitrogen. *Nat. Commun.* **2018,** *9*(1), 1–7.

Clausnitzer, V. Dragonfly communities in coastal habitats of Kenya: Indication of biotope quality and the need of conservation measures. *Biodiv. Conserv.* **2003,** *12*(2), 333–356.

Cock, M. J. W.; Biesmeijer, J. C.; Cannon, R. J. C.; Gerard, P. J.; Gillespie, D.; Jiménez, J. J.; Lavelle, P. M.; & Raina, S. K. The positive contribution of invertebrates to sustainable agriculture and food security. *CAB Rev.* **2012,** *7*, 1–27.

Critical Ecosystem Partnership Fund. CEPF fact sheet: Eastern Arc Mountains and coastal forests of Tanzania and Kenya, updated September **2007**. http://www.cepf.net/Documents/cepf.easternarc.factsheet.pdf [accessed 21 August 2022].

Critical Ecosystem Partnership Fund. *Ecosystem Profile: Eastern Arc Mountains & Coastal Forests of Tanzania & Kenya Biodiversity Hotspot*. Conservation International: Washington, DC, **2003**.

Daane, K. M.; Almeida, R. P. P.; Bell, V. A.; Walker, J. T. S.; Botton, M.; Fallahzadeh, M.; Mani, M.; Miano, J. L.; Sforza, R.; Walton, V. M.; Zanies, T. Biology and management of mealybugs in vineyards. In: Bostanian, N. J.; Vincent, C.; Isaacs, R. (Eds.), *Arthropod Management in Vineyards*. Springer Netherlands Press: Dordrecht, **2012,** pp. 271–307.

Doggart, N. Pande game reserve: A biodiversity survey. *TFCG Technical Paper* **2003,** (7), 1–100.

Döring, M. *English Wikipedia—Species Pages*. Wikimedia Foundation, **2020**

Dugan, P. J. *Wetland Conservation: A Review of Current Issues and Required Action*. IUCN: Gland, **1990**.

Erwin, T. L. The biodiversity question: How many species of terrestrial arthropods are there? In: Lowman, M. D.; Rinker, H. B. (Eds.), *Forest Canopies*. 2nd edn. Elsevier Academic Press: Burlington, MA, **2004,** pp. 259–269.

Forrest, J. R. Complex responses of insect phenology to climate change. *Curr. Opin. Insect Sci.* **2016,** *17*, 49–54.

Fraschetti, S.; Terlizzi, A.; Guarnieri, G.; Pizzolante, F.; D'ambrosio, P.; Maiorano, P.; Beqiraj, S.; Boero, F. Effects of unplanned development on marine biodiversity: A lesson from Albania (central Mediterranean Sea). *J. Coastal Res.* **2011,** (58), 106–115.

García, M.; Denno, B.; Miller, D. R.; Miller, G. L.; Ben-Dov, Y.; Hardy, N. B. ScaleNet: A literature-based model of scale insect biology and systematics. **2016**. [Online] URL: http://scalenet.info (accessed 15 October 2022).

Gereau, R. E.; Cumberlidge, N.; Hemp, C.; Hochkirch, A.; Jones, T.; Kariuki, M.; Lange, C. N.; Loader, S. P., Malonza, P. K.; Menegon, M.; Ndang'ang'a, P. K.; Rovero, F.; Shirk, P. Globally threatened biodiversity of the Eastern Arc Mountains and coastal forests of Kenya and Tanzania. *J. East African Nat. Hist.* **2016,** *105*(1), 115–201.

Grimaldi, D.; Engel, M. S. *Evolution of the Insects*. Cambridge University Press: New York, **2005**.

Halpern, B. S.; Walbridge, S.; Selkoe, K. A.; Kappel, C. V.; Micheli, F.; D'Agrosa, C. …Watson, R. A global map of human impact on marine ecosystems. *Science* **2008,** *319*, 948.

Hamilton, L. S.; Snedaker, S. C. *Handbook for Mangrove Area Management*. East-West Environment and Policy Institute: Honolulu, **1984**.

Hayes, L.; Mann, D. J.; Monastyrskii, A. L.; Lewis, O. T. Rapid assessments of tropical dung beetle and butterfly assemblages: Contrasting trends along a forest disturbance gradient. *Insect Conserv. Diver.* **2009,** *2*(3), 194–203.

Hemp, C. Annotated checklist of Orthoptera from Kazimzumbwi Forest Reserve, Tanzania with the description of new species and discussion of the biogeographic patterns of threatened species. *Zootaxa* **2017,** *4226*(2), zootaxa.4226.2.1.

Hickling, R.; Roy, D. B.; Hill, J. K.; Thomas, C. D. A northward shift of range margins in British Odonata. *Glob. Chang. Biol.* **2005,** *11*(3), 502–506.

Hodkinson, I. D.; Casson, D. A lesser predilection for bugs: Hemiptera (Insecta) diversity in tropical rain forests. *Biol. J. Linn. Soc.* **1991,** *43*, 101–109.

Hoekstra, J. M.; Boucher, T. M.; Ricketts, T. H.; Roberts, C. Confronting a biome crisis: Global disparities of habitat loss and protection. *Ecol. Lett.* **2005,** *8*(1), 23–29.

IUCN. The IUCN Red List of Threatened Species. **2015,** Version 2015.4. www.iucnredlist.org (accessed 2 October 2022).

Kanji, F.; Albano, G.; Macucule, A.; Soto, S; Massango, H. *The Coastal Forest of Mozambique*. WWF. Eastern Africa regional programme office. Eastern Africa coastal forest ecoregion programme. **1998,** p. 40.

Kroiss, J.; Lechner, K.; Strohm, E. Male territoriality and mating system in the European beewolf *Philanthus triangulum* F. (Hymenoptera: Crabronidae): Evidence for a "hotspot" lek polygyny. *J. Ethology* **2010,** *28*(2), 295–304. doi:10.1007/s10164-009-0185-5

Magnien, P.; Smets, K.; Pluot-Sigwalt, D.; Constant, J. A new species of Pygoplatys Dallas (Heteroptera, Tessaratomidae) from the Damar agroforests in Sumatra: Description, immatures and biology. *Nouvelle Revue d'Entomologie* **2008,** *24*(2), 99–112.

Majewski, M. S. *Pesticides in the Atmosphere: Distribution, Trends, and Governing Factors*. CRC Press, **2019**.

Martens, E. E. Coastal wetland utilisation. In: Crafter, S. A.; Njuguna, S. G.; Howard, G. W. (Eds.), *Wetlands of Kenya. Proceedings of the KWWG Seminar on Wetlands of Kenya, Nairobi, Kenya, 3–5July 1991*. IUCN: Gland, **1992,** pp. 115–134.

Matiku, P. *The Coastal Forests of Kenya: Forests data, threats, socio-economic issues, values, stakeholders, challenges, strategies, investment and enabling environment*—A national synthesis report for the development of the WWF-EARPO Eastern Africa Coastal Forests Ecoregion Programme. **2005**.

Mchenga, I. S.; Ali, A. I. Macro-fauna communities in a tropical mangrove forest of Zanzibar Island, Tanzania. *Global J. Bio-Sci. Biotech.* **2013,** *2*, 260–266.

McKelvey, J. J. Insects and human welfare. In: Pimentel, D. (Ed.), *Insects, Science and Society*. Academic Press: New York, **1975,** pp. 13–24.

Memon, P. R.; Panhwar, W. A. Study on Some Selected Species of Acrididae (Orthoptera) from District Dadu, Sindh-Pakistan, *Entomol. Zoo. Studies* **2021,** *9*(1), 6–11.

Miller, J. C. Insect natural history, multi-species interactions and biodiversity in ecosystems. *Biodiv. Conserv.* **1993,** *2*, 233–241.

Nartshuk, E. P.; Polevoi, A. V.; Przhiboro, A. A. A review of grass flies (Diptera, Chloropidae) of Karelia and Murmansk Province of Russia. *Fauna norvegica* **2020,** *40*, 47–92.

Ngongolo, K.; Mtoka, S. Using butterflies to measure biodiversity health in Wazo hill restored quarry. *J Entomol.d Zool. Studies* **2013,** *1*(4), 81–86.

Ngongolo, K.; Mtoka, S.; Rubanza, C. D. Floral visitors and pollinators of sesame (*Sesamum indicum* L.) from Kichi forest to the adjacent local communities' farms. *Entomol. Appl. Sci. Lett.* **2015,** *2*(2), 32–39.

Ngongolo, K.; Mtoka, S.; Mahulu, A.; Sigala, A. et al. The Abundance of Guinea fowl Butterflies (*Hamanumida daedalus*; Nymphalidae) in Kingupira Sector, Selous Game Reserve. *J. Entomol Appl Sci. Letter.* **2014b,** *1*(3).

Ngongolo. K.; Mtoka. S.; Mahulu. A. Wet season diversity of butterflies in restored mine of Wazo Hill Tanzania. *Intern. J. Fauna Biol. Studies* **2014a**, *1*(3), 01–03.

Ngumbau, V. M.; Luke, Q.; Nyange, M.; Wanga, V. O.; Watuma, B. M.; Mbuni, Y. M.; Munyao, J. N.; Oulo, M. A.; Mkala, E. M.; Kipkoech, S.; Itambo, M.; Hu, G. W.; Wang, Q. F. Annotated checklist of the coastal forests of Kenya, East Africa, *PhytoKeys* **2020,** *147*, 1–191, doi: 10.3897/phytokeys.147.49602.

Owens, A. C.; Lewis, S. M. The impact of artificial light at night on nocturnal insects: A review and synthesis. *Ecol. Evol.* **2018,** *8*(22), 11337–11358.

Panhwar, W. A. Biodiversity of Caelifera (Orthoptera) from Gorakh Hill Station, Dadu Sindh Pakistan. *J. Entomol. Zool. Stud.* **2018,** *6*(3), 372–378.

Patel, S.; Singh, R. Updated checklist and distribution of Mantidae (Mantodea: Insecta) of the World. *Intern. J. Res. Stud. Zool.* **2016,** *2*(4), 17–54.

Pearson, T. H.; Rosenberg, R. Macrobenthos succession in relation to organic enrichment and pollution in the marine environment. *Ocean. Mar. Bioi. Ann. Rev.* **1978,** *16*, 229–311.

Popov, G. B.; Fishpool, L. D.; Rowell, C. H. F. A review of the Acridinae s. str. (Orthoptera: Acridoidea: Acrididae) of eastern Africa with taxonomic changes and description of new taxa. *J Orthoptera Res.* **2019,** *28*(1), 37–105.

Rider, A. D. *Natural Enemies Index*. Department of Entomology, North Dakota State University, **2008**.

Robinet, C.; Roques, A. Direct impacts of recent climate warming on insect populations. *Integrative Zoology* **2010,** *5*(2), 132–142.

Rogo, L.; Odulaja, A. Butterfly populations in two forest fragments at the Kenya coast. *African J. Ecol.* **2001,** *39*(3), 266–275.

Saenger, P.; Heged, E. J.; Davie, J. D. S. *Global Status of Mangrove Ecosystems*. IUCN: Geneva, **1993**.

Salehe, J. *The Forests and Woodlands of the Coastal East Africa Region.* World Wide Fund for Nature (WWF), **2011**.

Scudder, G. G. The importance of insects. *Insect Biodiversity: Science and Society*, **2017,** *1*, 9–43.

Smit, J.; De Boer, R. *Philanthus Triangulum (Hymenoptera; Crabronidae) New for the Fauna of the Canary Islands* **2008**.

Stork, N. E. How many species are there? *Biodiversity and Conservation* **1993,** *2*, 215–232.

Stork, N. E. How many species of insects and other terrestrial arthropods are there on Earth? *Annual Rev. Entomol.* **2018,** *63*(1), 31–45. doi:10.1146/annurev-ento-020117-043348.

Stork, N. E.; Srivastava, D. S.; Eggleton, P.; Hodda, M.; Lawson, G.; Leakey, R. R. B.; Watt, A. D. Consistency of effects of tropical-forest disturbance on species composition and richness relative to use of indicator taxa. *Conserv. Biol.* **2017,** *31*(4), 924–933.

Tanzania Forest conservation Group (TFCG). *Coastal Forests of Kenya and Tanzania—Tanzania Regions (tfcg.org)*. Tanzania's Coastal Forests, **2006**.

Timberlake, J.; Goyder, D.; Crawford, F.; Burrows, J.; Clarke, G. P.; Luke, Q.; Matimele, H.; Müller, T.; Pascal, O.; de Sousa, C.; Alves, T. Coastal dry forests in northern Mozambique. *Plant Ecol. Evol.* **2011,** *144*(2), 126–137.

Wells, S. M. *Coral Reefs of the World*. 3 volumes. UNEP/IUCN: Gland, **1988**.

Whittaker, J. B. Presidential address: Insects and plants in a changing atmosphere. *J. Ecol.* **2011,** *89*(4), 507–518.

Winterton, S. L.; Gupta, A. Review of the green lacewing genus *Apochrysa* Schneider (Neuroptera: Chrysopidae). *Zootaxa* **2020,** *4729*(3), 329–346.

Winterton, S. L.; Brooks, S. J. Phylogeny of the apochrysine green lacewings (Neuroptera: Chrysopidae: Apochrysinae). *Ann. Entomol. Soc. Amer.* **2002,** *95*(1), 16–28.

Yonow, T.; Kriticos J. D.; Ota, N. The potential distribution of cassava mealybug (*Phenacoccus manihoti*), a threat to food security for the poor. *PLoS One* **2017,** *12*(3), e0173265. doi: 10.1371/journal.pone.0173265

Younge, A.; Negussie, G.; Burgess, N. *Eastern Africa Coastal Forest Programme: Regional Workshop Report Feb 4–7 2002*. WWF Eastern African Programme Office: Nairobi, **2002**.

Zoltani, C. K. Cardiovascular toxicity. In: *Veterinary Toxicology*. Academic Press, **2018,** pp. 227–238.

CHAPTER 8

AMPHIBIAN SPECIES OF THE COASTAL ECOREGION OF EASTERN AFRICA: RICHNESS, DISTRIBUTION, AND CONSERVATION

WILIRK NGALASON

Department of Zoology and Wildlife Conservation, University of Dar es Salaam, Dar es Salaam, Tanzania

ABSTRACT

Literature on the current status of amphibian species of the Eastern African coastal ecoregion is limited. This study therefore reviews amphibian species richness, composition, and conservation status along the region. It is assumed that amphibian species richness, composition, and levels of endemism vary between countries due to extensive latitudinal gradient. Of most important habitats for conservation of biodiversity are the coastal forests that range from southern Somalia through Kenya, Tanzania to Mozambique. They harbor endemic and diverse species of plants and animals. Ninety-two amphibian species occur within the region including 87 anurans and five caecilians. Twelve species are endemic including eight that occur in Tanzania, two in Kenya and two in Somalia. The offshore islands of Zanzibar, Pemba, and Mafia have 50% of the endemic species. Of the 92 species, three are considered critically endangered, seven are endangered, two are vulnerable, two are not evaluated, and 83 species are categorized as least concern by the International Union for Conservation of Nature. Major threats to the amphibian fauna are forest loss and fragmentation associated with urbanization and human population increase. Best practices in conservation and management of forests are recommended.

8.1 INTRODUCTION

8.1.1 AMPHIBIAN EVOLUTION AND DIVERSIFICATION

Amphibia is a class of vertebrate animals that are characterized by moist, glandular, and thin skins without scales, feathers, or hairs. The word "amphibia" basically comes from two

Biodiversity Hotspot of the Coastal Forests of Eastern Africa. T. Pullaiah and Ally J. K. Nkwabi (Eds.)

DOI: 10.1201/9781998511099-8

Greek words *amphi* meaning double and *bios* meaning life (du Preez and Carruthers, 2009). They are thought to have evolved during the Palaezoic Era during the Devonian period about 370 million years ago. They are said to have evolved from the lobe-fined fish sarcopterygians, which were a group of bony fishes (Vitt and Caldwell, 2014). Today there are two closest living members of the sarcopterygians including the modern coelacanths and the lungfishes. The coelacanths include the West Indian Ocean coelacanth (*Latimeria chalumnae*) that is found in deep waters near the coast of Comoro Islands, Madagascar, Mozambique and South Africa, and the Indonesian coelacanth (*Latimeria menadoensis*) that live in deep waters of north Sulawesi and Papua coasts in Indonesia (Fricke and Hissmann, 2000). The first fossils assigned to amphibians include *Elginerpeton*, *Ventastega*, and the ichthyostegalians (*Ichthyostega* and *Acanthostega*) (Vitt and Caldwell, 2014). These earliest amphibians had features resembling the fossil lobe-fined fish in many ways including radial fin rays supporting a tail fin, a lateral line system, labyrinthodont teeth, fish branchial arches supporting internal gills, and vertebra outline. The process of emerging from water entailed physiological and behavioral changes such as the development of skeletal system capable of supporting the amphibian body on land, and the evolution of lungs for breathing oxygen from the atmosphere. The fins of fish evolved into limbs to walk on land and lungs developed from gas-filled bladders in primitive fish (Perry et al., 2001; Hoffman et al., 2016). The modern and living amphibians of today are all grouped in the amphibian subclass Lissamphibia with three orders including Anura or Salientia (frogs and toads), Caudata or Urodela (newts and salamanders), and Apoda or Gymnophiona (caecilians). Lissamphibians are thought to have originated from temnospondyl group that evolved from the ichthyostegalians (Vitt and Caldwell, 2014). Currently, there are 8930 amphibian species worldwide that are distributed in all the continents except Antarctica (Frost, 2025) although the amphibian web names only 8926 species (AmphibiaWeb, 2025).

Anura is the largest order of amphibians with currently more than 7875 species (or 88% of the amphibians). This group is well-represented in all continents including the Eastern African region. The only continent the group has not been reported is in the Antarctica (Vitt and Caldwell, 2014). The anurans are tailless as adults and are characterized by having front and hind legs of unequal size, and have the ability to vocalize particularly during the breeding seasons to attract mates and defend territories. The remaining two orders of amphibians including urodelans and apodans do not have significant vocal calls, they are voiceless. The second largest amphibian group is Urodela with currently 826 species (or 9% of all the amphibians). This group consists of newts and salamanders which are characterized by possessing tails that they have retained throughout the larval and adult stages. Unlike frogs and toads the front and hind legs of urodelans are of an equal size. Urodelans are mostly found in the Northern Hemisphere except one group of plethodontid lungless salamander that was able to cross to the Southern Hemisphere in South America. In Africa, urodelans are restricted to the northern tip of the continent and are not found south of Sahara (Channing and Howell, 2006). The third group of the extant amphibians includes members of the Order Apoda with currently 225 species worldwide or 3% of all the amphibian species. These include the caecilians that have elongated bodies making them look like worms and with reduced eyes, an adaptation for subterranean life. They are tropical in distribution occurring in the tropical forests of South America, Africa, and South

East Asia (AmphibiaWeb, 2025). Only two amphibian groups including the anurans (frogs and toads) and apodans (caecilians) occur in the coastal Eastern African region.

This region stretches from its northern limit in Somalia to its southern limit in Mozambique. The coastal region of Eastern Africa has a mosaic of environmental conditions ranging from arid, semiarid to tropical, and moist conditions with high habitat diversity. The Eastern Africa coastal region consists of different habitat types including forests, woodland, bushland, grassland, thickets, farmland, and swamp vegetation which is mostly associated with larger rivers. The coastal forests are particularly most important for biodiversity conservation (Burgess et al., 1998). These forests are recognized as the African centers of high species richness and endemism of both plants and animals (Burgess et al., 1998; Mittermeier et al., 1998). Most of the forests are small highly fragmented remnants, most being less than 20 km^2 in size (Barratt, 2017) distributed in the ecoregion from southern Somalia, Kenya, Tanzania, and Mozambique. Due to the high levels of species richness and endemism, the coastal forests of the Eastern Africa are considered as one of the globally important regions for biodiversity conservation, along with the forests of the Eastern Arc Mountains (EAMs) (Mittermeier et al., 1998, 1999; Myers et al., 2000).

8.1.2 AMPHIBIANS STUDIES ALONG THE COASTAL EASTERN AFRICA ECOREGION

Studies on the amphibian fauna of the coastal region of the Eastern Africa are limited. Studies that have been conducted show the region to be very rich in biodiversity and levels of endemism in all groups of organisms including plants, invertebrates, amphibians, reptiles, birds, and mammals (Burgess et al., 1998). Much of the known literature on the amphibians of the coastal ecoregion of Eastern Africa is contained in Lanza (1983, 1990) on the amphibians and reptiles of Somali; Schiøtz (1999) on the tree frogs of Africa; Channing (2001) on the amphibians of central and southern Africa; Channing and Howell (2006) on the amphibians of East Africa, Spawls et al. (2006) on the reptiles and amphibians of East Africa; Pickersgill (2007) on the frogs of Southern and Eastern Africa; Malonza et al. (2006, 2018) on the herpetofauna of Tana River and Shimba Hills, respectively, in Kenya; Håkansson (2009) on the reptiles and amphibians of the Kenyan coastal hinterland; du Preez and Carruthers (2009) on to the frogs of southern Africa; Harper et al. (2010) on the amphibians of the EAMs and coastal forests of Tanzania and Kenya; Barratt (2017) on the amphibians of coastal forests of Eastern Africa; Spawls et al. (2019) on the amphibians of Kenya, and Channing and Rödel (2019) on the frogs and other amphibians of Africa. Information on lowland amphibian fauna is also contained in a study by Poynton (2000) and Poynton et al. (2007) and some localized studies and checklists in some forest patches, for example, Msuya (2001), Howell and Msuya (2002), Msuya et al. (2004), Howell et al. (2012), Bwong et al., 2017, and Taylor et al. (2019).

In recent years, there has been an extensive taxonomic review of the amphibian species which has led to the discovery of new species, splitting of some groups, refining or separating some populations from synonyms or redefining the geographical boundaries of species.

Most of the available literature on the amphibians of Eastern African ecoregion has not been reviewed to incorporate these changes. In addition, the amphibian fauna of the coastal Eastern Africa is scantly studied and there is not much literature on it. This calls for a study to review the amphibian fauna of the coastal Eastern Africa. This chapter therefore attempts to assess species richness, composition and conservation status of amphibians along the coastal forests of Eastern Africa. This study therefore focuses on the amphibian species that occur in the coastal forests and its ecoregion from southern Somalia to southern Mozambique. The taxonomy of the amphibian species followed that of the amphibian species of the World (Frost, 2025).

8.2 AMPHIBIAN SPECIES RICHNESS, COMPOSITION, ENDEMISM, AND CONSERVATION STATUS ALONG THE COASTAL ECOREGION OF EASTERN AFRICA

8.2.1 SPECIES RICHNESS AND DISTRIBUTION

A total of 92 amphibian species occur along the coastal Eastern Africa ecoregion (Appendix 8.1). The Somali coast is the poorest in terms of amphibian species richness with only 24 species, followed Kenyan and Mozambican coast with 54 species each, and the Tanzania coastal region is the richest with 61 species.

The amphibian fauna belongs to two amphibian orders including Anura and Apoda. No single species of tailed amphibians (Order Caudata) is found in the region. The majority of the amphibians were anurans with 87 species and the remaining five species were the apodans (Appendix 8.1). These species are adapted to different ways of life within their environment including those living aquatic, arboreal, subterranean and terrestrial life style. The highly aquatic species include the African clawed frog *Xenopus muelleri* with its extensive webbing (Figure 8.1), an adaptation for swimming. These species spend most of their life in aquatic environment. Some species including members of the family Hyperoliidae such as those in the genus *Afrixalus*, *Hyperolius,* and *Chiromantis* have expanded digits on their toes and fingers. Many of them are brightly colored, and the expanded digits enable those climbing substrates to arboreal sites. Other forms of adaptations observed in the Eastern African coastal amphibians are subterranean and terrestrial life forms. Species that live subterranean life include members of the genus *Boulengerula*, *Scolecomorphus,* and *Schistometopum*. These are worm-like amphibians, with vestigial eyes and without limbs for easy penetration in the ground (Figure 8.2). A few frogs also spend a significant amount of time underground particularly during the dry periods. Such species including the snout-burrowers (genus *Hemisus*) have hard-pointed snouts and strong limbs to enable them push into the ground. Other amphibian species occurring in the region live terrestrial life either in grassland or in leaf litter. These include the grass frogs (genus *Ptychadena*) that are found in grassland near water bodies as well as squeakers (genus *Arthroleptis*) that live in leaf litter in forested areas. The hind limbs of most terrestrial frogs and toads, for example, grass frogs are significantly elongated to enable them jump long distances of up to about 3 m. During breeding seasons most anurans go back to water to breed and therefore they have webbed digits to enable them swim. Most species in the family Phrynobatrachidae, Ptychadenidae, Pyxicephalidae, Ranidae, and Bufonidae (Appendix 8.1) spend part of their life both in terrestrial and aquatic environments, and they have aquatic larva. A few species do not breed in water instead they are direct developers, meaning that the eggs hatch into nonfree-swimming larva that develop and assume an adult morphology. Such species include some members of the family Arthroleptidae and Brevicipitidae. Another reproductive strategy observed is the viviparous or ovoviviparous in which the eggs are retained and develop in the female body either with or without maternal nourishment until they give birth to live young. This strategy is found in members of the family Dermophiidae and Scolecomorphidae (Appendix 8.1). Studies suggest that laying eggs on land is a key in the evolution of terrestrial larvae and/or direct development, and the evolution of live-bearing in caecilians is thought to have its origin from direct development strategy (Liedtke et al., 2022).

All the amphibian species in the coastal forests of Eastern Africa occur in twelve anuran families including Arthroleptidae, Brevicipitidae, Bufonidae, Hemisotidae, Hyperoliidae, Microhylidae, Phrynobatrachidae, Pipidae, Ptychadenidae, Pyxicephalidae, Ranidae, and Rhacophoridae. Two families including Pipidae and Ranidae are represented by only one species in each, that is, *Xenopus muelleri* and *Amnirana galamensis* (Figure 8.3), respectively, that occur in the coastal region whereas some families are represented only by two species including Hemisotidae with two species *Hemisus guineensis* and *Hemisus marmoratus*; Microhylidae with *Phrynomantis bifasciatus* and *Phrynomantis somalicus* and Rhacophoridae with *Chiromantis petersii* and *C. xerampelina*. The richest anuran family within the region is Hyperoliidae with 29 species followed by Bufonidae with 14 species and Ptychadenidae with 12 species. The remaining families have species ranging from three, for example, Brevicipitidae to nine, for example, Arthroleptidae (Appendix 8.1).

The caecilians are represented by three families including Family Dermophiidae with one species *Schistometopum gregorii*, Herpelidae with three species, *Boulengerula changamwensis*, *B. denhardti* and *B. uluguruensis*, and Scolecomorphidae with only one species *Scolecomorphus vittatus*.

Only 10 amphibian species occur throughout the Eastern African coast from Somalia to Mozambique. These species include *Hemisus marmoratus*, *Afrixalus delicatus*, *Hyperolius argus*, *H. pussilus*, *Kassina senegalensis*, *Phrynomantis bifasciatus*, *Phrynobatrachus acridoides*, *Ptychadena anchietae*, *P. schillukorum*, and *Amnirana galamensis* (Appendix 8.1). Twenty-four amphibian species occur in three of the four countries, for example, *Sclerophrys steindachneri* (in Somalia, Kenya, and Tanzania), *Arthroleptis stenodactylus* (in Kenya, Tanzania, and Mozambique), and also 24 species occur in two of the four countries, for example, *Poyntonophrynus lughensis* (Somalia and Kenya) or *Mertensophryne loveridgei* (Tanzania and Mozambique) (Appendix 8.1). Thirty-four species occur in only one of the countries, for example, *Lanzarana largeni* that is endemic to Somalia, *Boulengerula denhardti*, that is, endemic to Kenya, *Hyperolius ruvuensis* that is endemic to Tanzania and *Leptopelis mossambicus* that is found from Mozambique southwards to South Africa (Appendix 8.1). A relatively large proportion (37%) of the amphibian species in the ecoregion occur singly in one of the countries.

8.2.2 SPECIES COMPOSITION

Amphibian species composition in the coastal ecoregion of the four countries as assessed by the Jaccard Similarity Index (S_J) (Ludwig and Reynolds, 1988) indicated species similarity to be higher between adjacent than distant countries. The index uses species presence/absence data, which is widely used to assess similarity of species communities. The similarity index values range from 0% when no one species is common to both communities to 100% when all species in one community are also found in the other community. A higher similarity index suggests greater similarity in the number of amphibian species between countries.

In this study, similarity in amphibian species between the four coastal Eastern African countries was 23% whereas the similarity between Somalia, Kenya, and Tanzania was 38% and that of between Kenya, Tanzania, and Mozambique was 42%. Pairwise comparison showed two adjacent countries Kenya and Tanzania to have higher amphibian species similarity (58%) than between Tanzania and Mozambique (48%) and between Somalia and Kenya (39%). The lowest scale of species similarity was observed between two adjacent countries Somalia and Kenya followed by Tanzania and Mozambique. Much of parts of the Somalia and northern Mozambique are unexploited for amphibian species (e.g., see Conradie et al., 2016) which may account for the difference. There was no an amphibian species shared only by the following country pairs: Somalia and Tanzania, Somalia and Mozambique, and between Kenya and Mozambique. These country pairs are distantly located which may account for the observed results. The coastal Eastern Africa forests range from southern Somalia at about 1° north to 27° south in Mozambique. This is a big latitudinal gradient which is expected to impact the numbers and composition of amphibian species. The decrease in the numbers of amphibian species with an increase in latitude has been a long accepted phenomenon (see Rosenzweig, 1995).

8.2.3 AMPHIBIAN ENDEMISM

The level of amphibian endemism in the coastal Eastern African forests is considerably high. At least 12% or 13% of all the amphibian species occurring in the ecoregion are endemic. In Somalia alone, two amphibian species occurring on the southern coastal areas are endemic including *Lanzarana largeni* and *Ptychadena gansi*. In Kenya, two species including *Hyperolius rubrovermiculatus* and *Boulengerula denhardti* are endemic. In Tanzania, 61 species occur on the coast and of these eight species are endemic including six species that are endemic to the Islands of Pemba, Zanzibar, and Mafia, that is, *Mertensophryne howelli* (endemic to Zanzibar and Mafia Islands), *Hyperolius puncticulatus* (endemic to Zanzibar), *Hyperolius watsonae* (endemic to Ngezi Forest Reserve in Pemba), *Kassina jozani* (endemic to Jozani Forest in Zanzibar, Figure 8.4), *Phrynobatrachus pakenhami* (endemic to Pemba), *Afrixalus stuhlmanni* (endemic to both Zanzibar and Kilombero floodplains, Figure 8.5), *H. ruvuensis* (endemic to Ruvu South Forest Reserve, Barratt et al., 2017), and *Boulengerula uluguruensis*, endemic to the coastal Kazimzumbwi Forest Reserve and some other EAMs of Tanzania (see Figure 8.2). None of the amphibian species occurring on the coastal Mozambique is endemic. The decrease of level of endemism of amphibian species southward toward Mozambique coincides with the well-known phenomenon of the decline of species toward higher latitudes (Rosenzweig, 1995; Willig et al., 2003) and perhaps attributed to limited studies in Mozambique (Burgess et al., 1998; Conradie et al., 2016). Of the four Eastern African countries, Somalia and Mozambique are less studied in terms of amphibian fauna. Intensive studies in the coastal forests of these countries may reveal peculiar species. For example, a survey of herpetofauna conducted in the Afromontane forests in northern Mozambique came out with a country record and a putative new amphibian species (Conradie et al., 2016).

The findings from this study show the coastal lowland areas including the forests in the islands of Pemba, Zanzibar, and Mafia to possess relatively high number of endemic amphibian species. Some of the endemic species are restricted to only one locality, for example, *H. watsonae* is found only in Ngezi Forest Reserve in Pemba, whereas a few species, for example, *Boulengerula uluguruensis* are found in more than one locality. Areas where species diversity and endemism occur in much higher concentrations are often considered centers of endemism and the Eastern African coastal ecoregion is one of those areas (Burgess et al., 1998; Timberlake et al., 2011).

FIGURE 8.1 *Xenopus muelleri* Müller's Clawed Frog (John V. Lyakurwa).

FIGURE 8.2 *Boulengerula uluguruensis* Pink Caecilian (Joseph M. Christopher).

FIGURE 8.3 *Amnirana galamensis* Galam White-lipped Frog (John V. Lyakurwa).

FIGURE 8.4 *Kassina jozani* Running Frog from Jozani, Zanzibar (Charles A. Msuya).

FIGURE 8.5 *Afrixalus stuhlmanni* Stuhlimann's Spiny Reed Frog (John V. Lyakurwa).

8.2.4 CONSERVATION STATUS

The International Union for Conservation of Nature (IUCN) sets criteria for assessing extinction risk of animal, fungus, and plant species, which are used as indicators of the health of the biodiversity (IUCN, 2025). Criteria used to classify species include population size, rate of decline, area of geographical distribution, degree of population fragmentation, and the degree of distribution fragmentation (Mace et al., 2008). Based on the criteria, the IUCN red list divides species into nine categories including Not Evaluated (NE), Data Deficient, Least Concern (LC), Near Threatened, Vulnerable (VU), Endangered (EN), Critically Endangered (CR), Extinct in the Wild, and Extinct. Based on these categories, the coastal Eastern Africa amphibians were grouped into five categories (Appendix 8.1). Three species including *Boulengerula denhardt*, *H. ruvuensis,* and *H. watsonae* are considered CR. These species are facing an extremely high risk of extinction including reduction of the population size and the decline of the geographical range (IUCN, 2025). The risk of extinction for the species is more than 50% within over three generations. The second category includes seven amphibian species that are considered EN including *M. howelli*, *H. puncticulatus*, *H. rubrovermiculatus*, *Kassina jozani*, *Phyrnobatrachus pakenhami*, *P. ungujae,* and *Boulengerula changamwensis*. These species are also facing reduction in population size, reduction in the geographical range and the probability of extinction is at least 20% within 20 years up to a maximum of 100 years. Based on IUCN criteria, the area of occurrences is estimated to be less than 5000 km^2 and the area of occupancy is less than 500 km^2. Two amphibian species *Leptopelis grandiceps* and *Afrixalus sylvaticus* fall under VU status suggesting a population decline, extinction probability of at least 10% within 100 years, and the area of occurrences and occupancy is relatively bigger than that assigned to the CR or EN categories.

Extinction risks are largely caused by human impacts both on climate and on the loss of biodiversity. Habitat loss is a major threat to the amphibians of the coastal ecoregion mainly caused by deforestation, changing land-use patterns, urbanization, pollution, expansion of agricultural activities, lack of legal frameworks and effective policies, and management structures for conservation (UNEP, 2001; Younge et al., 2002; WWF, 2011). There are two species in the region, *Leptopelis broadleyi* and *Hyperolius microps* that have not been evaluated by the IUCN and thus are categorized as NE. Most of the amphibian species (78 species) are assigned LC category (Appendix 8.1). They are not a focus of conservation efforts because they are plenty in the wild and basically are not threatened with extinction.

8.3 CONCLUSIONS AND RECOMMENDATIONS

8.3.1 CONCLUSIONS

This study assesses the amphibian species of the coastal ecoregion of Somalia, Kenya, Tanzania, and Mozambique. Contained within the ecoregion are coastal forests ranging from southern Somalia to Mozambique. These forests are known for their high diversity and endemism in animal and plant species. Ninety-two amphibian species occur in the region including 87 species of anurans and five caecilian species. Sixty-one species occur on the coastal Tanzania, 54 species occur both in Kenya and Mozambique, and 24 species in the southern coastal ecoregion of Somalia. Twelve amphibian species are endemic including two species in Somalia, two in Kenya, and eight species in Tanzania. Three species are considered CR, seven species EN, two species VU, two species NE, and 78 species are categorized as LC based on the IUCN red list criteria. The observed findings reflect differences in habitat and latitudinal variability.

8.3.2 RECOMMENDATIONS

This study is a review of the amphibian species known to occur in the Eastern Africa coastal region. Some localities in the ecoregion including the northern portion of Mozambique and some forests have remained inadequately sampled and there have not been any recent surveys. Investigations in these areas may reveal interesting observation as well as the current status of the amphibians. Major threats to the amphibian fauna of the coastal region include habitat destruction, alteration and expansion of human settlements. Some amphibian species occurring in the region including the endemic ones such as *M. howelli*, *H. puncticulatus*, *H. rubrovermiculatus*, *H. ruvuensis*, *H. watsonae*, *Kassina jozani*, *Phrynobatrachus pakenhami*, *P. ungujae*, and *Boulengerula changamwensis* are restricted to only one or a few localities. Additionally, these species are categorized as EN or CR by the IUCN. This calls for improved and best practices in management and conservation of the habitats in which they occur for their continued survival.

ACKNOWLEDGMENTS

I am grateful to Professor Thammineni Pullaiah of the University of Sri Krishnadevaraya and Dr. Ally J. K. Nkwabi of the Tanzania Wildlife Research Institute for their valuable comments. I am grateful to Mr. John V. Lyakurwa and Mr. Joseph M. Christopher of the University of Dar es Salaam for contributing pictures of amphibians used in the document.

KEYWORDS

- **amphibian richness**
- **coastal forests**
- **Eastern African**
- **endemic**
- **conservation status**

REFERENCES

AmphibiaWeb. University of Califonia, Berkeley, CA, USA. **2025**. https://amphibiaweb.org. Accessed 29 September 2025.

Barratt, C. D. *Biodiversity Patterns and Conservation of the Coastal Forests of Eastern Africa*. PhD Dissertation, University of Basel: Switzerland, **2017**.

Barratt, C. D.; Lawson, L. P.; Bittencourt-Silva, G. B.; Doggart, N.; Morgan-Brown, T.; Nagel, P.; Loader, S. P. A new, narrowly distributed, and critically endangered species of spiny-throated reed frog (Anura: Hyperoliidae) from a highly threatened coastal forest reserve in Tanzania. *Herpetol. J.* **2017,** *27*, 13–24.

Burgess, N. D.; Clarke, G. P.; Rodgers, W. A. Coastal forests of eastern Africa: Status, endemism patterns and their potential causes. *Biol. J. Linn. Soc.* **1998,** *64*, 337–367.

Bwong, B. A.; Nyamache, J. O.; Malonza, P. K.; Wasonga, D. V.; Ngwava, J. M.; Barratt, C. D.; Nagel, P.; Loader, S. P. Amphibian diversity in Shimba Hills National Reserve, Kenya: A comprehensive list of specimens and species. *J. East Afr. Nat. Hist.* **2017,** *106*, 19–46, doi.org/10.2982/028.106.0104.

Channing, A. *Amphibians of Central and Southern Africa*, Protea Bookhouse: Pretoria, South Africa, **2001**.

Channing, A.; Howell, K. M. *Amphibians of East Africa*. Cornell University Press: Ithaca, New York, **2006**.

Channing, A.; Rödel, M. O. *Field Guide to the Frogs and Other Amphibians of Africa*. Struik Nature: Cape Town, South Africa, **2019**.

Conradie, W.; Bittencourt-Silva, G. B.; Engelbrecht, H.; Loader, S. P.; Menegon, M.; Nanvonamuquitxo, C.; Scott, M.; Tolley, K. Exploration into the hidden world of Mozambique's sky islands forests: New discoveries of reptiles and amphibians. *Zoosystematics Evol.* **2016,** *92*, 163–180, https://doi.org/10.3897/zse.92.9948.

du Preez, L.; Caruthers, V. *A Complete Guide to the Frogs of Southern Africa*. Stuik Nature: Cape Town, South Africa, **2009**.

Fricke, H.; Hissmann, K. Feeding ecology and evolutionary survival of the living coelacanth *Latimeria chalumnae*. *Mar. Biol.* **2000,** *136*, 379–386, https://doi.org/10.1007/s002270050697.

Frost, D. R. Amphibian Species of the World: An Online Reference. Version 6.2, American Museum of Natural History, New York, USA, **2025,** Electronic Database Accessible at https://amphibiansoftheworld.amnh.org/index.php. doi.org/10.5531/db.vz.0001, Accessed 29 September 2025.

Håkansson, N. T. Reptiles and amphibians from the Kenyan coastal hinterland. *Herpetol. Bull.* **2009,** *107*, 30–39.

Harper, E. B.; Measey, G. J.; Patrick, D. A.; Menegon, M.; Vonesh, J. R. *Field Guide to the Amphibians of the Eastern Arc Mountains and Coastal Forests of Tanzania and Kenya*. Camerapix Publishers International: Nairobi, Kenya, **2010**.

Hoffman, M.; Taylor, B. E.; Harris, M. B. Evolution of lung breathing from a lungless primitive vertebrate. *Respir. Physiol. Neurobiol.* **2016,** *224*, 11–16, https://doi.org/10.1016/j.resp.2015.09.016

Howell, K. M.; Msuya, C. A.; Mligo, C.; Werema, C.; Kihaule, P.; Honorati, M. K.; Suleiman, H. O. *Biodiversity Surveys of Poorly Known Coastal Forests of Southeastern Tanzania and Zanzibar.* A Consultancy Report of the Department of Zoology and Wildlife Conservation, University of Dar es Salaam, Tanzania to WWF, **2012**.

Howell, K. M.; Msuya, C. A. *Biodiversity studies in Kilwa and Lindi Districts: Reptiles and Amphibians*. Consultancy Report of the Department of Zoology and Wildlife Conservation, University of Dar es Salaam to Ornis Consult: Dar es Salaam, Tanzania, **2002**.

IUCN. The IUCN Red List of Threatened Species, Version 2022-1. https://www.iucnredlist.org. Accessed on 25th October 2022.

Lanza, B. Amphibians and reptiles of the Somali Democratic Republic: Checklist and biogeography. *Biogeographia*, **1990,** *14*, 407–465, doi.org/10.21426/B614110318.

Lanza, B. List of the Somali amphibians and reptiles, Monitore Zoologico Italiano. *Supplemento*, **1983,** *18*, 193–247, doi.org/10.1080/00269786.1983.11758573.

Liedtke, H. C.; Wiens, J. J.; Gomez-Mestre, I. The evolution of reproductive modes and life cycles in amphibians. *Nat. Commun.* **2022,** *13*, 7039, doi.org/10.1038/s41467-022-34474-4.

Ludwig, J. A.; Reynolds, J. F. *Statistical Ecology: A Primer on Methods and Computing*. John Wiley: New York, **1988**.

Mace, G. M.; Collar, N. J.; Gaston, K. J.; Hilton, T. C.; Akcakaya, H. R.; Leader, W. N.; Millner-Gulland, E. J.; Stuart, S. N. Quantification of extinction risk: IUCN's system for classifying threatened species. *Conserv. Biol.* **2008,** *22*, 1424–1442.

Malonza, P. K.; Mulwa, D. M.; Nyamache, J. O.; Jones, G. Biogeography of the Shimba Hills ecosystem herpetofauna in Kenya. *Zool. Res.* **2018,** *39*, 97–104, doi.org/10.24272%2Fj.issn.2095-8137.2017.048.

Malonza, P. K.; Wasonga, V. D.; Muchai, V.; Rotich, D.; Bwong, B. A.; Bauer, A. M. Diversity and biogeography of herpetofauna of the Tana River Primate National Reserve, Kenya. *J. East Afr. Nat. Hist.* **2006,** *95*, 95–109, doi.org/10.2982/0012- 8317(2006)95[95:DABOHO]2.0.CO;2

Mittermeier, R. A.; Myers, N.; Gil, P. R.; Mittermeier, C. G. *Hotspots: Earth's Biologically Richest and most Endangered Terrestrial Ecoregions.* CEMEX: Mexico City, **1999**.

Mittermeier, R. A.; Myers, N.; Thompsen, J. B.; da Fonseca, G. A. B.; Olivieri, S. Global biodiversity hotspots and major tropical wilderness areas. *Conserv. Biol.* **1998,** *12*, 516-520.

Msuya, C. *Habitats, Distribution and Feeding of Amphibians in Zaraninge Forest Reserve, Tanzania. A Case Study of the Ecology of a Community of Amphibians in Coastal Forest Ecosystem in Bagamoyo District, Tanzania.* PhD Thesis, University of Dar es Salaam, **2001**.

Msuya, C. A.; Sangu, G. M.; Kayombo, C. K.; Howell, K. M.; Ndangalasi, H. J.; Seleman, H.; Ngalason, W.; Kihaule, P. M. *Biological Diversity Survey of the Mbarawala and Ruawa-Likonde Plateaus in Lindi Region, Tanzania*. A Consultancy Report to the Royal Danish Embassy, Dar es Salaam, Tanzania, **2004**.

Myers, N.; Mittermeier, R. A.; Mittermeier, C. G.; Fonseca, G. A. B.; Kent, J. Biodiversity hotspots for conservation priorities. *Nature*, **2000,** *403*, 853–858.

Perry, S. F.; Wilson, R. J.; Straus, C.; Harris, M. B.; Remmers. J. E. Which came first, the lung or the breath? *Comp. Biochem. Physiol. A. Mol. Integr. Physiol.* **2001,** *129*, 37–47.

Pickersgill, M. *Frog Search. Results of Expedition to Southern and Eastern Africa*, Edition Chimaira, Frankfurt am Main: Germany, **2007**.

Poynton, J. C. Amphibians. In: Burgess, N. D.; Clarke, G. P. (Eds.), *Coastal Forests of Eastern Africa*. IUCN: Gland, Switzerland and Cambridge, UK, **2000,** p. 443.

Poynton, J. C.; Loader, S. P.; Sherrat, E.; Clarke, B. C. Amphibian diversity in East African biodiversity hotspots: Altitudinal and latitudinal patterns, *Biodivers. Conserv.* **2007,** *16*, 1103–1118, doi 10.1007/s10531-006-9074-1.

Rosenzweig, M. L. *Species Diversity in Space and Time*. Cambridge University Press: Cambridge, United Kingdom, **1995**.

Schiøtz, A. *Tree Frogs of Africa*. Edition Chimaira. Frankfurt am Main: Germany, **1999**.

Spawls, S.; Howell, K. M.; Drewes, R. C. *Pocket Guide to the Reptiles and Amphibians of East Africa.* A and C. Black: London, UK, **2006**.

Spawls, S.; Wasonga, D. V.; Drewes, R. C. The amphibians of Kenya, available on the web at www.kenyareptileatlas.com/onewebmedia/Amphibians%20of%20Kenya%20%20Book%20August%202019%20%20Downloadable%20Version.pdf, **2019**. Accessed 1 October 2022.

Taylor, R.; Malonza, P.; Spawls, S.; Ochieng, J.; Jackson, C.; Navarro, R. *A Checklist of the Reptiles and Amphibians of Arabuko-Sokoke Forest, Kenya, A Rocha Kenya.* Checklist dataset, **2019,** doi.org/10.15468/tcgw7t accessed via GBIF.org on 2022-12-16. Accessed 16 December 2022.

Timberlake, J.; Goyder, D.; Crawford, F.; Burrows, J.; Clarke, P.; Luke, Q.; Matimele, H.; Müller, T.; Pascal, O.; de Sousa, C.; Alves, T. Coastal dry forests in northern Mozambique. *Plant Ecol. Evol.* **2011,** *144*, 126–137, https://doi.org/10.5091/plecevo.2011.539.

United Nations Environment Programme (UNEP). *Eastern Africa Atlas of Coastal Resources.* United Nations Environment Programme: Nairobi, Kenya, **2001**.

Vitt, L. J.; Caldwell, J. P. *Herpetology: An Introductory Biology of Amphibians and Reptiles*, 4th Edition. Academic Press: San Diego, CA, USA, **2014**.

Willig, M. R. Kaufman, D. M.; Stevens, R. D. Latitudinal gradients of biodiversity: Pattern, process, scale, and synthesis. *Annu. Rev. Ecol. Evol. Syst.* **2003,** *34*, 273–309, https://doi.org/10.1146/annurev.ecolsys.34.012103.144032.

World Wide Fund for Nature (WWF). *Threatened Spaces, Disappearing Species: The Forests and Woodlands of the Coastal East Africa Region.* Coastal East Africa Initiative Report, Dar es Salaam, Tanzania, **2011,** https://d1qkbvpvihpfr5.cloudfront.net/downloads/wwf_cea_ni_species_report_final.pdf. Accessed 24 October 2022.

Younge, A.; Negussie, G.; Burgess, N. Eastern African Coastal Forest Programme, WWF—EARPO Regional Workshop Report, Nairobi, Kenya February 4–7, **2002**.

APPENDIX 8.1 Amphibian Species of the Coastal Ecoregion of Eastern African Countries (SO = Somalia, KE = Kenya, TZ = Tanzania, MZ = Mozambique, x = present, - = no record on the coastal ecoregion)

Sr. No.	Order, Family, Species	Common Name	Country				Conservation status
			SO	KE	TZ	MZ	
	Order Anura, Family Arthroleptidae						
1	*Arthroleptis stenodactylus* Pfeffer, 1893	Long-Fingered Squeaker	–	x	x	x	Least Concern
2	*Arthroleptis xenodactyloides* Hewitt, 1933	Dwarf Squeaker	–	x	x	x	Least Concern
3	*Leptopelis argenteus* (Pfeffer, 1893)	Silvery Tree Frog	–	x	x	x	Least Concern
4	*Leptopelis broadleyi* (Poynton, 1985)	Broadley's Tree Frog	–	–	–	x	Not Evaluated
5	*Leptopelis concolor* Ahl, 1929	Uniform Tree Frog	x	x	x	–	Least Concern
6	*Leptopelis flavomaculatus* (Günther, 1864)	Yellow-Spotted Tree Frog	–	x	x	x	Least Concern
7	*Leptopelis grandiceps* Ahl, 1929	Wide-headed Free Frog	–	x	–	–	Vulnerable
8	*Leptopelis mossambicus* Poynton, 1985	Mozambique Tree Frog	–	–	–	x	Least Concern
9	*Leptopelis parbocagii* Poynton and Broadley, 1987	Cryptic Tree Frog	–	–	–	x	Least Concern
	Order Anura, Family Brevicipitidae						
10	*Breviceps adspersus* Peters, 1882	Bushveld Rain Frog	–	–	–	x	Least Concern
11	*Breviceps mossambicus* Peters, 1854.	Mozambique Rain Frog	–	–	x	x	Least Concern
12	*Callulina kreffti* `Nieden, 1911	Krefft's Warty Frog	–	x	–	–	Least Concern
	Order Anura, Family Bufonidae						
13	*Mertensophryne howelli* (Poynton and Clarke, 1999)	Howell's Forest Frog	–	–	x	–	Endangered
14	*Mertensophryne lindneri* (Mertens, 1955)	Lindner's Forest Frog	–	–	x	x	Least Concern
15	*Mertensophryne loveridgei* (Poynton, 1991)	Loveridge's Forest Frog	–	–	x	x	Least Concern
16	*Mertensophryne micranotis* (Loveridge, 1925)	Woodland Forest Frog	–	x	x	–	Least Concern
17	*Mertensophryne taitana* (Peters, 1878)	Taita forest frog	–	x	x	–	Least Concern
18	*Poyntonophrynus beiranus* (Loveridge, 1932)	Beira Pygmy Toad	–	–	–	x	Least Concern
19	*Poyntonophrynus fenoulheti* (Hewitt and Methuen, 1912)	Fenoulhet's Pygmy Toad	–	–	–	x	Least Concern
20	*Poyntonophrynus lughensis* (Loveridge, 1932).	Lugh Pygmy Toad	x	x	–	–	Least Concern
21	*Schismaderma carens* (Smith, 1848)	Red Toad	–	–	x	x	Least Concern

APPENDIX 8.1 *(Continued)*

Sr. No.	Order, Family, Species	Common Name	Country				Conservation status
			SO	KE	TZ	MZ	
22	*Sclerophrys garmani* (Meek, 1897)	Garman's Toad	–	–	–	x	Least Concern
23	*Sclerophrys gutturalis* (Power, 1927)	Guttural Toad	–	x	x	x	Least Concern
24	*Sclerophrys pusilla* (Mertens, 1937)	Merten's Striped Toad	–	x	x	x	Least Concern
25	*Sclerophrys steindachneri* (Pfeffer, 1893).	Steindachner's Toad	x	x	x	–	Least Concern
26	*Sclerophrys xeros* (Tandy, Tandy, Keith and MacKay, 1976).	Desert Toad	x.	x	x	–	Least Concern
	Order Anura, Family Hemisotidae						
27	*Hemisus guineensis* Cope, 1865	Guinea Piglet Frog	–	–	–	x	Least Concern
28	*Hemisus marmoratus* (Peters, 1854).	Marbled Piglet Frog	x	x	x	x	Least Concern
	Order Anura, Family Hyperoliidae						
29	*Afrixalus aureus* Pickersgill, 1984	Golden Spiny Reed Frog	–	–	–	x	Least Concern
30	*Afrixalus brachycnemis* (Boulenger, 1896)	Lesser Spiny Reed Frog	–	–	x	x	Least Concern
31	*Afrixalus crotalus* Pickersgill, 1984	Snoring Spiny Reed Frog	–	–	–	x	Least Concern
32	*Afrixalus delicatus* Pickersgill, 1984	Delicate Spiny Reed Frog	x	x	x	x	Least Concern
33	*Afrixalus fornasini* (Bianconi, 1849)	Fornasini's Spiny Reed Frog	–	x	x	x	Least Concern
34	*Afrixalus stuhlmanni* (Pfeffer, 1893)	Stuhlimann's Spiny Reed Frog	–	–	x	–	Least Concern
35	*Afrixalus sylvaticus* Schiøtz, 1974	Forest Spiny Reed Frog	–	x	x	–	Vulnerable
36	*Hylambates maculatus* Duméril, 1853	Red-legged Wot-wot	–	x	x	x	Least Concern
37	*Hyperolius argus* Peters, 1854	Argus Reed Frog	x	x	x	x	Least Concern
38	*Hyperolius glandicolor* Peters, 1878	Peters' Reed Frog	x	x	x	–	Least Concern
39	*Hyperolius mariae* Barbour and Loveridge, 1928	Mary's Reed Frog	–	x	x	–	Least Concern
40	*Hyperolius marmoratus* Rapp, 1842	Marbled Reed Frog	–	–	–	x	Least Concern
41	*Hyperolius microps* Günther, 1864	Sharp-headed Reed Frog	–	x	x	x	Not Evaluated
42	*Hyperolius mitchelli* Loveridge, 1953	Mitchell's Reed Frog	–	–	x	–	Least Concern
43	*Hyperolius parkeri* Loveridge, 1933	Parker's Reed Frog	–	x	x	–	Least Concern

APPENDIX 8.1 *(Continued)*

Sr. No.	Order, Family, Species	Common Name	Country				Conservation status
			SO	KE	TZ	MZ	
44	*Hyperolius poweri* Loveridge, 1938	Power's Reed Frog	–	–	–	x	Least Concern
45	*Hyperolius puncticulatus* (Pfeffer, 1893)	Zanzibar Reed Frog	–	–	x	–	Endangered
46	*Hyperolius pusillus* (Cope, 1862)	Water Lily Reed Frog	x	x	x	x	Least Concern
47	*Hyperolius rubrovermiculatus* Schiøtz, 1975	Red-Lined Reed Frog	–	x	–	–	Endangered
48	*Hyperolius ruvuensis* Barratt, Lawson, and Loader, 2017.	Ruvu Reed Frog	–	–	x	–	Critically Endangered
49	*Hyperolius semidiscus* Hewitt, 1927	Yellow-striped Reed Frog	–	–	–	x	Least Concern
50	*Hyperolius substriatus* Ahl, 1931	Mainland Reed Frog	–	x	x	x	Least Concern
51	*Hyperolius tuberilinguis* Smith, 1849	Tinker Reed Frog	–	x	x	x	Least Concern
52	*Hyperolius viridiflavus* (Duméril and Bibron, 1841)	Common Reed Frog	x	x	x	–	Least Concern
53	*Hyperolius watsonae* Pickersgill, 2007	Watson's Reed Frog	–	–	x	–	Critically Endangered
54	*Kassina jozani* Msuya, Howell, and Channing, 2007	Jozani Running Frog	–	–	x	–	Endangered
55	*Kassina maculifer* (Ahl, 1924)	Northern Running Frog	x	x	–	–	Least Concern
56	*Kassina senegalensis* (Duméril and Bibron, 1841)	Senegal Running Frog	x	x	x	x	Least Concern
57	*Kassina somalica* Scortecci, 1932	Somali Running Frog	x	x	x	–	Least Concern
	Order Anura, Family Microhylidae						
58	*Phrynomantis bifasciatus* (Smith, 1847)	Banded Rubber Frog	x	x	x	x	Least Concern
59	*Phrynomantis somalicus* (Scortecci, 1941)	Somali Rubber Frog	x	x	–	–	Least Concern
	Order Anura, Family Phrynobatrachidae						
60	*Phrynobatrachus acridoides* (Cope, 1867)	East African Puddle Frog	x	x	x	x	Least Concern
61	*Phrynobatrachus mababiensis* FitzSimons, 1932	Mababe Puddle Frog	–	x	x	x	Least Concern
62	*Phrynobatrachus natalensis* (Smith, 1849)	Natal Puddle Frog	–	–	x	x	Least Concern
63	*Phrynobatrachus pakenhami* Loveridge, 1941	Pakenham's Puddle Frog	–	–	x	–	Endangered
64	*Phrynobatrachus scheffleri* (Nieden, 1911)	Scheffler's Puddle Frog	–	x	x	–	Least Concern
65	*Phrynobatrachus ungujae* Pickersgill, 2007	Unguja Puddle Frog	–	x	x	–	Endangered

APPENDIX 8.1 *(Continued)*

Sr. No.	Order, Family, Species	Common Name	Country				Conservation status
			SO	KE	TZ	MZ	
	Order Anura, Family Pipidae						
66	*Xenopus muelleri* (Peters, 1844)	Müller's Clawed Frog	–	x	x	x	Least Concern
	Order Anura, Family Ptychadenidae						
67	*Hildebrandtia macrotympanum* (Boulenger, 1912).	Northern Ornate Frog	x	x	–	–	Least Concern
68	*Hildebrandtia ornata* (Peters, 1878)	Common Ornate Frog	–	–	x	x	Least Concern
69	*Lanzarana largeni* (Lanza, 1978)	Lanza's Frog	x	–	–	–	Least Concern
70	*Ptychadena anchietae* (Bocage, 1868).	Anchieta's Grass Frog	x	x	x	x	Least Concern
71	*Ptychadena gansi* Laurent, 1965	Gan's Grass Frog	x	–	–	–	Least Concern
72	*Ptychadena guibei* Laurent, 1954	Guibé's Grass Frog	–	–	–	x	Least Concern
73	*Ptychadena mascareniensis* (Duméril and Bibron, 1841). Note: it is a complex of many species (Channing and Rödell, 2019)	Mascarene Grass Frog	–	–	x	x	Least Concern
74	*Ptychadena mossambica* (Peters, 1854)	Mozambique Grass Frog	–	x	x	x	Least Concern
75	*Ptychadena nilotica* (Seetzen, 1855)	Nile Grass Frog	–	x	x	–	Least Concern
76	*Ptychadena oxyrhynchus* (Smith, 1849)	Sharp-nosed Grass Frog	–	x	x	x	Least Concern
77	*Ptychadena schillukorum* (Werner, 1908)	Schilluk Grass Frog	x	x	x	x	Least Concern
78	*Ptychadena taenioscelis* Laurent, 1954	Southern Dwarf Grass Frog	–	–	x	x	Least Concern
	Order Anura, Family Pyxicephalidae						
79	*Pyxicephalus angusticeps* Parry, 1982	Beira Bull Frog	–	x	x	x	Least Concern
80	*Pyxicephalus edulis* Peters, 1854	Edible Bull Frog	–	x	x	x	Least Concern
81	*Tomopterna krugerensis* Passmore and Carruthers, 1975	Knocking Sand Frog	–	–	–	x	Least Concern
82	*Tomopterna marmorata* (Peters, 1854)	Marbled Sand Frog	–	–	–	x	Least Concern
83	*Tomopterna natalensis* (Smith, 1849)	Natal Sand Frog	–	–	–	x	Least Concern

APPENDIX 8.1 *(Continued)*

Sr. No.	Order, Family, Species	Common Name	Country				Conservation status
			SO	KE	TZ	MZ	
	Order Anura, Family Ranidae						
84	*Amnirana galamensis* (Duméril and Bibron, 1841	Galam White-lipped Frog	x	x	x	x	Least Concern
	Order Anura, Family Rhacophoridae						
85	*Chiromantis kelleri* Boettger, 1893	Keller's Foam-nest Frog	x	x	–	–	Least Concern
86	*Chiromantis xerampelina* Peters 1854	Southern Foam-nest Frog	–	x	x	x	Least Concern
87	*Chiromantis petersii* Boulenger, 1882	Peters' Foam-nest Frog	x	x	–	–	Least Concern
	Order Apoda, Family Herpelidae						
88	*Boulengerula changamwensis* Loveridge, 1932	Changamwe Caecilian	–	x	–	–	Endangered
89	*Boulengerula denhardti* Nieden, 1912	Denhardt's Caecilian	–	x	–	–	Critically Endangered
90	*Boulengerula uluguruensis* Barbour and Loveridge, 1928	Pink caecilian	–	–	x	–	Least Concern
	Order Apoda, Family Dermophiidae						
91	*Schistometopum gregorii* (Boulenger, 1894)	Gregory's Caecilian	–	x	x	–	Least Concern
	Order Apoda, Family Scolecomorphidae						
92	*Scolecomorphus vittatus* (Boulenger, 1895)	Banded Caecilian	–	x	–	–	Least Concern
	Total number of species		**24**	**54**	**61**	**54**	

CHAPTER 9

REPTILE SPECIES OF THE COASTAL ECOREGION OF EASTERN AFRICA: RICHNESS, DISTRIBUTION, AND CONSERVATION

WILIRK NGALASON* and JOHN V. LYAKURWA

Department of Zoology and Wildlife Conservation, University of Dar es Salaam, Dar es Salaam, Tanzania

*Corresponding author

ABSTRACT

Literature on the current status of the reptilian fauna of the coastal Eastern African ecoregion, including Somalia, Kenya, Tanzania, and Mozambique, is limited. This study therefore reviews reptilian species richness, composition, and conservation status along the region. It is assumed that reptilian species richness, composition, and endemism vary between the countries due to extensive latitudinal gradients. Furthermore, the study assesses the conservation status of the reptiles that occur in the region. The study reveals 293 species including one crocodile, 16 testudines, and 276 squamates. Of the 293 species, 79 (27% of the species) are endemic to the region, with coastal Tanzania having the highest numbers (26 species), followed by Somalia with 13 species, Mozambique with six species, and coastal Kenya with five species. The threat status categories set by the International Union for Conservation of Nature identify two species that are Critically Endangered, four species that are Endangered, seven species that are Vulnerable, and five species that are Near Threatened. Habitat destruction is one of the major threats to the species. This calls for best practices in conservation and management for the continued existence of the reptile species.

9.1 INTRODUCTION

9.1.1 REPTILE EVOLUTION AND RICHNESS

Reptiles belong to a class of vertebrate animals comprising crocodiles in the Order Crocodylia; terrapins, tortoises, and turtles in the Order Testudines; worm lizards, lizards,

Biodiversity Hotspot of the Coastal Forests of Eastern Africa. T. Pullaiah and Ally J. K. Nkwabi (Eds.)

DOI: 10.1201/9781998511099-9

and snakes in the Order Squamata, and tuataras in the Order Rhynchocephalia. Only the first three orders have representatives in the coastal Eastern African ecoregion, whereas the tuataras are restricted to New Zealand (Vitt and Caldwell, 2014). Those four orders are the living groups of reptiles, although there are several extinct groups, including the dinosaurs that dominated the earth more than 65 million years ago. Reptiles are characterized by a dry scaly skin or scutes with neither feathers as in birds, nor hairs as in mammals. They have their embryos covered in an amniotic membrane; thus, together with birds and mammals, they are collectively termed amniotes (Vitt, 2016). The amniotic membrane encloses the embryo, whereas other membranes, including the allantois, deal with embryo's waste and gaseous management, the yolk sac that provides nutrients to the developing embryo, chorion that encloses the embryo, and the egg shell that protects the entire contents (Vitt and Caldwell, 2014). Reproductively, two types of modes are recognized, including those that lay eggs (oviparous) and live-bearing (viviparous). They have internal fertilization, which in most species results into an amniotic egg that is laid on the ground, in cavities, or in underground nests. In this mode the embryo develops outside the maternal body, being nourished by the contents of the egg. One introduced species in Eastern Africa coast the Flower pot blind snake *Indotyphlops braminus,* has deviated from this mode of reproduction, as it is parthenogenetic, females being able to reproduce without males, enabling a single individual to establish a population (Spawls et al., 2018). The other mode involves the development of the embryo within the maternal body after fertilization has taken place. In this mode, the embryo obtains its nutrients from the contents of the egg and/ or from the mother until they give birth to a live new one which is a miniature of an adult. In the Eastern African coastal ecoregion, both modes are recognized whereas the majority of the reptiles are oviparous including all members of the order Crocodylia and Testudines (Spawls et al., 2018). Most squamates are oviparous with few being viviparous. Species that give live birth include some snakes, such as the puff adder *Bitis arietans* and the sea snake *Hydrophis platurus*, and some lizards for example, variable skink *Trachylepis varia* (Spawls et al., 2018). Reptiles mostly obtain their heat from the surroundings, thus termed ectotherms. Very little heat is generated from muscular activities but since they do not have any insulation ability that heat is dissipated to the environment. Heat regulation is mainly achieved through behavioral means. They reach their preferred body temperature by absorbing heat from the environment mainly through basking and will retreat into shade/ burrow when it becomes too hot or cold.

Reptiles are thought to have evolved about 315 million years ago during the Carboniferous period from one lineage of amphibians that was more terrestrial than other amphibians (Branch, 1998; Vitt and Caldwell, 2014). The first unquestionable fossil assigned to reptile was *Hylonomus lyelli*. The evolution and radiation of reptiles from amphibians involved modification in the anatomy, physiology, and behavior including the development of impermeable skin that resists desiccation, amniotic eggs, which reduced the need for water during reproduction, movement and support on land without the support of water, and well-developed lungs to support respiration on land (Vitt and Caldwell, 2014). The tremendous radiation of reptiles after their evolution led to many species, including the extant groups as well as the extinct groups, such as dinosaurs, which dominated the earth during the Mesozoic era. The dinosaurs became extinct about 65 million years ago (Spawls et al., 2018).

Modern reptiles are very much diversified in life forms, leading to species that live terrestrial, fossorial, arboreal, or aquatic life. These various life forms are all represented in the coastal Eastern Africa ecoregion. Reptiles that live in aquatic environment, including crocodiles, sea snakes, and sea turtles, have features that enable them to live their way of life, such as laterally flattened tails for swimming (in crocodiles and sea snakes) and limbs that are modified into flippers (in turtles). The fossorial species, such as some skinks and worm lizards, lack limbs and have rudimentary eyes and compact skulls, being modified for burrowing life (Figure 9.1). Arboreal species such as chameleons have prehensile tails and limb digits fused into opposable bundles (Figure 9.2), whereas the tree geckoes have expanded digits with setae to make them excellent climbers (Figure 9.3). Big variation in body morphology is also observed in terrestrial species, including snakes, lizards, and tortoises. Worldwide there are at least 12,643 reptile species including 27 species of crocodiles, 366 species of testudines, one species of tuataras and 12,249 squamates (Uetz et al., 2025). The squamates are further divided into snakes Suborder Serpentes (4187 species), lizards Suborder Sauria (7859 species), and the worm lizards Suborder Amphisbaenia (203 species) (Uetz et al., 2025). More than 2088 subspecies are also recognized although some authors reject the idea of subspecies that they synonymize with parent species or consider them valid species (Uetz et al., 2025). The distribution of reptiles is mainly attributed to environmental factors. Ambient temperature is the key factor since they are ectotherms. They have successfully invaded most of the world, including oceans and deserts, except where it is very cold, like in the Antarctica.

9.1.2 STUDIES ON REPTILES ALONG THE COASTAL EASTERN AFRICA ECOREGION

The history of reptile studies along the east African coast conforms to that of the entire East African herpetology, for example, see Loveridge (1957), Broadley and Howell (1991) and Howell (2000). It began with specimen collections during late 1800s which set the foundation for herpetological studies in the coastal region. The collections made led to the description of several new species that occur on the coastal Eastern Africa, for example, *Brachyophis revoili, Epacrophis boulengeri, Afrotyphlops platyrhynchus, Letheobia pembana,* and *Chirindia ewerbecki.* Most of the information on the reptile species of the coastal Eastern Africa ecoregion is patchy and can be found in Loveridge (1957), Lanza (1983, 1990), Broadley (1990, 1992), Broadley and Howell (1991, 2000), Howell (1980, 1993), Branch (1998), Tilbury (2010), Pascal (2011), Conradie et al. (2016), Malonza et al. (2006, 2018), Håkansson (2009), Spawls et al. (2002, 2006, 2018), Weinell et al. (2017), Verburgt et al. (2018), Bates (2019), Branch et al. (2019), Taylor et al. (2019), and Hallermann and Hawlitscheck (2025). There are also several localized studies, both published and unpublished, for some specific species and habitats within the region, for example, Bayliss (1993), Clarke and Dickinson (1995), Howell and Msuya (2002), Msuya et al. (2004), Wegner et al. (2009), and Howell et al. (2012) who added to our knowledge on the reptiles of the ecoregion. Despite these few studies, much is still not known about the taxonomy, ecology, the diversity, and the distribution of reptiles in the

FIGURE 9.1 Lineolate Blind Snake (*Afrotyphlops lineolatus*) a fossorial blind snake in the coastal eastern Africa region. Note the reduced eyes (*John V. Lyakurwa*).

FIGURE 9.2 Flap-nosed Chameleon (*Kinyongia tenuis*) one of the forest-dependent species in the coastal eastern Africa region. Note the digits fused into opposable bundles (*John V. Lyakurwa*).

FIGURE 9.3 Dull Green Day Gecko (*Phelsuma dubia*) expanded digits with setae (*John V. Lyakurwa*).

coastal Eastern Africa ecoregion. The most recent publication encompassing the reptiles of the coastal forests of the whole of Eastern Africa region was by Broadley and Howell (2000). Several taxonomic studies and changes have occurred regarding the reptile fauna of the ecoregion since then. These involved splitting of species that were known to occur in the region, making it difficult to establish the distributional limits of the newly described or elevated species. Some of the species affected by recent taxonomic changes includes the testudines *Pelomedusa subrufa,* the lizards *Hemidactylus mrimaensis*, *Lygodactylus tsavoensis, Melanoseps pygmaeus, Panaspis wahlbergi,* and *Scolecoseps broadleyi;* and the snakes *Afrotyphlops nanus, Letheobia pembana, L. swahilica, Leptotyphlops howelli, L. mbanjensis, Myriopholis ionides, M. tanae, Dispholidus pembae, Thelotornis usambaricus,* and *Naja ashei* (Appendix 9.1). In addition, there has been extensive review of the reptile groups, for example, at family level, which has also led to nomenclatural changes. This study assesses the reptile species richness, composition, and conservation status along the coastal Eastern Africa ecoregion from southern Somalia to southern Mozambique. In addition to the literature review, we also queried the Vertnet database (http://vertnet.org/) using the combination of words "Reptilia," "Coastal" and Tanzania/Somalia/Kenya/

Mozambique for information on specimens collected in the region. The taxonomy used in this chapter followed Spawls et al. (2018) and Uetz et al. (2025).

9.2 REPTILE SPECIES RICHNESS, COMPOSITION, ENDEMISM AND CONSERVATION STATUS ALONG THE COASTAL ECOREGION OF EASTERN AFRICA

9.2.1 SPECIES RICHNESS AND DISTRIBUTION

This study recorded 293 reptile species that occur along the coastal Eastern Africa ecoregion. The species belong to the three Orders of reptiles including Crocodylia (one species), Testudines (16 species), and Squamata (276 species). Order Rhychocephalia, which contains only one species, *Sphenodon punctatus,* does not occur in the region. More species were recorded in the Tanzanian coastal ecoregion than in the other three countries, with 168 species, followed by Kenya with 145 species, Mozambique with 141, and Somalia with 125 species (Appendix 9.1).

Only one species of crocodile *Crocodylus niloticus* is found in the coastal region of the four countries (Appendix 9.1). This species is mostly found along big rivers that empty their water in the Indian Ocean, including Juba in Somalia, Tana in Kenya, Rufiji in Tanzania, and Zambezi in Mozambique. This species is also found in other relatively bigger rivers with perennial water that traverse the coastal ecoregion, including Pangani, Wami, Ruvu, and Ruvuma.

Four families of Testudines occur in the region, including the land tortoises Family Testudinidae, fresh water terrapins Family Pelomedusidae, soft shelled turtles Family Trionychidae, and the sea turtles Family Cheloniidae (Appendix 9.1). Each family has five species except Trionychidae that has only one species *Trionyx triunguis* in Somalia. Members in the Order Testudines occur in different habitats, including on terrestrial habitats, for example, the leopard tortoise *Stigmochelys pardalis*, in fresh water (the terrapins), for example, *Pelusios castanoides* and in the marine environment (the sea turtles), for example, *Chelonia mydas*.

Order Squamata is the largest group, with 276 reptile species that occur along the coastal region of Eastern Africa. The group includes three suborders: Sauria (the typical lizards) with 132 species, Amphisbaenia (worm lizards) with 13 species, and suborder Serpentes (Ophidia) with 131 species. The squamata group is very diverse, with species adapted to different ways of life, including the arboreal species, such as chameleons, geckos, and some snakes; fossorial species, including the worm lizards, blind or thread snakes, and the limbless skinks, as well as aquatic species, for example, the sea snake *Hydrophis platurus*.

At least 12 reptilian species occurring in the region are confined to the coastal forests. These represent 4% of all the species occurring in the region, including the forest gecko *Ancylodactylus barbouri*, the dwarf gecko *Lygodactylus broadleyi*, *L. conradti*, *L. viscatus,* and *L. williamsi*. Other species include the skinks *Sepsina tetradactyla*, *Melanoseps rondoensis,* and *Scolecoseps litipoensis*. The snakes *Thelotornis usambaricus*, *Natriciteres sylvatica,* and *Elapsoidea nigra* are other forest-associated species that are found within the region. Several other reptile species are found both in forests and other habitats, for example, *Kinyongia tenuis* and *Rieppeleon brevicaudatus* that are found in forests, woodland, and farmlands (Spawls et al., 2018).

Three out of the 293 reptile species known to occur on the coastal Eastern Africa are non-native including the Indian garden lizard *Calotes versicolor*, the Flower pot blind snake *Indotyphlops braminus,* and the Aldabran tortoise *Aldabrachelys gigantea* (Spawls et al., 2018). The former two species occur in Asia and are thought to have been transported in cargo (Spawls et al., 2018). The latter species originates from Aldabra Island in Seychelles and are kept in enclosures (Spawls et al., 2018).

Coastal Eastern African reptile species richness presented in this chapter is likely to change in near future due to some unresolved taxonomic issues and the continuing exploration and description of new species. For example, Trape and Mediannikov (2016) split several species from the genus *Boaedon* and restricted *B. fuliginosus* to West Africa. Trape and Mediannikov (2016) suggest the population in the Eastern African region to corresponds to other species. Since this is not yet achieved, we retain the name *B. fuliginosus* for the population in the coastal Eastern Africa ecoregion. Taxonomic problems are also observed in other species complex groups such as *Ancylodactylus*, *Dipsadoboa, Lygodactylus, Philothamnus,* and *Panaspis*. *Dipadoboa shrevei* complex was previously known to have representatives in Tanzania (Spawls et al., 2018), that is, *D. kageleri* and *D. shrevei*. The former is now elevated to fully species and the latter is restricted to Mozambique (Branch et al., 2019). The southeastern Tanzania population (Spawls et al., 2018) is well isolated from *D. shrevei* (Branch et al., 2019) but its taxonomic situation is not yet resolved. A number of species are also poorly known, only from a handful of specimens, for example, *Hemidactylus tanganicus*, *Lygodactylus inexpectatus,* and *Scolecoseps litipoensis*.

9.2.2 SPECIES COMPOSITION

Of the 293 species known to occur in the coastal Eastern Africa, only 43 species were found throughout the coastal ecoregion from southern Somalia to Mozambique, 37 species occurred only in the Somalian coast, 12 species only in the Kenyan coast, 37 species only in the Tanzania coast, and 44 species in the Mozambican coast (Appendix 9.1). Twelve species occur in the coastal areas of Somalia, Kenya, and Tanzania but not in coastal Mozambique; three species in the coastal Somalia, Tanzania, and Mozambique but not in coastal Kenya and 22 species in the coastal Kenya, Tanzania, and Mozambique but not in coastal Somalia (Appendix 9.1). Pairwise comparison indicated that 81 of the 293 species were shared by coastal Somalia and Kenya; 59 species by coastal Somalia and Tanzania; 48 species by coastal Somalia and Mozambique; 105 species by coastal Kenya and Tanzania; 68 species by coastal Kenya and Mozambique, and 91 species were shared by the coastal areas of Tanzania and Mozambique. Some of the species might be absent in the coastal region of one country but occur in other parts of the country, for example, the Black-necked spitting cobra (*Naja nigricollis*) reaches the coastal areas in Tanzania but not in Kenya, and the Banded velvet gecko (*Homopholis fasciata*) reaches coastal Kenya but not in coastal Tanzania, although they occur in those countries.

Reptilian species composition in the coastal ecoregion of the four countries Somalia, Kenya, Tanzania, and Mozambique was assessed by the Jaccard Similarity Index (S_J) (Ludwig and Reynolds, 1988). As one would expect, countries adjacent to each other had higher species similarity than the distantly located countries. The overall species similarity index between the four countries was only 24.7%; between the three countries Somalia, Kenya, and Tanzania was 39.3%; between Somalia, Kenya, and Mozambique was 31.4%; between Somalia, Tanzania, and Mozambique was 27.5%, and between Kenya, Tanzania, and Mozambique was 41.1%. The index between two countries Somalia and Kenya was 62.8%; between Somalia and Tanzania was 44%; between Somalia and Mozambique was 36.4%; between Kenya and Tanzania was 69.1%; between Kenya and Mozambique was 55.2%, and between Tanzania and Mozambique was 52.3%. The differences in similarity indices between countries may be accounted by a big latitudinal gradient between the countries resulting in different environmental conditions and habitats that favor different species of reptiles. The coastal Eastern Africa ecoregion ranges from southern Somalia at about 1° North to 27° South in Mozambique.

9.2.3 ENDEMISM

The level of endemism of reptilian fauna in the coastal Eastern African ecoregion is considerably high. At least 79 species or 27% of all the reptiles occurring in the coastal region are endemic. These include species that are only restricted to one country or shared only by two or more counties. Tanzanian coastal region has the highest numbers with 26 endemic species, followed by Somalia with 13 species, Mozambique with six species, and Kenya with five species (Appendix 9.1). Twenty-nine (29) coastal Eastern Africa endemic species are shared between countries (mostly between two countries), for example, *Atractaspis engdahli*, *Mochlus mabuiiformis*, *Crotaphopeltis braestrupi*, *Geocalamus acutus*, *Hemidactylus barbouri*, *Lygodactylus broadleyi Lygodactylus kimhowelli, Lygodactylus mombasicus, Lygodactylus tsavoensis*, *Acontias percivali, Mochlus pembanum*, *Gastropholis prasina*, *Gastropholis vittata*, *Rieppeleon brevicaudatus*, *Kinyongia tenuis*, *Letheobia pembana*, *Letheobia swahilica*, *Leptotyphlops howelli, Elapsoidea nigra* and *Thelotornis usambaricus* (Appendix 9.1). The highest level of endemism in Tanzania was contributed by the existence of the offshore Islands of Pemba and Zanzibar that have at least seven endemic species, including *Phelsuma parkeri*, *Trachylepis albotaeniata*, *Letheobia pembana*, *Leptotyphlops pembae*, *Lycophidion pembanum*, *Dispholidus pembae,* and *Natriciteres pembana.* Most of the endemic species in the region are mainly restricted to the coast and are found nowhere else outside the belt, for example, most amphisbaenians, *Scolecoseps boulengeri*, *Lygodactylus inexpectatus*, *Melanoseps rondoensis, Prosymna semifasciata*, *Letheobia swahilica*, *Hemidactylus barbouri*, and *Letheobia lumbriciformis.* Merging of the coastal and the Eastern Arc Mountain habitats at the foothills may also have contributed to the higher levels of endemism in the Tanzanian coastline. This is well exemplified by Kimboza Nature Forest Reserve that is in transition between the Eastern Arc Mountains and coastal forest (Lyakurwa et al., 2024). These mountains are known for their higher endemism in many taxonomic groups (Dinesen et al., 2001; Gereau et al., 2016). Areas with the highest species diversity and endemism are often considered centers of endemism, including the Eastern African coastal ecoregion (Burgess et al., 1998; Timberlake et al., 2011).

9.2.4 CONSERVATION STATUS

The International Union for Conservation of Nature (IUCN) sets criteria for assessing extinction risk of animal, fungus, and plant species, which are used as indicators of the health of the biodiversity (IUCN, 2025). Criteria used to classify species include population size, rate of decline, area of geographical distribution, degree of population fragmentation, and the degree of distribution fragmentation (Mace et al., 2008). Based on the criteria, the IUCN red list divides species into nine categories, including Not Evaluated (NE), Data Deficient (DD), Least Concern (LC), Near Threatened (NT), Vulnerable (VU), Endangered (EN), Critically Endangered (CR), Extinct in the Wild, and Extinct. Based on these categories the coastal Eastern Africa reptiles were grouped into seven categories (Appendix 9.1).

Two species, including the Hawksbill turtle *Eretmochelys imbricata* and the turquoise dwarf gecko *Lygodactylus williamsi* (Figure 9.4) are assigned CR status. These species are facing extremely high risk of extinction due to habitat loss and illegal harvesting. The second category includes four species that are considered EN, including two turtles *Cycloderma frenatum* and *Chelonia mydas*, the chameleon *Kinyongia tenuis* (Figure 9.2) and the snake *Prosymna semifasciata*. These species are also facing reduction in population size, reduction in the geographical range, and the probability of extinction is high (IUCN, 2025). Seven species, including the turtles *Caretta caretta*, *Lepidochelys olivacea*, *Dermochelys coriacea,* and *Trionyx triunguis* as well as the snakes *Aparallactus werneri*, *Thelotornis usambaricus* (Figure 9.5), and *Bitis gabonica* are assigned the VU status. Five species are assigned NT status, including *Lygodactylus viscatus*, *Cryptoblepharus boutonii*, *Scolecoseps broadleyi*, *Gastropholis prasina*, and *Python sebae*. Other categories set by the IUCN in which reptiles within the ecoregion fall include NE category with 19 species, DD category with 39 species, and the LC category with 217 species or 74% of all the species (Appendix 9.1). The species in the LC category are currently not considered of conservation concern because they are widely spread, have sufficient populations and are not threatened with extinction. However, some species which are in this category are thought to be species complex (Malonza et al., 2016; Spawls et al., 2018; Malonza and Bauer 2022) and geographical ranges of some species were overestimated because species delimitations were poorly characterized (Gereau et al., 2016).

9.3 CONCLUSIONS AND RECOMMENDATIONS

9.3.1 CONCLUSIONS

This study reviews the reptilian fauna occurring along the coastal Eastern Africa ecoregion. This region ranges from southern Somalia through Kenya and Tanzania to southern

FIGURE 9.4 Turquoise Blue Dwarf (*Lygodactylus williamsi*) one of the Tanzanian endemic and critically endangered reptile species (*John V. Lyakurwa*).

FIGURE 9.5 Usambara Dwarf Gecko (*Thelotornis usambaricus*) one of the eastern Africa coastal region snakes that is considered Vulnerable by IUCN (*John V. Lyakurwa*).

Mozambique. At least 293 reptile species occur in the region, including one crocodile species *Crocodylus niloticus*, 16 testudines, and 276 squamates. Species richness varied between countries, with coastal Tanzania having 168 species, followed by Kenya with 145 species, Mozambique with 141 species, and lastly Somalia with 125 species. Some species were distributed all along the coastal region, whereas others were endemic to a certain country or locality. The level of endemism was highest along the coastal Tanzania with 26 endemic species followed by Somalia with 13 species, Mozambique with six species and Kenya with five species. The level of reptilian species similarity between the countries was

24% indicating differences in species composition among the four countries. The majority of the species (74%) in the region are assigned a LC category by the IUCN suggesting that they are not threatened with extinction although the other species fall under the DD, NT, or threat categories of CR, EN, and VU. The big latitudinal gradient and habitat variability between countries perhaps may have contributed to the findings observed.

9.3.2 *RECOMMENDATIONS*

Three non-native reptilian species occur on the coast. There is a need to monitor their distribution and population size to enable deducing their possible ecological impact to the native species and environment. Furthermore, this study reveals some reptile species that are endemic to the region, whereas others are categorized as threatened species. Most of these species fall outside the protected area network. The continued existence of the species depends very much on the implementation of best practices in conservation and management of their habitats by the responsible authorities and individuals. Additionally, some species in the coastal Eastern Africa have some taxonomic problems and are considered complex groups. Some localities within the region have not been adequately surveyed, and also some species were described only from a few specimens. Further studies are recommended to resolve the issues.

ACKNOWLEDGMENTS

We thank Professor Thammineni Pullaiah of the Sri Krishnadevaraya University and Dr. Ally J. K. Nkwabi of the Tanzania Wildlife Research Institute for their valuable comments.

KEYWORDS

- **reptile species**
- **coastal ecoregion**
- **Eastern Africa**
- **endemism**
- **conservation status**
- **Zanzibar-Inhambane**

REFERENCES

Bates, M. F. Catalogue of reptiles from Mozambique in the collection of the National Museum, Bloemfontein, South Africa, Animal and Plant Systematics Articles, **2019.** https://nationalmuseumpublications.co.za/

catalogue-of-reptiles-from-mozambique-in-the-collection-of-the-national-museum-bloemfontein-south-africa-2/. Accessed 22 December 2022.

Bayliss, J. Preliminary biological investigation into Kimboza Forest Reserve, Morogoro Region, Morogoro District, Tanzania. Unpublished report. **1993**.

Branch, B. 1998. *Field Guide to Snakes and other Reptiles of Southern Africa*, 3rd edn. Stuik: Cape Town, South Africa, **1998**.

Branch, W. R.; Bayliss, J.; Bittencourt-Silva, G. B.; Loader, S. P.; Menegon, M.; Nanvonamuquitxo, C.; Tolley, K. A. A new species of tree snake (*Dipsadoboa*, Serpentes: Colubridae) from 'Sky Island' forests in northern Mozambique, with notes on other members of the *Dipsadoboa werneri* group. *Zootaxa* **2019,** *4646*, 541–563.

Broadley, D. G. Reptiles and amphibians from the Bazaruto Archipelago, Mozambique. *Arnoldia Zimbabwe* **1992,** *9*, 539–548.

Broadley, D. G. The herpetofaunas of the islands off the coast of South Mozambique. *Arnoldia Zimbabwe* **1990,** *9*, 469–493.

Broadley, D. G.; Howell, K. M. A checklist of the reptiles of Tanzania, with synoptic keys. *Syntarsus* **1991,** *1*, 1–70.

Broadley, D. G.; Howell, K. M. Reptiles. In: Burgess, N. D.; Clarke, G. P. (Eds.), *Coastal Forests of Eastern Africa*. IUCN: Gland, Switzerland and Cambridge, UK, **2000,** p. 443.

Burgess, N. D.; Clarke, G. P.; Rodgers, W. A. Coastal forests of eastern Africa: Status, endemism patterns and their potential causes. *Biol. J. Linn. Soc*. **1998,** *64*, 337–367.

Clarke, G. P.; Dickinson, A. *Status Reports for 11 Coastal Forests in Coast Region, Tanzania.* Frontier-Tanzania Technical Report No. 17. The Society for Environmental Exploration, UK/The University of Dar es Salaam: Tanzania, **1995**.

Conradie, W.; Bittencourt-Silva, G. B.; Engelbrecht, H.; Loader, S. P.; Menegon, M.; Nanvonamuquitxo, C.; Scott, M.; Tolley, K. Exploration into the hidden world of Mozambique's sky islands forests: New discoveries of reptiles and amphibians. *Zoosystematics Evol*. **2016,** *92*, 163–180. https://doi.org/10.3897/zse.92.9948.

Dinesen, L.; Lehmberg, T.; Rahner, M. C.; Fjeldså, J. Conservation priorities for the forests of the Udzungwa Mountains, Tanzania, based on primates, duikers and birds. *Biol. Conserv*. **2001,** *99*, 223–236.

Gereau, R. E.; Cumberlidge, N.; Hemp, C.; Hochkirch, A.; Jones, T.; Kariuki, M.; Lange, C. N.; Loader, S. P.; Malonza, P. K.; Menegon, M.; Ndang'ang'a, P. K.; Rovero, F.; Shirk, P. Globally threatened biodiversity of the Eastern Arc Mountains and coastal forests of Kenya and Tanzania. *J. East Afr. Nat. Hist.* **2016,** *105*, 115–201.

Hallermann, J.; Hawlitscheck, O. Two new species of Boaedon from Ethiopia and Somalia, with a review of the species of East Africa. *Zootaxa*. **2025**, 5569, 119–137.

Håkansson, N. T. Reptiles and amphibians from the Kenyan coastal hinterland. *Herpetol. Bull*. **2009,** *107*, 30–39.

Howell, K. M. An overview of East African amphibian studies, past, present and future: A view from Tanzania, *Afr. J. Herpetol.* **2000,** *49*, 147–164.

Howell, K. M. Herpetofauna of the Eastern African forests. In: Lovett, J. C.; Wasser, S. K. (Eds.), *Biogeography and Ecology of the Rain Forests of Eastern Africa*. Cambridge University Press: UK, **1993**.

Howell, K. M. Pugu forest reserve: Biological values and development, *Afr. J. Herpetol*. **1980,** *19*, 73–82.

Howell, K. M.; Msuya, C. A.; Mligo, C.; Werema, C.; Kihaule, P.; Honorati, M. K.; Suleiman, H. O. *Biodiversity Surveys of Poorly Known Coastal Forests of Southeastern Tanzania and Zanzibar*. A Consultancy Report of the Department of Zoology and Wildlife Conservation, University of Dar es Salaam: Tanzania to WWF, **2012**.

Howell, K. M.; Msuya, C. A. *Biodiversity studies in Kilwa and Lindi Districts: Reptiles and Amphibians.* Consultancy Report of the Department of Zoology and Wildlife Conservation, University of Dar es Salaam to Ornis Consult: Dar es Salaam, Tanzania, **2002**.

IUCN. The IUCN red list of threatened species, Version 2025-1, **2025**. https://www.iucnredlist.org, ISSN 2307–8235. Accessed 17th June 2025.

Lanza, B. Amphibians and reptiles of the Somali Democratic Republic: Checklist and biogeography. *Biogeographia* **1990,** *14*, 407–465, https://doi.org/10.21426/B614110318.

Lanza, B. List of the Somali amphibians and reptiles. *Monitore Zoologico Italiano, Supplemento* **1983,** *18*, 193–247, https://doi.org/10.1080/00269786.1983.11758573.

Loveridge, A. Check list of the reptiles and amphibians of East Africa: Uganda, Kenya, Tanganyika, Zanzibar. *Bull. Mus. Comp. Zool*. **1957,** *117*, 153–362.

Ludwig, J. A.; Reynolds, J. F. *Statistical Ecology: A Primer on Methods and Computing*. John Wiley & Sons: New York, **1988**.

Lyakurwa, J. V.; Loader, S. P.; Ngalason, W.; Gumbs, R.; Ofori-Boateng, C.; Liedtke, H. C. Kimboza, a small lowland forest with an outstanding herpetofauna diversity in East Africa. *Ecol. Evol.* **2024**, 14, 1–14.

Mace, G. M.; Collar, N. J.; Gaston, K. J.; Hilton, T. C.; Akcakaya, H. R.; Leader, W. N.; Millner-Gulland, E. J.; Stuart, S. N. Quantification of extinction risk: IUCN's system for classifying threatened species. *Conserv. Biol.* **2008,** *22*, 1424–1442.

Malonza, P.; Bauer, A. Resurrection of the African gecko genus *Ancylodactylus* Müller, 1907 (Squamata: Gekkonidae) and description of six new species from Kenya. *Zootaxa* **2022,** *5141*, 101–139.

Malonza, P. K.; Granthon, C.; Williams, D. A. A new species of dwarf gecko in the genus *Lygodactylus* (squamata: Gekkonidae) from central Kenya. Zootaxa **2016**, 4061, 418–428. https://doi.org/10.11646/zootaxa.4061.4.6

Malonza, P. K.; Mulwa, D. M.; Nyamache, J. O.; Jones, G. Biogeography of the Shimba Hills ecosystem herpetofauna in Kenya. *Zool. Res.* **2018,** *39,* 97–104, https://doi.org/10.24272%2Fj.issn.2095-8137.2017.048.

Malonza, P. K.; Wasonga, V. D.; Muchai, V.; Rotich, D.; Bwong, B. A.; Bauer, A. M. Diversity and biogeography of herpetofauna of the Tana River Primate National Reserve, Kenya. *J. East Afr. Nat. Hist.* **2006,** *95*, 95–109, https://doi.org/10.2982/0012- 8317(2006)95[95:DABOHO]2.0.CO;2

Msuya, C. A.; Sangu, G. M.; Kayombo, C. K.; Howell, K. M.; Ndangalasi, H. J.; Seleman, H.; Ngalason, W.; Kihaule, P. M. *Biological Diversity Survey of the Mbarawala and Ruawa-Likonde Plateaus in Lindi Region, Tanzania.* A Consultancy Report to the Royal Danish Embassy: Dar es Salaam, Tanzania, **2004**.

Pascal, O. *The Coastal Forests of Northern Mozambique, 2008-2009 Expeditions. Our Planet Reviewed Program Report no. 1*. Pro-Natura International/Museum national d'Histoire naturelle: Paris, **2011**.

Spawls, S.; Howell, K. M.; Drewes, R. C. *Pocket Guide to the Reptiles and Amphibians of East Africa*. A and C. Black: London, UK, **2006**.

Spawls, S.; Howell, K. M.; Drewes, R. C.; Ashe, J. *A Field Guide to the Reptiles of East Africa*. A and C Black: London, UK, **2002**.

Spawls, S.; Howell, K. M.; Hinkel, H.; Menegon, M. *Field Guide to East African Reptiles*. 2nd Ed. Bloomsbury: London, UK, **2018**.

Taylor, R.; Malonza, P.; Spawls, S.; Ochieng, J.; Jackson, C.; Navarro, R. A checklist of the reptiles and amphibians of Arabuko-Sokoke Forest, Kenya, A Rocha Kenya. Checklist dataset, **2019,** https://doi.org/10.15468/tcgw7t accessed via GBIF.org on 2022-12-16. Web visited 16th December 2022.

Tilbury, C. R. *Chameleons of Africa, An Atlas Including the Chameleons of Europe, the Middle East and Asia*. Edition Chimaira/Serpent's Tale NHBD: Frankfurt, Germany, **2010**.

Timberlake, J.; Goyder, D.; Crawford, F.; Burrows, J.; Clarke, P.; Luke, Q.; Matimele, H.; Müller, T.; Pascal, O.; de Sousa, C.; Alves, T. Coastal dry forests in northern Mozambique. *Plant Ecol. Evol.* **2011,** *144*, 126–137, https://doi.org/10.5091/plecevo.2011.539.

Trape, J.-F.; Mediannikov, O. Cinq serpents nouveaux du genre *Boaedon* Duméril, Bibron and Duméril, 1854 en Afrique centrale. *Bull. Soc. Herpétol. Fr.* **2016,** *159*, 61–111.

Uetz, P.; Freed, P.; Aguilar, R.; Reyes, F.; Kudera, J; Hošek, J. (Eds). The Reptile Database. **2025**. http://www.reptile-database.org. Accessed 17th June 2025.

Verburgt, L.; Verburgt, U. K.; Branch, R. B. A new species of *Scolecoseps* (Reptilia: Scincidae) from coastal north-eastern Mozambique. *Afr. J. Herpetol* **2018,** *67*, 86–98.

Vitt, L. J. Reptile diversity and history. In: Dodd, C. K. Jr. (Ed.), *Reptiles Ecology and Conservation: A Handbook of Techniques*. Oxford University Press: UK, **2016**, p. 462.

Vitt, L. J.; Caldwell, J. P. *Herpetology: An Introductory Biology of Amphibians and Reptiles*, 4th Edition. Academic Press: San Diego, CA, USA, **2014**.

Wegner, G.; Howell, K. M.; Davenport, T. R. B.; Burgess, N. D. The forgotten 'coastal forests' of Mtwara, Tanzania: A biologically impoverished and yet important ecosystem. *J. East Afr. Nat. Hist.* **2009,** *98*, 167–209.

Weinell, J. L.; Portik, D. M.; Bauer, A. M. Noteworthy records of amphibians and reptiles from Mozambique. *AHN Surveys* **2017,** *65*, 3–10.

APPENDIX 9.1 Reptile Species of the Coastal Ecoregion of Eastern African Countries

Sr. No.	Order, Suborder, Family, Species	Common Name	Country				Conservation Status and Endemism
			SO	KE	TZ	MZ	
	Order Crocodylia, Family Crocodylidae, Crocodiles						
1	*Crocodylus niloticus* Laurenti, 1768	Nile crocodile	x	x	x	x	Least concern
	Order Testudines, Family Testudinidae, Tortoises						
2	*Stigmochelys pardalis* (Bell, 1828)	Leopard tortoise	x	x	x	x	Least concern
3	*Kinixys belliana* Gray, 1831	Bell's hingeback tortoise	x	x	x	x	Not evaluated
4	*Kinixys spekii* Gray, 1863	Speke's hingeback tortoise	–	x	–	x	Not evaluated
5	*Kinixys zombensis* Hewitt, 1931	Eastern hinged tortoise	–	x	x	x	Not evaluated
6	*Aldabrachelys gigantea* (Schweigger, 1812)	Aldabran tortoise	–	x	x	–	Not evaluated, introduced
	Order Testudines, Family Pelomedusidae						
7	*Cycloderma frenatum* Peters, 1854	Zambezi flap-shelled turtle/Zambezi soft-shelled turtle	–	–	x	x	Endangered
8	*Pelomedusa neumanni* Petzold, Vargas-Ramírez, Kehlmaier et al., 2014	Neumann's marsh terrapin	–	x	x	–	Not evaluated
9	*Pelusios castanoides* Hewitt, 1931	Yellow-bellied hinged terrapin	–	x	x	x	Least concern
10	*Pelusios sinuatus* (Smith, 1838)	Serrated hinged terrapin	x	x	x	x	Not evaluated
11	*Pelusios subniger* (Bonnaterre, 1789)	Pan hinged terrapin	–	–	–	x	Not evaluated
	Order Testudines, Family Cheloniidae, Sea Turtles						
12	*Caretta caretta* (Linnaeus, 1758)	Loggerhead turtle	x	x	x	x	Vulnerable
13	*Chelonia mydas* (Linnaeus, 1758)	Green turtle	x	x	x	x	Endangered
14	*Eretmochelys imbricata* (Linnaeus, 1766)	Hawksbill turtle	x	x	x	x	Critically endangered
15	*Lepidochelys olivacea* (Eschscholtz, 1829)	Olive ridley turtle	x	x	x	x	Vulnerable
163	*Dermochelys coriacea* (Vandelli, 1761)	Leatherback turtle	x	x	x	x	Vulnerable
	Order Testudines, Family Trionychidae, Soft-shelled Turtles						
17	*Trionyx triunguis* (Forskål, 1775)	Nile soft-shelled turtle	x	–	–	–	Vulnerable
	Squamata, Sauria, Gekkonidae, Geckoes						
18	*Ancylodactylus africanus* (Werner, 1895)	Usambara Forest gecko African gecko	–	x	x	–	Least concern
19	*Ancylodactylus barbouri* (Perret, 1986)	Uluguru Forest gecko	–	–	x	–	Least concern, Tanzania endemic, forest

APPENDIX 9.1 *(Continued)*

Sr. No.	Order, Suborder, Family, Species	Common Name	Country				Conservation Status and Endemism
			SO	KE	TZ	MZ	
20	*Ancylodactylus uzungwae* (Perret, 1986)	Udzungwa Forest gecko	–	–	x	–	Data deficient, Tanzania endemic, forest
21	*Hemidactylus albopunctatus* (Loveridge, 1947)	White-spotted gecko	x	–	–	–	Least concern
22	*Hemidactylus angulatus* Hallowell, 1854	East African house gecko	–	x	x	–	Least concern
23	*Hemidactylus barbouri* Loveridge, 1942	Barbour's gecko	–	x	x	–	Not evaluated
24	*Hemidactylus granchii* Lanza, 1978	Granchi's leaf-toed gecko	x	–	–	–	Data deficient, Somali endemic
25	*Hemidactylus frenatus* Duméril and Bibron, 1836	Pacific gecko	x	x	–	–	Least concern
26	*Hemidactylus klauberi* Scortecci, 1948	–	x	–	–	–	Data deficient, Somali endemic
27	*Hemidactylus mabouia* (Moreau De Jonnès, 1818)	Tropical house gecko	x	x	x	x	Least concern
28	*Hemidactylus macropholis* Boulenger, 1896	Boulenger's gecko	x	–	–	–	Least concern
29	*Hemidactylus mercatorius* Gray, 1842	–	x	–	–	x	Least concern
30	*Hemidactylus modestus* (Günter, 1894)	Tana river gecko	–	x	–	–	Data deficient, Kenya endemic
31	*Hemidactylus mrimaensis* Malonza and Bauer, 2014	Kaya gecko	–	x	–	–	Not evaluated, Kenya endemic
32	*Hemidactylus platycephalus* Peters, 1854	Tree gecko	x	x	x	x	Least concern
33	*Hemidactylus puccionii* Calabresi, 1927	Somali plain gecko	x	–	–	–	Data deficient, Somali endemic
34	*Hemidactylus robustus* Heyden, 1827	Heyden's gecko	x	–	–	–	Least concern
35	*Hemidactylus ruspolii* Boulenger, 1896	Prince Ruspoli's gecko	x	–	–	–	Least concern
36	*Hemidactylus squamulatus* Tornier, 1896	Nyika gecko	x	x	x	–	Least concern
37	*Hemidactylus tanganicus* Loveridge, 1929	Dutumi gecko	–	–	x	–	Data deficient, Tanzania endemic
38	*Hemidactylus tropidolepis* Mocquard, 1888	Ogaden gecko	x	x	–	–	Least concern
39	*Holodactylus africanus* Boettger, 1893	Mocquard's leaf-toed gecko	x	–	–	–	Least concern
40	*Holodactylus cornii* Scortecci, 1930	Somalia-Krallengecko	x	–	–	–	Data deficient

APPENDIX 9.1 *(Continued)*

Sr. No.	Order, Suborder, Family, Species	Common Name	Country				Conservation Status and Endemism
			SO	KE	TZ	MZ	
41	*Lygodactylus angolensis* Bocage, 1896	Angolan dwarf gecko	–	–	x	x	Least concern
42	*Lygodactylus angularis* Günther, 1893	Angulate dwarf gecko	–	x	x	x	Least concern
43	*Lygodactylus broadleyi* Pasteur, 1995	Broadley's dwarf gecko	–	x	x	–	Least concern, forest
44	*Lygodactylus capensis* (Smith, 1849)	Cape dwarf gecko	–	–	x	x	Least concern
45	*Lygodactylus conradti* Matschie, 1892	Conradt's dwarf gecko	–	–	x	–	Data deficient, Tanzania endemic, forest
46	*Lygodactylus grotei* Sternfeld, 1911	Grote's dwarf gecko	–	–	x	x	Least concern
47	*Lygodactylus viscatus* Vaillant, 1873	Copal/Howell's dwarf gecko	–	–	x	–	Near threatened, Tanzania endemic, forest
48	*Lygodactylus inexpectatus* Pasteur, 1965	Dar es salaam Dwarf gecko	–	–	x	–	Data deficient, Tanzania endemic
49	*Lygodactylus keniensis* Parker, 1936	Kenya dwarf gecko	x	–	–	–	Least concern
50	*Lygodactylus kimhowelli* Pasteur, 1995	Kim Howell's/zebra dwarf gecko	–	x	x	–	Least concern
51	*Lygodactylus mombasicus* Loveridge, 1935	White-headed dwarf gecko	–	x	x	–	Least concern
52	*Lygodactylus picturatus* (Peters, 1870)	Yellow-headed dwarf gecko	x	x	x	x	Least concern
53	*Lygodactylus scorteccii* Pasteur, 1959	Scortecci's dwarf gecko	x	x	–	–	Data ceficient
54	*Lygodactylus somalicus* Loveridge, 1935	Somali dwarf gecko	x	–	–	–	Least concern
55	*Lygodactylus tsavoensis* Malonza, Bauer, Granthon et al., 2019	Tsavo dwarf gecko	–	x	–	–	Not evaluated
56	*Lygodactylus williamsi* Loveridge, 1952	Turquoise/electric blue dwarf gecko	–	–	x	–	Critically endangered, Tanzania endemic, forest
57	*Chondrodactylus turneri* (Gray, 1864)	Turner's thick-toed gecko	–	–	x	x	Least concern
58	*Elasmodactylus tetensis* (Loveridge, 1952)	Tete thick-toed gecko	–	–	x	x	Least concern
59	*Elasmodactylus tuberculosus* Boulenger, 1895	Tuberculate thick-toed gecko	–	–	x	x	Least concern
60	*Pachydactylus maculatus* Gray, 1845	Spotted thick-toed gecko	–	–	–	x	Least concern
61	*Homopholis fasciata* (Boulenger, 1890)	Banded velvet gecko	x	x	–	x	Least concern
62	*Homopholis wahlbergii* (Smith, 1849)	Wahlberg's velvet gecko	–	–	–	x	Least concern

APPENDIX 9.1 *(Continued)*

Sr. No.	Order, Suborder, Family, Species	Common Name	Country				Conservation Status and Endemism
			SO	KE	TZ	MZ	
63	*Phelsuma parkeri* Loveridge, 1941	Pemba day gecko	–	–	x	–	Least concern, endemic to Pemba, Tanzania
64	*Phelsuma dubia* (Boettger, 1881)	Dull green day gecko	–	x	x	x	Least concern
65	*Pristurus crucifer* (Valenciennes, 1861)	Valenciennes rock gecko	x	–	–	–	Least concern
66	*Pristurus phillipsii* Boulenger, 1895	Somai rock gecko	x	–	–	–	Least concern
67	*Pristurus simonettai* (Lanza and Sassi, 1968)	Coastal rock gecko	x	–	–	–	Least concern, Somali endemic
	Order Squamata, Suborder Sauria, Family Scincidae, Skinks						
68	*Acontias aurantiacus* (Peters, 1854)	Golden blind legless skink	–	–	–	x	Least concern
69	*Acontias percivali* Loveridge, 1935	Percival's legless skink	–	x	–	–	Least concern
70	*Acontias plumbeus* Bianconi, 1849	Giant legless skink	–	–	–	x	Least concern
71	*Chalcides ocellatus* (Forskål, 1775)	Ocellated skink	x	–	–	–	Least concern
72	*Scelotes arenicola* (Peters, 1854)	Zululand dwarf burrowing skink	–	–	–	x	Least concern
73	*Scelotes mossambicus* (Peters, 1882)	Mozambique dwarf burrowing skink	–	–	–	x	Least concern
74	*Scelotes duttoni* Broadley, 1990	Dutton's dwarf burrowing skink	–	–	–	x	Least concern, Mozambique endemic
75	*Trachylepis albotaeniata* (Boettger, 1913)	Pemba speckle-lipped skink	–	–	x	–	Least concern, Tanzania endemic
76	*Trachylepis boulengeri* (Sternfeld, 1911)	Boulenger's skink	–	–	x	x	Least concern
77	*Trachylepis brevicollis* (Wiegmann, 1837)	Short-necked skink	x	x	–	–	Least concern
78	*Trachylepis depressa* (Peters, 1854)	Eastern Coastal skink	–	–	–	x	Least concern
79	*Trachylepis ferrarai* (Lanza, 1978)	Ferrara's Mabuya	x	–	–	–	Data deficient, Somali endemic
80	*Trachylepis hildebrandtii* (Peters, 1874)	Hildebrandt's skink	x	–	–	–	Least concern
81	*Trachylepis maculilabris* (Gray, 1845)	Speckle-lipped skink	x	x	x	x	Least concern
82	*Trachylepis margaritifera* (Peters, 1854)	Rainbow skink	–	–	x	x	Least concern
83	*Trachylepis megalura* (Peters, 1878)	Grass top/long-tailed skink	–	x	x	x	Least concern
84	*Trachylepis planifrons* (Peters, 1878)	Tree skink	x	x	x	–	Least concern

APPENDIX 9.1 *(Continued)*

Sr. No.	Order, Suborder, Family, Species	Common Name	Country				Conservation Status and Endemism
			SO	KE	TZ	MZ	
85	*Trachylepis striata* (Peters, 1844)	Striped skink	x	x	x	x	Least concern
86	*Trachylepis varia* (Peters, 1867)	Variable skink	–	x	x	x	Least concern
87	*Trachylepis variegata* (Peters, 1870)	Variegated skink	–	–	–	x	Least concern
88	*Mochlus laeviceps* (Peters, 1874)	Common writhing skink	x	–	–	–	Data deficient
89	*Mochlus lanceolatum* (Broadley, 1990)	Broadley's writhing skink	–	–	–	x	Not evaluated, Mozambique endemic
90	*Mochlus mabuiiform*e (Loveridge, 1935)	Mabula-like writhing skink	x	x	–	–	Not evaluated
91	*Mochlus mafianum* (Broadley, 1994)	Mafia writhing skink	–	–	x	–	Least concern, Tanzania endemic
92	*Mochlus pembanu*m (Boettger, 1913)	Pemba island writhing skink	–	x	x	–	Least concern
93	*Mochlus productum* (Boulenger, 1909)	Boulenger's writhing skink	x	–	–	–	Least concern, Somali endemic
94	*Mochlus simonettai* (Lanza, 1979)	Simonetta's writhing skink	x	–	–	–	Data deficient, Somali endemic
95	*Mochlus sundevalli* (Smith, 1849)	Sundevall's writhing skink	x	x	x	x	Least concern
96	*Mochlus somalicus* (Parker, 1942)	Somali writhing skink	–	x	x	–	Not evaluated
97	*Mochlus tana*e (Loveridge, 1935)	Tana river writhing skink	x	x	–	–	Least concern
98	*Mochlus vinciguerrae* (Parker, 1932)	Vinciguerra's writhing skink	x	–	–	–	Least concern
99	*Cryptoblepharus africanus* (Sternfeld, 1918)	Coral rag skink	x	x	x	x	Least concern
100	*Cryptoblepharus boutonii* (Des Jardin, 1831)	Bouton's skink	–	–	–	x	Near threatened
101	*Panaspis massaiensi*s (Angel, 1924)	Maasai snake-eyed skink	–	x	x	–	Not evaluated
102	*Panaspis wahlbergi* (Smith, 1849)	Wahlberg's snake-eyed skink	–	–	–	x	Least concern
103	*Sepsina tetradactyla* Peters, 1874	Four-fingered skink	–	–	x	–	Least concern, forest
104	*Melanoseps loveridgei* Brygoo and Roux-Estève, 1982	Loveridge's limbless skink	–	–	x	–	Least concern
105	*Melanoseps pygmaeus* Broadley, 2006	Pigmy limbless skink	–	x	x	–	Least concern
106	*Melanoseps rondoensis* Loveridge, 1942	Rondo limbless skink	–	–	x	–	Data deficient, Tanzania endemic, forest

APPENDIX 9.1 *(Continued)*

Sr. No.	Order, Suborder, Family, Species	Common Name	Country				Conservation Status and Endemism
			SO	KE	TZ	MZ	
107	*Scolecoseps acontias* (Werner, 1913)	Legless sand skink	–	–	x	–	Data deficient, Tanzania endemic
108	*Scolecoseps boulengeri* Loveridge, 1920	Boulenger's limbless skink	–	–	–	x	Data deficient, Mozambique endemic
109	*Scolecoseps broadleyi* Verburgt, Verburgt and Branch, 2018	–	–	–	–	x	Near threatened, Mozambique endemic
110	*Scolecoseps litipoensis* Broadley, 1995	Litipo sand skink	–	–	x	–	Data deficient, Tanzania endemic, forest
	Suborder Sauria, Family Lacertidae, Typical Lizards						
111	*Gastropholis prasin*a Werner, 1904	Green keel-bellied lizard	–	x	x	–	Near threatened
112	*Gastropholis vittat*a Fischer, 1886	Striped keel-bellied lizard	–	x	x	x	Least concern
113	*Holaspis laevis* Werner, 1896	Eastern blue-tailed gliding lizard	–	x	x	x	Least concern
114	*Nucras boulengeri* Neumann, 1900	Boulenger's scrub lizard	–	x	x	–	Least concern
115	*Nucras ornata* (Gray, 1864)	Ornate scrub lizard	–	–	x	x	Not evaluated
116	*Latastia boscai* Bedriaga, 1884	Bosca's long-tailed lizard	x	–	–	–	Least concern
117	*Latastia petersiana* (Mertens, 1938)	Peters' longtail lizard	x	–	–	–	Data deficient, Somalia endemic
118	*Latastia johnstonii* Boulenger, 1907	Johnston's/Malawi long-tailed lizard	–	–	x	–	Least concern
119	*Latastia longicaudata* (Reuss, 1834)	Southern long-tailed lizard	x	x	–	–	Least concern
120	*Heliobolus neumanni* (Tornier, 1905)	Neumann's sand lizard	–	x	x	–	Data deficient
121	*Heliobolus spekii* (Günther, 1872)	Speke's sand lizard	x	x	x	–	Least concern
122	*Ichnotropis capensis* (Smith, 1838)	Cape rough-scaled lizard	–	–	–	x	Least concern
123	*Meroles squamulosa* (Peters, 1854)	Mozambique rough-scaled lizard	–	–	x	x	Not evaluated
124	*Philochortus hardeggeri* (Steindachner, 1891)	Hardegger's orangetail lizard	x	–	–	–	Least concern
125	*Pseuderemias brenneri* (Peters, 1869)	Brenner's sand racer	x	–	–	–	Least concern
126	*Pseuderemias striatus*(Peters, 1874)	Peter's sand lizard	x	x	–	–	Data deficient
127	*Pseuderemias smithii* (Boulenger, 1895)	Smith's sand lizard	x	–	–	–	Least concern

APPENDIX 9.1 *(Continued)*

Sr. No.	Order, Suborder, Family, Species	Common Name	Country				Conservation Status and Endemism
			SO	KE	TZ	MZ	
	Suborder Sauria, Family Cordylidae, Girdled Lizards						
128	*Platysaurus maculatus* Broadley, 1965	Spotted flat lizard	–	–	x	x	Least concern
129	*Cordylus tropidosternum* (Cope, 1869)	Tropical Girdled lizard	–	x	x	x	Least concern
	Suborder Sauria, Family Gerrhosauridae, Plated Lizards						
130	*Broadleysaurus major* (Duméril, 1851)	Great plated lizard	x	x	x	x	Least concern
131	*Gerrhosaurus flavigularis* Wiegmann, 1828	Yellow-throated plated lizard	x	x	x	x	Least concern
132	*Gerrhosaurus nigrolineatus* Hallowell, 1857	Black-lined plated lizard	–	x	x	x	Least concern
	Suborder Sauria, Family Agamidae, Agamas						
133	*Acanthocercus atricollis* (Smith, 1849)	Blue-headed tree agama	–	–	–	x	Least concern
134	*Acanthocercus gregorii* (Günther, 1894)	Blue-throated agama	–	x	x	–	Least concern
135	*Agama armata* Peters, 1855	Peter's ground agama	–	–	–	x	Least concern
136	*Agama lionotus* Boulenger, 1896	Kenya red-headed rock agama	–	x	x	–	Least concern
137	*Agama mossambica* Peters, 1854	Mozambique agama	–	–	x	x	Least concern
138	*Agama robecchii* Boulenger, 1891	Robecchi's agama	x	–	–	–	Least concern
139	*Agama rueppelli* Vaillant, 1882	Ruppell's agama	x	–	–	–	Least concern
140	*Calotes versicolor* (Daudin, 1802)	Indian garden lizard	–	x	–	–	Least concern, Introduced species from Asia
	Suborder Sauria, Family Chamaeleonidae, Chameleons						
141	*Rieppeleon brachyurus* (Günther, 1893)	Beardless pygmy chameleon	–	–	x	x	Least concern
142	*Rieppeleon brevicaudatu*s (Matschie, 1892)	Bearded pygmy chameleon	–	x	x	–	Least concern, forest and outside forests
143	*Rieppeleon kerstenii* (Peters, 1868)	Kenya pygmy chameleon	x	x	x	–	Least concern
144	*Chamaeleo dilepis* Leach, 1819	Flap-necked chameleon	x	x	x	x	Least concern
145	*Chamaeleo gracilis* Hallowell, 1844	Slender chameleon	x	x	–	–	Least concern
146	*Kinyongia tenui*s (Matschie, 1892)	Usambara soft-horned/flap-nosed chameleon	–	x	x	–	Endangered, forest
147	*Trioceros melleri* (Gray, 1865)	Giant one-horned chameleon	–	–	x	x	Least concern

APPENDIX 9.1 *(Continued)*

Sr. No.	Order, Suborder, Family, Species	Common Name	Country				Conservation Status and Endemism
			SO	KE	TZ	MZ	
	Suborder Sauria, Family Varanidae, Monitor Lizards						
148	*Varanus albigularis* (Daudin, 1802)	White-throated savanna monitor	x	x	x	x	Least concern
149	*Varanus niloticus* (Linnaeus, 1766)	Nile/water monitor	x	x	x	x	Least concern
	Suborder Amphisbaenia, Family Amphisbaenidae, Worm Lizards						
150	*Geocalamus acutus* Sternfeld, 1912	Voi wedge-snouted worm lizard	–	x	–	–	Least concern
151	*Loveridgea ionidesii* (Battersby, 1950)	Liwale round-snouted worm lizard	–	–	x	–	Least concern, Tanzania endemic
152	*Chirindia ewerbecki* Werner, 1910	Mbanja worm lizard	–	–	x	–	Least concern, Tanzania endemic
153	*Chirindia rondoensis* (Loveridge, 1941)	Rondo round-headed worm lizard	–	–	x	–	Data deficient, Tanzania endemic
154	*Chirindia swynnertoni* Boulenger, 1907	Swynnerton's round-headed worm lizard	–	–	x	x	Least concern
155	*Ancylocranium barkeri* Loveridge 1946	Barker's sharp-snouted worm lizard	–	–	x	–	Data deficient, Tanzania endemic
156	*Ancylocranium ionidesi* Loveridge, 1955	Ionides' sharp-snouted worm lizard	–	–	x	–	Data deficient, Tanzania endemic
157	*Zygaspis vandami* (Fitzsimons, 1930)	Van Dam's round-headed worm lizard	–	–	–	x	Least concern
158	*Zygaspis violacea* (Peters, 1854)	Violet round-headed worm lizard	–	–	–	x	Least concern, Mozambique endemic
159	*Zygaspis quadrifrons* (Peters, 1862)	Kalahari round-headed worm lizard	–	–	–	x	Least concern
160	*Monopeltis sphenorhynchus* (Peters, 1879)	Slender spade-snouted worm lizard	–	–	–	x	Least concern
	Suborder Amphisbaenia, Family Trogonophidae						
161	*Agamodon anguliceps* Peters, 1882	Angled worm lizard	x	–	–	–	Least concern, Somali endemic
162	*Agamodon compressus* Mocquard, 1888	Flat worm lizard	x	–	–	–	Data deficient, Somalia endemic
	Suborder Ophidia (Serpentes), Family Typhlopidae, Blind Snakes						
163	*Afrotyphlops fornasinii* (Bianconi, 1849)	Fornasini's blind snake	–	–	–	x	Least concern

APPENDIX 9.1 *(Continued)*

Sr. No.	Order, Suborder, Family, Species	Common Name	Country				Conservation Status and Endemism
			SO	KE	TZ	MZ	
164	*Afrotyphlops lineolatus* (Jan, 1864)	Lineolate blind snake	–	–	x	–	Least concern
165	*Afrotyphlops mucruso* (Peters, 1854)	Zambezi giant blind snake	–	x	x	x	Least concern
166	*Afrotyphlops nanus* Broadley and Wallach, 2009	Kenyan dwarf blind snake	–	x	–	–	Data deficient, Kenya endemic,
167	*Afrotyphlops platyrhynchus* (Sternfeld, 1910)	Tanga blind snake	–	–	x	–	Data deficient, Tanzania endemic
168	*Afrotyphlops rondoensis* (Loveridge, 1942)	Rondo blind snake	–	–	x	–	Data deficient, Tanzania endemic
169	*Afrotyphlops schlegelii* (Bianconi, 1849)	Schlegel's beaked blind snake	–	–	–	x	Least concern
170	*Indotyphlops braminus* (Daudin, 1803)	Flower pot blind snake	x	x	x	x	Least concern, Exotic/ Introduced species
171	*Letheobia lumbriciformis* (Peters, 1874)	Worm-like beaked snake /Zanzibar gracile blind snake	–	x	x	–	Least concern
172	*Letheobia pallida* Cope, 1868	Pallid gracile blind snake	–	x	x	–	Data deficient
173	*Letheobia pembana* Broadley and Wallach, 2007	Pemba gracile blind snake	–	–	x	–	Data deficient, Tanzania endemic in Pemba
174	*Letheobia swahilica* (Broadley and Wallach, 2007)	Swahili gracile blind snake	–	x	x	–	Least concern
175	*Rhinotyphlops unitaeniatus* (Peters, 1878)	Yellow-striped blind snake	x	x	x	–	Least concern
	Suborder Ophidia (Serpentes), Family Leptotyphlopidae, Worm/Thread Snakes						
176	*Epacrophis boulengeri* (Boettger, 1913)	Lamu worm snake	–	x	–	–	Data deficient, Kenya endemic
177	*Leptotyphlops distanti* (Boulenger, 1892)	Distant's thread snake	–	–	–	x	Least concern
178	*Leptotyphlops incognitus* (Broadley and Watson, 1976)	Incognito thread snake	–	–	–	x	Least concern
179	*Leptotyphlops howelli* Broadley and Wallach, 2007	Kim Howell's worm snake	–	–	x	–	Data deficient
180	*Leptotyphlops macrops* (Broadley and Wallach, 1996)	Large-eyed worm snake	–	x	x	–	Least concern
181	*Leptotyphlops mbanjensis* Broadley and Wallach, 2007	Mbanja worm snake	–	–	x	–	Data deficient, Tanzania endemic

APPENDIX 9.1 *(Continued)*

Sr. No.	Order, Suborder, Family, Species	Common Name	Country				Conservation Status and Endemism
			SO	KE	TZ	MZ	
182	*Leptotyphlops merkeri* (Werner, 1909)	Merker's worm snake	–	–	x	–	Least concern
183	*Leptotyphlops pembae* Loveridge, 1941	Pemba worm snake	–	–	x	–	Least concern, Tanzania endemic in Pemba
184	*Leptotyphlops telloi* Broadley and Watshon, 1976	Tello's thread snake	–	–	–	x	Not evaluated
185	*Myriopholis braccianii* (Scortecci, 1929)	Scortecci's worm snake	x	x	–	–	Least concern
186	*Myriopholis ionidesi* (Broadley and Wallach, 2007)	Ionides' worm snake	–	–	x	x	Least concern
187	*Myriopholis longicauda* (Peters, 1854)	Long-tailed thread snake	x	–	–	x	Least concern
188	*Myriopholis tanae* (Broadley and Wallach, 2007)	Tana worm snake	x	x	–	–	Data deficient
	Suborder Ophidia (Serpentes), Family Booidae, Boas						
189	*Eryx colubrinus* (Linnaeus, 1758)	Kenya sand boa	x	x	–	–	Least concern
	Suborder Ophidia (Serpentes), Family Pythonidae, Pythons						
190	*Python natalensis* Smith, 1840	Southern African rock python	–	x	x	x	Least concern
191	*Python sebae* (Gmelin, 1789)	Central African rock python	x	x	x	–	Near threatened
	Suborder Ophidia (Serpentes), Family Lamprophiidae, African Snakes						
192	*Boaedon fuliginosus* (Boie, 1827)	Brown house snake	x	x	x	x	Least concern
193	*Boaedon capensis* Duméril, Bibron and Duméril, 1854	Brown house snake	–	x	–	x	Least concern
194	*Lamprophis fuscus* Boulenger, 1893	Yellow-bellied house snake	–	–	–	x	Least concern
195	*Lycodonomorphus obscuriventris* Fitzsimons 1963	Floodplain water snake	–	–	–	x	Data deficient
196	*Lycophidion capense* (Smith, 1831)	Cape wolf snake	x	x	x	x	Least concern
197	*Lycophidion depressirostre* Laurent, 1968	Flat-snouted wolf snake	x	x	x	–	Least concern
198	*Lycophidion meleagre* Boulenger, 1893	Specked wolf snake	–	x	x	–	Least concern
199	*Lycophidion pembanum* Laurent, 1968	Pemba wolf snake	–	–	x	–	Data deficient, Tanzania endemic in Pemba
200	*Gracililima nyassae* (Günther, 1888)	Dwarf/black file snake	x	x	x	x	Least concern
201	*Limaformosa capensis* (Smith, 1847)	Cape file snake	–	–	x	x	Least concern
202	*Limaformosa chanleri* (Stejneger, 1893)	Chanler's file snake	x	x	–	–	Least concern

APPENDIX 9.1 *(Continued)*

Sr. No.	Order, Suborder, Family, Species	Common Name	Country				Conservation Status and Endemism
			SO	KE	TZ	MZ	
203	*Limaformosa savorgnani* (Mocquard, 1887)	Congo file snake	x	–	–	–	Least concern
	Suborder Ophidia (Serpentes), Family Pseudoxyrhophiidae, Elapoid snakes						
204	*Duberria variegata* (Peters, 1854)	Variegated or spotted slug eater	–	–	–	x	Least concern
	Suborder Ophidia (Serpentes), Family Prosymnidae						
205	*Prosymna janii* Bianconi, 1862	Mozambique shovel-snout	–	–	–	x	Least concern
206	*Prosymna lineata* (Peters, 1871)	Lined shovel-snout	–	–	–	x	Least concern
207	*Prosymna pitmani* Battersby, 1951	Pitman's shovel-snout	–	–	x	–	Least concern
208	*Prosymna ruspolii* (Boulenger, 1896)	Prince Ruspoli's shovel-snout	x	x	–	–	Least concern
209	*Prosymna semifasciata* Broadley, 1995	Banded shovel-snout	–	x	x	–	Endangered, forest
210	*Prosymna sundevalli* (Smith, 1849)	Sundevall's shovel-snout	–	–	–	x	Least concern
211	*Prosymna stuhlmanni* (Pfeffer,1893)	East African shovel-snout	x	x	x	x	Least concern
	Suborder Ophidia (Serpentes), Family Psammophiidae						
212	*Hemirhagerrhis hildebrandtii* (Peters, 1878)	Kenya bark snake	x	x	x	–	Least concern
213	*Hemirhagerrhis kelleri* Boettger, 1893	Striped bark snake	x	x	–	–	Least concern
214	*Hemirhagerrhis nototaenia* (Günther, 1864)	Southeastern bark snake	x	x	x	x	Least concern
215	*Psammophis angolensis* (Bocage, 1872)	Dwarf sand snake	–	–	x	x	Least concern
216	*Psammophis biseriatus* Peters, 1881	Link-marked sand snake	x	x	–	–	Least concern
217	*Psammophis mossambicus* Peters, 1882	Olive/hissing sand snake	–	x	x	x	Least concern
218	*Psammophis orientalis* Broadley, 1977	Eastern stripe-bellied sand snake	–	x	x	x	Least concern
219	*Psammophis sudanensis* Werner, 1919	Northern stripe-bellied sand snake	–	–	x	–	Least concern
220	*Psammophis punctulatus* Duméril, Bibron and Duémril, 1854	Specked sand snake	x	x	–	–	Least concern
221	*Psammophis tanganicus* Loveridge, 1940	Tanganyika sand snake	x	x	–	–	Least concern
222	*Psammophylax tritaeniatus* (Günther, 1868)	Southern striped skaapsteker	–	–	–	x	Least concern
223	*Rhamphiophis rostratus* Peters, 1854	Rufous beaked snake	x	x	x	x	Least concern
224	*Rhamphiophis rubropunctatus* (Fischer, 1884)	Red-spotted beaked snake	x	x	x	–	Least concern

APPENDIX 9.1 *(Continued)*

Sr. No.	Order, Suborder, Family, Species	Common Name	Country				Conservation Status and Endemism
			SO	KE	TZ	MZ	
	Suborder Ophidia (Serpentes), Family Atractaspididae, Burrowing asps						
225	*Aparallactus capensis* Smith, 1849	Cape centipede eater	–	x	x	x	Least concern
226	*Aparallactus guentheri* Boulenger, 1895	Black centipede eater	–	x	x	x	Least concern
227	*Aparallactus jacksonii* (Günther, 1888)	Jackson's centipede eater	x	x	x	–	Least concern
228	*Aparallactus lunulatus* (Peters, 1854)	Plumbeous centipede eater	x	x	x	x	Least concern
229	*Aparallactus nigriceps* (Peters, 1854)	Mozambique centipede eater	–	–	–	x	Data deficient, Mozambique endemic
230	*Aparallactus turneri* Loveridge, 1935	Malindi centipede eater	–	x	–	–	Least concern, Kenyan endemic
231	*Aparallactus werneri* Boulenger 1895	Usambara centipede eater	–	–	x	–	Vulnerable, Tanzanian endemic
232	*Amblyodipsas katangensis* De White and Laurent, 1942	Ionides' purple-glossed snake	–	–	x	x	Least concern
233	*Amblyodipsas microphthalma* (Bianconi 1852)	Eastern purple-glossed snake	–	–	–	x	Least concern
234	*Amblyodipsas polylepis* (Bocage, 1873)	Common purple-glossed snake	x	x	x	x	Least concern
235	*Atractaspis bibronii* Smith, 1849	Bibron's burrowing Asp	x	x	x	x	Least concern
236	*Atractaspis engdahli* Lönberg and Anderson, 1913		x	–	–	–	Least concern
237	*Atractaspis fallax* Peters, 1867	Eastern small-scaled burrowing Asp	x	x	x	–	Least concern
238	*Brachyophis revoili* Mocquard, 1888	Revoil's short snake	x	–	–	–	Data deficient, Somali endemic
239	*Chilorhinophis butleri* Werner, 1907	Butler's black-and-yellow burrowing-snake	–	–	x	x	Least concern
240	*Xenocalamus bicolor* Günther, 1868	Bicolored quil-snouted snake	–	–	–	x	Least concern
	Suborder Ophidia (Serpentes), Family Micrelapsidae						
241	*Micrelaps bicoloratus* Sternfeld, 1908	Two-colored snake	–	x	–	–	Least concern
242	*Micrelaps vaillanti* (Mocquard, 1888)	Desert black-headed snake	x	x	–	–	Least concern
	Suborder Ophidia (Serpentes), Family Colubridae, Colubrid Snakes						
243	*Platyceps brevis* (Boulenger, 1895)	Smith's racer	x	x	–	–	Least concern
244	*Meizodon semiornatus* (Peters, 1854)	Semiornate snake	x	x	x	x	Least concern

APPENDIX 9.1 *(Continued)*

Sr. No.	Order, Suborder, Family, Species	Common Name	Country				Conservation Status and Endemism
			SO	KE	TZ	MZ	
245	*Meizodon plumbiceps* (Boettger, 1893)	Black-headed smooth snake	x	–	–	–	Least concern
246	*Philothamnus angolensis* Bocage, 1882	Angola green snake	–	–	–	x	Least concern
247	*Philothamnus battersbyi* Loveridge, 1951	Battersby's green snake	x	x	x	–	Least concern
248	*Philothamnus hoplogaster* (Günther, 1863)	Southeastern green snake	–	x	x	x	Least concern
249	*Philothamnus macrops* (Boulenger, 1895)	Usmbara green snake	–	x	x	x	Least concern
250	*Philothamnus natalensis* (Smith, 1848)	Eastern green snake	–	–	–	x	Least concern
251	*Philothamnus punctatus* Peters, 1867	Speckled green snake	x	x	x	x	Least concern
252	*Philothamnus semivariegatus* (Smith, 1840)	Spotted bush snake/spotted wood snake	x	x	x	x	Least concern
253	*Dispholidus pembae* Hughes, 2021	–	–	–	x	–	Not evaluated, Tanzania endemic in Pemba
254	*Dispholidus typus* (Smith, 1828)	Boomslang	x	x	x	x	Least concern
255	*Thelotornis mossambicanus* (Bocage, 1895)	Eastern vine snake	x	–	x	x	Least concern
256	*Thelotornis capensis* Smith, 1849	Vine snake	–	–	–	x	Least concern
257	*Thelotornis usambaricus* Broadley, 2001	Usambara vine snake	–	x	x	–	Vulnerable, forest
258	*Dasypeltis medici* Bianconi, 1859	Rufous/East African egg eater	x	x	x	x	Least concern
259	*Dasypeltis scabra* (Linnaeus, 1758)	Common or Rhombic egg eater	x	x	x	x	Least concern
260	*Telescopus obtusus* (Reuss, 1834)	Large-eyed snake	x	x	–	–	Least concern
261	*Telescopus semiannulatus* Smith, 1849	Tiger snake	–	x	x	x	Least concern
262	*Crotaphopeltis braestrupi* Rasmussen, 1985	Tana Herald snake	x	x	–	–	Least concern
263	*Crotaphopeltis hotamboeia* (Laurenti, 1768)	White-Lipped snake	x	x	x	x	Least concern
264	*Crotaphopeltis tornieri* (Werner, 1908)	Tornier's cat snake	–	–	x	–	Least concern
265	*Dipsadoboa aulica* (Günther, 1864)	Marbled tree-snake	x	–	x	x	Least concern
266	*Dipsadoboa flavida* (Broadley and Stevens, 1971)	Cross-Barred tree-snake	x	x	x	x	Least concern
267	*Scaphiophis albopunctatus* Peters, 1870	Hook-nosed snake	–	x	x	–	Least Concern
268	*Natriciteres olivacea* (Peters, 1854)	Olive marsh snake	x	x	x	x	Least concern
269	*Natriciteres pembana* (Loveridge, 1935)	Pemba marsh snake	–	–	x	–	Least concern, Endemic to Tanzania, Pemba Island

APPENDIX 9.1 *(Continued)*

Sr. No.	Order, Suborder, Family, Species	Common Name	Country				Conservation Status and Endemism
			SO	KE	TZ	MZ	
270	*Natriciteres sylvatica* Broadley, 1966	Forest marsh snake	–	–	x	x	Least concern, a forest species
	Suborder Ophidia (Serpentes), Family Elapidae, Cobras, Mambas and Relatives						
271	*Elapsoidea boulengeri* Boettger, 1895	Boulenger's garter snake	–	–	x	x	Least concern
272	*Elapsoidea broadleyi* Jacobsen, 1997	–	x	–	–	–	Data deficient, Somali endemic
273	*Elapsoidea chelazzii* Lanza, 1979	Somali garter snake	x	–	–	–	Data deficient, Somali endemic
274	*Elapsoidea loveridgei* Parker, 1949	East African garter snake	x	–	–	–	Least concern
275	*Elapsoidea nigra* Günther, 1888	Usambara garter snake	–	x	x	–	Least concern, forest species
276	*Elapsoidea semiannulata* Bocage, 1882	Half-banded garter snake	–	–	–	x	Least concern
277	*Elapsoidea sundervallii* Smith, 1848	Sundevalls' garter snake	–	–	–	x	Not evaluated
278	*Naja annulifera* Peters, 1854	Snouted cobra	–	–	–	x	Least concern
279	*Naja ashei* Wüster and Broadley, 2007	Ashe's spitting cobra	x	x	–	–	Least concern
280	*Naja mossambica* Peters, 1854	Mozambique spitting cobra	x	–	x	x	Least concern
281	*Naja nigricollis* Reinhardt, 1843	Black-necked spitting cobra	x	–	x	–	Least concern
282	*Naja pallida* Boulenger, 1896	Red spitting cobra	x	x	–	–	Least concern
283	*Naja subfulva* Laurent, 1955	Brown forest cobra	–	x	x	x	Least concern
284	*Dendroaspis angusticeps* (Smith, 1849)	Green mamba	–	x	x	x	Least concern
285	*Dendroaspis polylepis* (Günther, 1864)	Black mamba	x	x	x	x	Least concern
286	*Hydrophis platurus* (Linnaeus, 1766)	Yellow-bellied sea snake	x	x	x	x	Least concern
287	*Aspidelaps scutatus* (Smith, 1849)	Shield-nosed snake	–	–	–	x	Least concern
	Suborder Ophidia (Serpentes), Family Viperidae, Vipers						
288	*Causus defilippii* (Jan, 1863)	Snouted night adder	–	x	x	x	Least concern
289	*Causus resimus* (Peters, 1862)	Velvety-green night adder	x	x	–	–	Least concern
290	*Causus rhombeatus* (Lichtenstein, 1823)	Rhombic night adder	–	x	–	x	Least concern

APPENDIX 9.1 *(Continued)*

Sr. No.	Order, Suborder, Family, Species	Common Name	Country				Conservation Status and Endemism
			SO	KE	TZ	MZ	
291	*Bitis arietans* (Merrem, 1820)	Puff adder	x	x	x	x	Least concern
292	*Bitis gabonica* (Duméril, Bibron and Duméril, 1854)	Gaboon viper	–	–	x	x	Vulnerable
293	*Proatheris superciliaris* (Peters, 1855)	Lowland swamp viper	–	–	–	x	Least concern
	Species richness		125	145	168	141	
	Endemic species		13	5	26	6	

KE = Kenya; MZ = Mozambique; SO = Somalia; TZ = Tanzania, x = present, – = no record in the coastal ecoregion.

CHAPTER 10

BIRD DIVERSITY OF THE COASTAL FORESTS OF EASTERN AFRICA BIODIVERSITY HOTSPOT

JASSON JOHN

Department of Zoology and Wildlife Conservation, College of Natural and Applied Sciences, University of Dar es Salaam, Dar es Salaam, Tanzania

ABSTRACT

The coastal forests of Eastern Africa are home to five endemic forest birds and several nonforest species of the East Coast Biome utilizing or found in adjacent and associated habitats including offshore islands. The Pemba Island in Tanzania hosts four endemic nonforest birds, and by itself an Endemic Bird Area. The inseparable connectivity of the Coastal Forests with adjoining forest ecosystems is evidenced by a huge number of forest dependent bird species shared among them especially with the Eastern Arc Mountains. Moreover, the lowland forests serve as corridor habitats linking forests in the north with those in extreme southeastern Tanzania; the East African Spotted Ground Thrush race, *fischeri*, breeds in Rondo forest, southern Tanzania and winters in the coastal forests in northern Tanzania and Kenya coast. Despite their significance for conservation of endemic, rare and threatened species, these forests are threatened by encroachment, urbanization, wildfire, infrastructure development, and unsustainable use of forest resources. This chapter outlines several aspects to be included in any conservation strategy to sustainably manage the coastal forests and their characteristic avifauna.

10.1 INTRODUCTION

The Coastal Forests (CFs) of Eastern Africa region (hereafter, termed as "Coastal Forests") are among the few global unique areas for the conservation of avifauna. Apart from hosting endemic and near-endemic bird species, it has many forest dependent bird species. Adjoining habitats within the coastal belt, the grassland, and woodlands, most of which might have arisen due to anthropogenic effects such as frequent fires, host some nonforest species that are restricted to this region. Transition areas between CFs and grassland and woodland are occupied by degraded forests and thickets, which preserve some characteristics of the

Biodiversity Hotspot of the Coastal Forests of Eastern Africa. T. Pullaiah and Ally J. K. Nkwabi (Eds.)

DOI: 10.1201/9781998511099-10

previous forest communities. These nonforest bird species partially or solely benefit from the conservation of CFs. The microclimate along river valleys just before estuaries is also important component in the CFs that add value to its ornithological importance.

The CFs overlap with another biodiversity forest hotspot area, the Eastern Arc Mountains (EAMs) of Kenya and Tanzania (Burgess et al., 2007). In reality, no hard boundary exists between these two forest types (Lovett et al., 2001) and in some mountain blocks there is a continuum between the Eastern Arc and CF types. The avifauna of the two forest types are connected and especially due to altitudinal and intra-African migration (Dowsett-Lemaire, 1989; Baker and Baker, 1992; Burgess and Mlingwa, 2000; Werema, 2014). Within this general area, two important Endemic Bird Areas (EBA) exist; Pemba and East African CFs hosting many restricted-range species (Stattersfield et al., 1998). A large part of the EAMs falls within the African Highland Biome. In additional to EBAs, the East Coast Biome holds many of the bird species characteristic of the CFs (Baker and Baker, 2002). Thus, several sites hosting endemic (including populations that define an EBA and biome-restricted species) and/or populations of threatened species have been identified as critical for bird conservation, and popularly known as Important Bird and Biodiversity Areas (IBAs).

A first detailed compilation of the ornithological importance of the CFs is documented in Mlingwa et al. (2000). This book chapter in Burgess and Clarke (2000) characterizes the CFs avifauna in terms of richness, endemism, and threats. The publication provides an in-depth literature on the avifauna of this region from 1930 to 1990. However, the discussion is more focused on Somalia, Kenya, and Tanzania and little attention is put on Mozambique CFs and on the small areas of lowland forest in the eastern margins of Malawi and Zimbabwe, which have their strongest affinities to the CFs. Understandably, earlier work on rare birds for three countries' CFs had been compiled in: Vincent (1933, 1934), Mozambique; Dowsett-Lemaire (1989), Malawi; and Irwin (1994), Zimbabwe. Since this publication, over 20 years ago, a lot has happened especially in scientific knowledge on the avifauna of this area; few taxonomic revisions made, species conservation status changed, and threats have also intensified necessitating a new review.

The approach for this chapter is to focus on all species with restricted distributional ranges within the CFs of Eastern Africa, which largely falls within the Zanzibar-Inhambane regional mosaic and Swahilian region mosaic (White, 1983)/Swahilian regional center of endemism (Mlingwa et al., 2000). However, special attention is put on the current status of the endemic avifauna and an assessment on the geographical distribution of the wide-ranging forest birds found in the CFs is made. Taxonomy and nomenclature follow the Handbook of the Birds of the World and Birdlife taxonomic checklist version 7 (HWB and Birdlife International, 2022).

10.2 DIVERSITY AND DISTRIBUTION OF FOREST BIRDS

The presence of forest dependent species is a strong reason to why the CFs region is considered a global hotspot for bird conservation. These species are sensitive to the alterations of the forest habitats and the efforts put in the region by different scholars at different period (Bennun et al., 1996; Mlingwa et al., 2000; Jensen et al., 2005) allow ornithologists to assign many of the CF species to different forest dependence categories. There are three major categories of forest birds:

- Forest Specialists (FF) are typical of the forest interior and likely to disappear when the forest is modified.
- Forest Generalists (F) occur in undisturbed forest but can exist at the forest edge or in modified and fragmented forests but continue to depend upon forest for some of their resources, such as nesting sites.
- Forest Visitors (f) occur in forests but are more typical of other habitats, especially moist woodlands and thickets. They do not depend on forests and their presence can be an indication of disturbance.

This section discusses FF and F species reflecting on the importance of the CFs. Nonforest and forest visitors unique to the CF region are discussed in subsequent sections.

Appendix 10.1 lists 86 forest birds along the CFs; Somalia (36 species), Kenya (78 species), Tanzania (84 species), Mozambique (71 species), Zimbabwe (59 species), and Malawi (63 species). Of 86 forest dependent species, 38 (44%) are FF and 48 (56%) are forest generalist (F). Only eight species are Globally Threatened; five Endangered (EN), three Vulnerable (VU) while another eight species are Near-threatened (NT). The threat category is based on the definitions provided in Collar et al. (1994) and Mace and Stuart (1994); EN: facing a very high risk of extinction in the wild with 20% chance of extinction in 20 years, VU: facing a high risk of extinction in the wild with 10% chance of extinction in 100 years, NT: species very close to VU. Globally Threatened Species are those found in the three categories: Critically Endangered (CR), Endangered, and Vulnerable.

Most species are found throughout the CFs in the six countries (some in fact extending into other biomes) and with few country endemics (see a section on endemism and threat below). Some of the species in Appendix 10.1 are not typical CF birds—notably several are essentially montane, with small temporary or permanent populations (Mlingwa, 1996; Mlingwa et al., 2000; Baker and Baker, 2002): Swynnerton's Robin *Swynnertonia synnertoni*, White-chested Alethe *Alethe fulleborni*, White-starred Robin *Pogonocichla stellata*, Black-fronted Bush-shrike *Chlorophoneus nigrifrons*, Olive-headed Bulbul *Arizelocichla striifacies,* and White-winged Apalis *Apalis chariessa*. Many scholars have reported strong species link to other lowland forest types particularly those of the lowland EAMs (Stuart and Jensen, 1985; Evans et al., 1994) and further south in Malawi (Dowsett-Lemaire, 1989) and Zimbabwe (Irwin, 1994). Lowland CFs along Africa's east coast have served as a corridor linking CFs in the north with those in extreme southeastern Tanzania and northern Mozambique (Fjeldså and Bowie, 2021), and many species subscribe to this phenomenon. Lowland forests of EAMs of Tanzania have an avifauna which contains many of the species typical of both the Eastern Arc and CFs (Borghesio et al., 2008), and the distinction is even hard for the East Usambara, NE Tanzania, because of the continuation of lowland forests to montane habitats.

A patchy or restricted distribution of forest birds in CFs including within countries is reported in Mlingwa et al. (2000). This is due to the forest vegetation types and structure across the CF belt. The vegetation in the CFs is spatially heterogenous in terms of vegetation communities (Burgess and Clarke, 2000) and structure (Lowe and Clarke, 2000), influencing the presence and absence of species. Habitat structure can also determine species composition especially of understory bird communities. The distribution of species can also be influenced by biogeographical barriers such as rivers (Kingdon, 1990) and landscape spatial structures including those in mountains and valleys in separating forests (Bowie et al., 2023). Although, there are many rivers along the coastal belt of Eastern Africa opening into the Indian Ocean, it is only Rufiji River in southeast Tanzania which has been thought to limit the distribution of some local avifauna (Baker and Baker, 2002); for example, Reichenow's Batis *Batis reichenowi* is found south of the river while Forest Batis *B. mixta* and Little Yellow Flycatcher *Erythrocercus holochlorus* have their southern distribution limit north of Rufiji River. Generally, community differences among the forests and countries are due to biogeographic histories and local ecologies.

10.3 ENDEMIC BIRDS OF CFs AND ADJOINING HABITATS

Endemism is a natural phenomenon related to geographical limits of range of distribution of biodiversity, collectively, or taxon-wise. When a taxon is restricted in its range of distribution related to geographical unit such taxon is called an Endemic taxon to that limited unit and the latter unit is called an Endemic Area. Endemism not only signifies the restricted range but also the exclusiveness or peculiarities of the taxon in its form and function in the process of evolution. In the context of evolutionary phases, endemic species may reveal the development of new, entirely autochthonous or relictual forms. Due to the presence of range-restricted species, the CF of Eastern Africa is identified as an EBA (Stattersfield et al., 1998). This EBA is close to the Pemba EBA and has affinities to the Tanzania-Malawi mountains EBAs. Although, many species found in CFs have a wide distribution in Africa (Mlingwa et al., 2000), a few shows a disjunct distribution resulting into subspecies. Mackworth-Praed and Grant (1960), Mlingwa et al. (2000) and Jensen et al. (2005) list at least 10 subspecies of 12 species endemic to CFs (including "f" species): *zanzibaricus* (Fisher's Turaco, Zanzibar only), *woodwardia/hylophona* (Green Barbet, Lindi Forests, Tanzania), *debilis* (Lowland Tiny Greenbul, south of Rufiji to Mozambique), *rabai* (Lowland Tiny Greenbul, north of the Rufiji River), *puguensis* (Pale-breasted Illadopsis, Pugu Hills, and Zanzibar), *sokokensis* (East Coast Akalat) (see also Section 10.3.4), *fischeri* (Spotted Ground Thrush), *chariessa* (White-winged Apalis, Tana River, Kenya), *griseiceps* (Kretschmer's Longbill, Rufiji to Mozambique), *kretschmeri* (Kretschmer's Longbill), *pileata* (Bleating Camaroptera, Kenya to Mozambique), and *yokanae* (Plain-backed Sunbird). Nonetheless, recent molecular studies and improved field morphology and behavioral observations have helped to describe full species, for example, Tropical Boubou "*sublacteus*" is now a full species, East Coast Boubou *Laniarius sublacteus,* endemic to Coastal East Africa (Stevenson and Fanshawe, 2020), and *Batis reichenowi* recently described (del Hoyo et al., 2016) formerly treated as a subspecies of *B. mixta* (Fjeldså et al., 2006). Information for the endemic *nigricauda* (Gorgeous Bush-shrike) and *vosseleri* (Fraser's eagle-owl) are provided in Section 10.3.1.

Endemic or near endemic bird species to the CFs and associated environs are listed in Table 10.1. The species have been grouped into the following groups; Found only in CF, Unique to CFs but with population in EAMs (CF/EAM), nonforest species but endemic or near endemic to the E and SE CF belt and Zanzibar-Inhambane regional mosaic (ENF), Restricted in the forests of E and SE Africa but do not fall in (CF) and (CF/EAM) (ESE Africa), and those Endemic to Pemba Island, Zanzibar, Tanzania (EPI).

TABLE 10.1 Endemic and Near-Endemic to Coastal Forest belt of Eastern Africa. Threat categories are based on IUCN 2023 list (BirdLife International, 2023a)

Sr. No.	Common Name	Species	Endemism	Threat Categories
1	Sokoke Scops-owl	*Otus ireneae*	CF	EN
2	Sokoke Pipit	*Anthus sokokensis*	CF	EN
3	Spotted Ground-thrush	*Geokichla guttata fischeri**	CF	VU
4	Little Yellow Flycatcher	*Erythrocercus holochlorus*	CF	LC
5	Reichenow's Batis	*Batis reichenowi*	CF	NT
6	Gorgeous Bush-shrike	*Telophorus quadricolor nigricauda**	CF	LC
7	Clarke's Weaver	*Ploceus golandi*	CF	EN
8	Fischer's Turaco	*Tauraco fischeri*	CF/EAM	NT
9	Fraser's eagle-owl	*Bubo poensis vosseleri**	CF/EAM	LC
10	Mombasa Woodpecker	*Campethera mombassica*	CF/EAM	LC
11	Usambara Hyliota	*Hyliota usambarae*	CF/EAM	EN
12	East Coast Batis	*Batis soror*	CF/EAM	LC
13	Amani Sunbird	*Hedydipna pallidigaster*	CF/EAM	EN
14	Banded Sunbird	*Anthreptes rubritorques*	CF/EAM	VU
15	Uluguru Violet-backed Sunbird	*Anthreptes neglectus*	CF/EAM	LC
16	East Coast Boubou	*Laniarius sublacteus*	ENF	LC
17	Malindi Pipit	*Anthus melindae*	ENF	LC
18	Tana River Cisticola	*Cisticola restrictus*	ENF	DD
19	Coastal Cisticola	*Cisticola haematocephala*	ENF	LC
20	Lesser Seedcracker	*Pyrenestes minor*	ENF	LC
21	Violet-breasted Sunbird	*Cinnyris chalcomelas*	ENF	LC
22	Woodwards' Batis	*Batis fratrum***	ENF	LC
23	Mouse-colored Sunbird	*Cyanomitra verreauxii***	ENF	LC
24	Scaly Babbler	*Turdoides squamulata***	ENF	LC
25	Southern Banded Snake-eagle	*Circaetus fasciolatus***	ESE Africa	NT
26	Brown-headed Parrot	*Poicephalus cryptoxanthus***	ESE Africa	LC
27	Mangrove Kingfisher	*Halcyon senegaloides***	ESE Africa	LC
28	White-eared Barbet	*Stactolaema leucotis***	ESE Africa	LC
29	Eastern Green Tinkerbird	*Pogoniulus simplex*	ESE Africa	LC
30	Fisher's Greenbul	*Phyllastrephus fischeri*	ESE Africa	LC

TABLE 10.1 *(Continued)*

Sr. No.	Common Name	Species	Endemism	Threat Categories
31	Lowland Tiny Greenbul	*Phyllastrephus debilis*	ESE Africa	LC
32	Kretschmer's Longbill	*Macrosphenus kretschmeri*	ESE Africa	LC
33	East Coast Akalat	*Sheppardia gunning*	ESE Africa	NT
34	Chestnut-fronted Helmetshrike	*Prionops scopifrons***	ESE Africa	LC
35	Black-bellied Starling	*Notopholia corruscus***	ESE Africa	LC
36	Plain-backed Sunbird	*Anthreotes reichenowi***	ESE Africa	NT
37	Pemba Green-pigeon	*Treron pembaensis*	EPI	VU
38	Pemba Scops-owl	*Otus pembaensis*	EPI	VU
39	Pemba Sunbird	*Cinnyris pembae*	EPI	LC
40	Pemba White-eye	*Zosterops vaughani*	EPI	LC

*Endemic subspecies, **Near-endemic species (distribution extending beyond Coastal Forest range states, for example, South Africa, Eswatini, and Ethiopia).

EN = Endangered, VU = vulnerable, NT = Near-Threatened, DD=Data Deficiency, LC=Least Concern, IUCN=International Union for the Conservation of Nature.

10.3.1 CFs ENDEMIC SPECIES

In this category, there are five forest dependent species (Table 10.1, Figure 10.1) found in East Coast of Somalia, Kenya, and Tanzania. Little Yellow Flycatcher is endemic to coastal East Africa from Somalia to Tanzania, north of Rufiji River found in thick primary and secondary CFs. Reichenow's Batis is found south of the Rufiji River and Clarke's Weaver is endemic to coastal Kenya in Miombo woodland and adjacent forests particularly in Dakacha Woodland and Arabuko-Sokoke Forests. Sokoke Pipit is endemic to coastal Kenya (mainly in Arabuko-Sokoke and Shimba Hills) and northern Tanzania in CFs where it spends much time foraging in leaf litter and among root tangles whereas the Sokoke Scops-owl is found in CFs in Arabuko-Sokoke Forest and foothills of Usambaras, north-east Tanzania.

Two endemic subspecies of Spotted Ground-thrush (*fischeri*) and Gorgeous Bush-shrike (*nigricauda*), respectively, have been included here because of their significant populations in the region and threat. Spotted Ground-thrush found in DR Congo (from a single specimen from montane forest of Upemba), Kenya, Malawi, Mozambique, South Africa, South Sudan and Tanzania has five subspecies; *G. g. guttata, G. g. fischeri, G. g. belcheri, G. g. maxis* and *G. g. lippensi. G. g. fischeri,* previously estimated at *c.*200 pairs (Ndang'ang'a et al., 2005, 2008), but now believed to be close to extinction (BirdLife International, 2023b) is endemic to CFs of East Africa: winters in northern Tanzania and Kenya and breeds in Rondo Plateau (Holsten et al., 1991) in southern Tanzania. Gorgeous Bush-shrike is found in Kenya, Tanzania, Mozambique, South Africa, Eswatini, Zimbabwe, South Sudan, DRC Congo, Congo, Gabon, and Zambia. *Telophorus viridis* and *T. quadricolor* (Four-colored Bush-shrike) (Sibley and Monroe, 1990) have been lumped into *T. viridis* following del

Hoyo et al (2016), and *T. q. nigricauda* (Figure 10.2) is endemic to Coastal SE Kenya and E Tanzania found in dense tangled thickets.

FIGURE 10.1 Some endemic bird species to Coastal Forests (*Photos: Per Holmen*)

10.3.2 SPECIES UNIQUE TO CFs BUT WITH POPULATIONS IN EAMs

A total of seven species and one subspecies are listed in this category. Two species; Usambara Hyliota, Banded Sunbird, and the subspecies of Fraser's Eagle-owl *B. p. vosseleri* are endemic to Tanzania (Figure 10.3). Usambara Hyliota is found in low and mid altitudes in the Usambaras particularly the East Usambaras (Seddon et al., 1999), whereas Banded Sunbird is locally common in the Usambara Mountains. Elsewhere rare, with records in EAMs: Nguru, Nguu, Ulugurus, and Udzungwa (Seddon et al., 1999), in intact submontane forest, secondary forest clearings (Borghesio et al., 2008), and in the ecotone between submontane and miombo woodland (Hipkiss et al., 1994). Fraser's Eagle-owl *B. poensis* (del Hoyo et al., 2014) was previously split as *B. poensis* and *B. vosseleri* [Usambara Eagle-owl] following Sibley and Monroe (1990) and *B. p. vosseleri* (Figure 10.3) is restricted to CFs (Evans et al., 1994) and EAMs of Tanzania (Sibley and Monroe, 1990; Hunter et al., 1996) while the nominate species is found in west and central Africa.

FIGURE 10.2 Gorgeous Bush-shrike, subspecies *nigricauda*, endemic to East Africa Coastal forests. This picture is from Ruvu South Forest in Tanzania.

Photo Credit: Author

Fischer's Turaco is restricted in dense coastal lowland to submontane forests of southern Somalia, Kenya and northern Tanzania (Turner, 1977; Borghesio and Ndang'ang'a, 2003) but up to 1500 m in Nilo, East Usambara Mountains. An endemic race *zanzibaricus* in Zanzibar (Pakenham, 1938), which was rumored to be extinct (Britton, 1980; Mlingwa et al., 2000) was estimated at *c.* 1400 individuals in 2001 on Unguja Island (Borghesio and Ndang'ang'a, 2003). Mombasa Woodpecker and East Coast Batis

Batis soror occur in forest and coastal woodlands throughout their distributional ranges up to 1000m in East Usambaras. Amani Sunbird is endemic in CFs and montane forests in Kenya and Tanzania. Limited to one site in Kenya in the Arabuko-Sokoke forest, and two sites in Tanzania: the Udzungwas at 1350–1400 m (Dinesen et al., 1993) and in the East Usambaras up to 1000 m (Collar and Stuart, 1985; Hipkiss et al., 1994) where is common in forest, forest edges and smallholder farmlands sharing habitats with the Uluguru Violet-backed Sunbird (Borghesio et al., 2008). Within its range, the Uluguru Violet-backed Sunbird is also common in tropical and subtropical lowland moist forests.

Banded Sunbird   Usambara Eagle-owl

FIGURE 10.3 Some of the unique species of Coastal Forests with shared populations in the EAMs (*Photos: Per Holmen*).

Mlingwa et al. (2000) included the White-winged Apalis in this category but the only coastal nominate race *A.c. cheriessa* in Kenya has potentially already extinct. The lower Tana River forests, the home of this race, has been well studied recently (Owino et al., 2008) with no record of this species since 1961 (Stevenson and Fanshawe, 2020).

10.3.3 NONFOREST SPECIES ENDEMIC TO THE E AND SE OF CF BELT

Six species are endemic while three species are near endemic to the CF belt of which Zanzibar-Inhambane and Swahiliani regional mosaics are part; Malindi Pipit (Figure 10.4), Tana River Cisticola, and Coastal Cisticola are endemic to the coastal belt while the rest have wide distribution within the CFs. The pipit and cisticolas, unlike some of the Pemba Island endemics which evolved as forest species (see Section 10.3.5), have not been associated with forests, they occur in seasonally flooded areas, grasslands near swamps and coastal scrubs. The Tana River Cisticola is classified as data deficiency because it has not been sighted in its known range for many years now (BirdLife International, 2023c) and it is in fact considered extinct (Stevenson and Fanshawe, 2020). The Coastal Cisticola (Figure 10.4), endemic to the coast region of Somalia, Kenya and Tanzania, was formerly considered conspecific with Winding Cisticola *C. marginatus* (del Hoyo et al., 2016), a widely distributed species in central and west Africa. The East Coast Boubou *L. sublacteus*

(Figure 10.4) formerly considered conspecific with Tropical Boubou *L. aethiopicus*, is endemic to coastal East Africa. Inhabits forest undergrowth and coastal thickets in coastal lowlands but up to montane habitats in East Usambaras and Ulugurus in Tanzania.

Malindi Pipit East Coast Boubou Coastal Cisticola

FIGURE 10.4 Some of the nonforest endemic bird species to eastern coastal forests (*Photos: Per Holmen*).

The Lesser Seedcracker has a wider distribution across the coastal belt within the Zanzibar-Inhambane region occurring in dense secondary growth, thickets, and overgrown farmlands up to 900 m. Violet-breasted Sunbird is endemic to NE Africa in forests, scrub, and dry savanna habitats.

The rest of the species in this category are classified as near endemic (Table 10.1) because their ranges extend beyond the CFs into South Africa and Ethiopia. Woodwards' Batis is mostly found in Mozambique and its population in Malawi is found in Malawi Hills Forest Reserve and Lengwe National Park while in Zimbabwe it is found in Haron-Rusitu Junction and Botanical Reserve. Although Mouse-colored Sunbird was classified as endemic to CFs (Mlingwa et al., 2000), its distribution extends to South Africa into Tongaland-Pondoland regional mosaic (White, 1983) and is also endemic to East Coastal Biome (Fishpool and Evans, 2001). The nominate race of the Scaly Babbler is found in dense coastal bushland of Kenya (Owino et al., 2008) and Somalia while the race *corolinae* is found in Ethiopia (Stevenson and Fanshawe, 2020).

Although the role of the forests to these species is not well established, it is obvious that these forests influence the microclimate of surrounding habitats and act as a buffer from human pressure, both benefiting the nonforest bird species.

10.3.4 SPECIES RESTRICTED IN THE FORESTS OF E AND SE AFRICA BUT DO NOT FALL IN SECTION 10.3.1 AND 10.3.2

Twelve species are listed in this category (Table 10.1). All include forest dependent species (see Appendix 10.1), except the Mangrove Kingfisher which has not been classed into categories of forest dependent. However, it has been found to breed in CFs, woodland and riparian thickets (Davies et al., 2012) indication the value of these habitats to its survival, for this fact, it is therefore being included here. Nonetheless, this species is endemic to the East Coastal Biome (Fishpool and Evans, 2001).

At least seven species, including the Mangrove Kingfisher in this category extend their distribution into South Africa and some into montane habitats away from EAMs. For example, the White-eared Barbet is found in CFs and bush in South, but up to 2600 m in Northern Tanzania, whereas the Kretschmer's Longbill occurs in subtropical or tropical moist lowland forests and subtropical or tropical moist montane forests including Kilimanjaro Mountains (Britton, 1980). The East Coast Akalat occurs in three races: the nominate race *S. g. gunningi* is found in small, mainly coastal lowland forests around Beira in Mozambique. *S. g. bensoni* occurs at 475–1750 m in northern Malawi; and *S. g. sokokensis* is found up to 300 m in Kenya and up to 800 m in NE Tanzania, Pugu Hills, Rondo Plateau, Dondwe forest, and Zanzibar (Keith et al., 1992; Collar et al., 1994). However, Seddon et al. (1999), from biometrics of 20 individuals mist-netted, found resemblance of the montane Nguu (EAM, Tanzania) birds with that of *S.g. bensoni* in Malawi which may require further studies due to the distance between the two subpopulations.

Bird species in this category are widespread, and because of this most populations are not threatened and only three species are currently classified as globally Near-Threatened: Southern Banded Snake-eagle, East Coast Akalat and Plain-backed Sunbird.

10.3.5 PEMBA ISLAND (EPI) ENDEMIC BIRD SPECIES

Pemba is one of the two main Indian Ocean islands of Zanzibar in Tanzania, the other being Unguja (large island) (itself commonly referred to as "Zanzibar"). According to the theory of biogeography by MacArthur and Wilson (1963), a corollary of the theory of island biogeography is that the hallmarks of greater "oceanicity" (Peake, 1972) would then be expected in Pemba's biota relative to that of Unguja with a depauperate fauna skewed toward vagile species and with a greater degree of endemism. Pemba was found to be an important minor/subcenter of endemism for the region as reported by Burgess et al. (1998) and Burgess (2000); however, the island is affected by various human activities that may affect the habitat of the endemic species. Four bird species are endemic to Pemba (Catry et al., 2000): Pemba Green-pigeon, Pemba Scops-owl, Pemba Sunbird, and Pemba White-eye (Figure 10.5). The pigeon and owl although common in forest, woodland and plantations, and they probably evolved as forest-dependent species (Mlingwa et al., 2000). It is thought that the Pemba Island might have been covered by forests until a few hundred years ago (Greenway et al., 1988).

10.4 THREATS AND SPECIES CONSERVATION STATUS

The global threat status of bird species follows the IUCN red list of 2023 (BirdLife International, 2023a). The IUCN Red List is a critical indicator of the health of the world's biodiversity. Species are periodically reviewed by BirdLife International, the IUCN authority for birds, and depending on new knowledge and information new entries and down-listing are made. For those species that CFs are critical habitats (Table 10.1); their major threats include fire, encroachments, habitat lost, and degradation (Figure 10.6, see also Section 10.5.2).

Pemba Sunbird (Female) | Pemba Scops-owl

Pemba Green Pigeon | Pemba White-eye

FIGURE 10.5 Four Pemba Island endemic bird species (*Photos: Per Holmen*).

10.4.1 ENDANGERED SPECIES

- **Sokoke Scops-owl**—It has a very small, severely fragmented range, within which its population and the quality of its habitat is declining (Virani et al., 2010). In Tanzania the owl inhabits a CF that, while still extensive, is under some pressure from fires, expansion of exotic commercial tree plantations such as *Tectona grandis* and *Cedrela odorata* in East Usambaras, encroachment and tree-cutting (with potential nest cavities) and climate change (Monadjem et al., 2013).
- **Sokoke Pipit**—It has a very small range, within which the total area of its forest habitat is declining owing to the clearance and degradation of the species' forest habitat, mainly through charcoal burning, agricultural encroachment, logging, and pole cutting. Habitat deterioration has welcomed cattle grazers, and enhanced regeneration of exotic plant species in some of the CFs of the Kisarawe District in Tanzania. Furthermore, these forests are under pressure from frequent wildfire and infrastructure development such as rails, powerlines, and roads.
- **Clarke's Weaver**—It has a small and fragmented range, within which its highly threatened due to drought and its woodland habitat being cleared for cultivation.

There is still encroachment, selective and illegal logging in its primary habitats (Schürmanna et al., 2020); Dakacha Woodland and Arabuko-Sokoke forests.

Senna siamea, exotic plant, dominating eastern part of Pugu forests, Tanzania

Frequent wildfire has changed the vegetation of the northern part of Pugu forests

Active borrow pit in Pugu forests for roads and rail construction

Charcoal from coastal forests are major source of energy in neighborhood urban centers

FIGURE 10.6 Some of the major threats to the Coastal Forests: exotic plant species, wildfire, infrastructure developing projects, and unsustainable utilization of forest resources.

Photo Credit: Author

- **Usambara Hyliota**—It has extremely small range, very small population size, and because of its tendency to inhabit the canopy of mature trees (Seddon et al., 1999) which are target for logging especially in private lands, threats remain high. Additionally, its forest and forest-edge habitat quality in the lowland is declining due to encroachments and exotic tree plantations.
- **Amani Sunbird**—It has a very small and severely fragmented range. Away from its strongholds in Arabuko-Sokoke and the East Usambaras, it is found only very locally, even within suitable evergreen forest habitat (Seddon et al., 1999). Forest clearance and alteration are ongoing over much of its range (Borghesio et al., 2008), and its population and range are therefore supposedly on decline.

10.4.2 VULNERABLE SPECIES

- **Spotted Ground-thrush:** It has a small (patchy) population (Ndang'ang'a et al., 2005), which is undergoing a continuing decline because of destruction and degradation of its habitat. Forest blocks in some parts of its range, for example, at Mt Mulanje in Malawi and Mt Namuli in Mozambique have completely been destructed (Timberlake, 2017). The East Africa race, *fisheri*, although its known breeding site is now found in a newly gazetted Rondo Nature Forest Reserve, is probably CR, for the same threats facing Sokoke Pipit in much of its range especially passage and wintering habitats.
- **Banded Sunbird:** It has a small and fragmented distribution in which the quality and extent of its habitat is declining. Although it may be tolerant to some degree of habitat disturbance (Borghesio et al., 2008), it still requires tall, mature forest trees (Seddon et al., 1999).
- **Pemba Green-pigeon:** It has a rapidly declining population as a result of habitat loss and degradation, owing primarily to a local agricultural trend away from plantation crops and toward open farmland (Catry et al., 2000). The species is frugivorous feeding on the fruits of *Antidesma venosum*, *Croton sylvaticus*, *Ficus lutea*, *Ficus natalensis*, and Betel-nut Palm *Areca catechu* (Catry et al., 2000) and prefers areas with high canopy height and canopy cover, both declining on the island. Some of its feed fruits are overexploited commercially or medicinally (Nahonyo et al., 2005) threatening the survival of this species.
- **Pemba Scops-owl:** It has a single small population which is continuing to decline as a result of habitat loss and degradation. Conversion of areas that were formerly clove plantations to open farmlands is a major threat to this species (Catry et al., 2000).

10.4.3 NEAR-THREATENED SPECIES

- **Reichenow's Batis:** Recently, described species because formerly *B. mixta* and *B. reichenowi* (del Hoyo et al., 2016) were previously lumped as *B. mixta* (Fjeldså et al., 2006). It is restricted in south-east Tanzania, in the Mikindani and Lindi area. It occurs in dense vegetations such as riparian forest (Eriksen et al., 1994; Jensen et al., 2005). The population is suspected to be declining as a result of the loss of CF in Tanzania (Kashiagili et al., 2013).
- **Fischer's Turaco:** It has a moderately small population in CFs of Somalia, Kenya and Tanzania, which is suspected to be declining owing to the effects of habitat degradation across its range, mainly due to clearance for agriculture. In Somalia, a few pairs lives in lower Jubba valley, where up to 80% of the riverine forest has been cleared in recent past (Madgwick, 1986; Ash and Miskell, 1998) and no recent data; thus, there is little hope for its survival there.
- **Southern Banded Snake-eagle:** This species has a limited range and usually occurs at low densities (Brown et al., 1982; Seddon et al., 1999). Throughout its range,

human pressure, deforestation for plantations, settlements, and other anthropogenic development threaten its habitats.

- **East Coast Akalat:** It occurs as four subspecies in scattered forests in south-east Kenya, eastern Tanzania, northern Malawi and Mozambique. It inhabits closed, moist forest in the Tanzanian mountains (Fjeldså et al., 2000), lowland and mid-altitude humid forest in Malawi (Dowsett-Lemaire, 1989) and Mozambique (Spottiswoode et al., 2008), and tracts of mixed woodland and open, dry forest in coastal/lowland areas (Nemeth and Bennun, 2000). It prefers forest areas with dead logs and moss, which it uses for nesting material. Arabuko-Sokoke Forest (Kenya) has probably the largest and stable population in Kenya (Matiku et al., 1999; Banks et al., 2012). In Tanzania, the species is under high risk: the largest population is thought to occur in Ruvu South Forest Reserve (Baker and Baker, 2002), although this is now under severe pressure for charcoal production, fire, and cattle grazing (Author's recent field observation); the population in Jozani forest on Zanzibar occupies a tiny habitat. Elsewhere in Mozambique and lowland Malawi, the species' habitat is under severe human pressure.
- **Plain-Backed Sunbird:** It has a disjunct distribution, with one subpopulation in the coastal lowlands of Kenya and northeastern Tanzania, and another in Mozambique and Zimbabwe. It occurs in dense CF and ironwood *Androstachys* forests in Mozambique (Clancey, 1996). In the CF in Kilwa, Tanzania, it may prefer large forest patches which are constantly being reduced in size, but smaller patches elsewhere may hold sparser populations (Maclean et al., 2012). Its population in eastern Zimbabwe is uncertain. Although it is widely distributed, it remains little-known and nowhere common. It is therefore classed as Near-Threatened because it is thought to have a moderately small population, which is suspected to be in decline owing to deforestation throughout its range.

10.5 IMPORTANT SITES FOR BIRD CONSERVATION IN CFs

10.5.1 KEY BIRDS AND BIODIVERSITY AREAS, THE PROCESS

Bird species unique to the CFs have not reached a level of *ex-situ* conservation and the best option would be to target their habitats, an approach which is less expensive and sustainable. It is however interesting that many of the forest blocks in the CFs have a certain level of protection (Ndang'ang'a et al., 2009); some fall within the national reserve/parks such as Shiba Hills in Kenya, Nyerere and Saadani in Tanzania, Quirimbas and Gorongosa in Mozambique, and many others are forest reserves. However, because many of the unique species to this region are range-restricted species, thus most of their habitats fall within Key Biodiversity/Bird Important Areas (KBAs/IBAs). Sites qualify as global KBAs if they meet one or more of 11 criteria, clustered into five higher level categories (IUCN, 2016) namely: A. Threatened biodiversity (Criteria A1 and A2), B. Geographically restricted biodiversity (Criteria B1, B2, B3, and B4), C. Ecological integrity (Criterion

C), D. Biological processes (Criteria D1, D2, and D3), E. Very high irreplaceability, as determined through quantitative analysis (Criterion E).

Furthermore, because there is a significant overlap among the criteria used for both KBA and IBA, many of the sites qualify for both categories. IBAs form a subset of KBAs that have been embedded in various international agreements and serve as indicators for the biodiversity policy framework of the Convention on Biological Diversity and the Sustainable Development Goals. The identification of IBAs (formerly called Important Bird Areas until 2013) has a long history (Donald et al., 2019). In general, the regular presence of a Critical or Endangered bird species, irrespective of population size, at a site may be enough for a site to qualify as an IBA. For vulnerable species, the presence of more than threshold numbers at a site is necessary to trigger selection.

The selection of IBAs is achieved through the application of quantitative ornithological criteria, grounded in up-to-date knowledge of the sizes and trends of bird populations. The criteria ensure that the sites selected as IBAs have true significance for the international conservation of bird populations and provide a common currency that all IBAs adhere to, thus creating consistency among, and enabling comparability at global levels using criteria listed below (Box 1). Additional criteria for regional and subregional criteria are discussed in Donald et al. (2019). Sites should ideally be assessed against all relevant criteria for which data are available, but meeting thresholds under any of the criteria or subcriteria is enough for a site to be recognized as an IBA, provided that supporting documentation requirements are properly met. A main aim of the IBA Programme of BirdLife International is to secure the long-term conservation of sites that are significant for birds and biodiversity.

Box 1: Major four criteria used to identify Important Bird and Biodiversity Areas
A1: Globally threatened Species Criterion: the site is known or thought regularly to hold significant numbers of a Globally Threatened species.
A2: Restricted Range Species Criterion: the site is known or thought to hold a significant population of at least two range-restricted species.
A3: Bioregion-restricted assemblages Criterion: the site is known or thought to hold a significant component of the group of species whose distributions are largely or wholly confined to one biome-realm.
A4: Congregations Criterion: the site is known or thought to hold congregations of $\geq$1% of the global population of one or more species on a regular or predictable basis.

10.5.2 KEY THREATENED IBAs IN CFs

The entire CF region is threatened with degradation and fragmentation (Burgess and Muir, 1994; del Hoyo et al., 1994; Ndang'ang'a et al., 2009) as a result of the extraction of wood for use as timber, charcoal, poles, and firewood. Slash and burn agriculture threaten the forest habitat and the loss of cultivated land and plantations has created local range declines of some key species (Catry et.al., 2000; Cooper, 2015). Most of the lowland CFs have been cleared (Clancey, 1996; Kashiagili et al., 2013). Habitat degradation in the CFs is increasing due to the high demands of forest products from the increasing human population. Forest fires and invasive plant species are among the major threats to the remaining forests. For example, forests near large urban centers such as Malindi, Mombasa, Tanga, Dar es Salaam, Beira, and Maputo are under serious threat from waves of forest destruction for charcoal production (Ahrends et al., 2010). In general, more than 80% of the Indian Ocean CF belt has already been lost and fragmented due to these pressures, whilst the remaining habitat is projected to undergo significant changes due to climate change (Colyn et al., 2020). This region is home for many range restricted bird species, and the endemic species (Table 10.2) which are the most vulnerable to extinction should these forests disappear.

TABLE 10.2 Important Bird and Biodiversity Areas for Coastal Forest Endemic Avifauna (including Pemba Endemics)

Sr. No.	Endemic Bird Species	Important Bird and Biodiversity Areas
1	Little Yellow Flycatcher	**Kenya:** Arabuko-Sokoke Forest, Diani Forest, Gede Ruins National Monument, Marenji Forest, Mrima Hill Forest, Shimba Hills, Lower Tana River Forests, and Boni and Dodori National Reserves.
		Somalia: Aangole-Farbiito, Far Waamo, and Laag Badaana.
		Tanzania: Selous Game Reserve, Bagamoyo District Coastal Forests, Kisarawe District Coastal Forests, Pande Game Reserve and Dondwe Coastal Forests, Handeni District Coastal Forests, Pangani District Coastal Forests, and East Usambara Mountains.
2	Reichenow's Batis	**Tanzania:** Coastal Forests of Lindi and Mtwara districts.
3	Clarke's Weaver	**Kenya:** Arabuko-Sokoke Forest and Dakatcha Woodland.
4	Sokoke Pipit	**Kenya:** Arabuko-Sokoke Forest, Dakatcha Woodland, Dzombo Hill Forest, Gede Ruins National Monument, Kaya Gandini, Marenji Forest, Shimba Hills.
		Tanzania: Bagamoyo District Coastal Forests, Kisarawe District Coastal Forests, Pande Game Reserve and Dondwe Coastal Forests, Muheza District Coastal Forests.
5	Sokoke Scops-owl	**Kenya:** Arabuko-Sokoke Forest and Dakatcha Woodland.
		Tanzania: East Usambara Mountains.
6	Pemba Green-pigeon	**Tanzania:** Pemba Island
7	Pemba Scops-owl	**Tanzania:** Pemba Island
8	Pemba Sunbird	**Tanzania:** Pemba Island
9	Pemba White-eye	**Tanzania:** Pemba Island

A total of 26 IBAs (Table 10.2) seems to be important for the conservation of endemic species within the CF region. Many of these sites are small and highly fragmented (Mlingwa et al., 2000; Soi, 2018). Some of these endemic species such as Reichenow's Batis, Clarke's Weaver, Pemba Green-pigeon, Pemba Scops-owl, Pemba Sunbird, and Pemba White-eye are restricted in one or two sites. Despite being classified as IBA in danger (BirdLife International, 2020), Arabuko-Sokoke is among the most important forest for bird conservation in CFs. Dakatcha Woodland in Kenya is another IBA in danger, which hold substantial populations of Sokoke Pipit and Clarke's Weaver. The Pemba Island is unique EBA (Stattersfield et al., 1998). In addition to the Pemba Island, the East Usambaras and the two IBAs in Danger (in Kenya) are probably the most important sites for protection CF endemic avifauna. Lindi and Mtwara district CFs host Reichenowi's Batis and a subspecies of Green Barbet. These sites should be treated as flagship areas not only for the conservation of endemic birds but also for other species characteristic of the CFs.

In order to achieve any conservation plans put forward for these critical sites, a long-term monitoring for the *pressure*, *state*, and *response*, is needed. The state score measures the condition of KBA trigger species and their habitats; the *pressure* score measures the threats facing the site; and the *response* score measures the actions being taken to conserve the sites (Ndang'ang'a et al., 2009). Trigger species are the birds for which the IBA was designated (Gacheru et al., 2022). Fortunately, such mechanisms are already in place in Kenya where Nature Kenya, a nongovernmental organization and a BirdLife International country partner, has a robust and well-coordinated IBA monitoring program of any country in Africa (Bennun et al., 2005; Gacheru et al., 2022). This program was established in 2004; and since then at least 17 annual reports have been produced and all can be found on the Nature Kenya website (KBA Status & Trends Reports—Nature Kenya). If a similar program is adopted in other countries along the CF biodiversity hotspot, the status of these IBAs could be updated periodically and specific actions undertaken.

10.5.3 IBAs AS LOCAL CENTERS OF ENDEMISM

Burgess (2000) lists several locations as subcenters of endemism within the CFs. The identification of these sites was based on several endemic taxa: plants, vertebrates and invertebrates (especially butterfly) similar to the classification made by White (1993) who defined areas with more than 100 endemic species as local center of endemism. Burgess (2000) defined minor center of endemism as the one with at least one endemic vertebrate species, and marginal center of endemism based on the concentration of endemic species. Some of the IBAs in the CFs are found within the areas defined by Burgess (2000). The centers of endemism include: Pemba Islands, for the four endemic birds (Tables 10.1 and 10.2); Usambara-Kwale, ranging from Shimba Hills in SE Kenya to the lowland of East Usambaras in NE Tanzania which is important for bird species characteristic of the Kenyan and Tanzanian CF EBAs (Burgess and Muir, 1994; Bennun and Njoroge, 2000; Fishpool and Evans, 2001; Soi, 2018); Lindi District CFs for it hosts Reichenowi's Batis and a subspecies of Green Barbet, *hylophona*; Lower Tana River forests host the Tana River Cisticola and the *chariessa* (subspecies of White-winged Apalis, probably locally extinct now). These areas are not only critical for the conservation of bird species but for a variety of other taxa endemic in CFs (Burgess and Clarke, 2000).

10.6 THE WAY FORWAD

CFs biodiversity hotspot continues to be threatened by habitat loss, and the increasing human population impacts the region through intensified use of the forest's natural resources, encroachments for settlement and agriculture, and infrastructure development projects such as roads, rails, and hydropower plants. Fires have become a primary means of clearing land for cultivation and when uncontrolled, which is often the case, become wildfire burning large areas of CFs annually. Due to urbanization, fragmentation and deforestation, the CFs are now reduced to an archipelago of forested islands. In fact, there is evidence for near-complete loss of forest cover in many sites, illustrating urgent threats facing them, particularly when they lack some form of protection (Ndang'ang'a et al., 2009). It is therefore worrying that the conservation of the endemic avifauna will not be sustainable unless restoration and connectivity are considered. Priority should be given to the threatened critical sites (IBAs) for the conservation of CFs endemic avifauna: Arabuko-Sokoke, Dakatcha Woodlands, Mtwara and Lindi CFs, and the East Usambara forests. Although all the aforementioned sites are under anthropogenic pressure, Arabuko-Sokoke and Dakatcha are at high risk and have been named as IBA in danger by the BirdLife International.

The future of the CFs will depend on effective joint conservation efforts among the stakeholders, and any conservation plans should prioritize but not limited to the following:

- Raise public awareness of the extent and consequences of forest loss and degradation, as well as the potential for forest landscape restoration. The importance of using native plant species in restoring degraded forests should be emphasized (Outwater et al., 2019) to avoid impacts of invasive species that may arise when using exotic species. CFs can easily regenerate if current threats are reduced.
- Enhance ecosystem connectivity by effectively managing riparian areas and increasing vegetal cover along river systems, general lands between forests, and coastal settlements.
- Promote community forestry and implement legislation and guidelines for collaborative natural resources management with communities and increase budgetary allocations to support conservation efforts especially environmentally friendly income-generating activities such as agroforestry and beekeeping programs by local communities.
- Ensure appropriate environmental safeguards (e.g., Environmental and Social Impact Assessments, Environmental Audit, Strategic Environmental Assessments) and best practices (e.g., decommissioning, adaptive management) are applied by all infrastructure development projects.
- Tap tourism opportunities including birdwatching (Figure 10.7) as this will provide jobs and other benefits to the local population and reduce unsustainable use of the forest resources. There are already some ecotourism initiatives near some of the IBAs in coastal Kenya and Pugu Hills in Tanzania which need to be strengthened and used as case studies for the rest of sites.
- Strengthen local awareness on the importance of biodiversity especially endemic species of the CFs and the opportunities they bring. Develop and implementing target-site and species-level-based conservation initiatives.
- Conduct research and monitoring of the IBAs and adjoining habitats. This is critical especially in the CFs in Somalia, Malawi, Mozambique, Zimbabwe, and Tanzania

where current information for many sites and species is lacking. Research and monitoring will provide current data upon which immediate and adaptive conservation actions can be considered.

FIGURE 10.7 Birdwatching is among the popular ecotourism activities in some coastal forests.
Photo Credit: Author

ACKNOWLEDGMENTS

I acknowledge my family for being around during the write up of this chapter. Edigar Apolinary helped with compilation of the species list.

KEYWORDS

- **coastal forests**
- **endemic bird area**
- **forest dependency**
- **important bird and biodiversity area**
- **species conservation status**

REFERENCES

Ahrends, A.; Burgess, N. D.; Milledge, S. A. H.; Bulling, M. T.; Fisher, B.; Smart, J. C. R.; Clarke, G. P.; Mhoro, B. E.; Lewis, S. L. Predictable waves of sequential forest degradation and biodiversity loss spreading from an African city. *Proc. Natl. Acad. Sci. U.S.A.* **2010,** *107*, 14556–14561.

Ash, J. S.; Miskell, J. E. *Birds of Somalia*. Pica Press: Robertsbridge, UK, **1998**.

Baker, N.; Baker, E. *Important Bird Areas in Tanzania: A first inventory*. Wildlife Conservation Society of Tanzania: Dar es Salaam, Tanzania, **2002**.

Baker, N. E.; Baker, E. M. Four afrotropical migrants on the East African coast: Evidence for a common origin. *Scopus* **1992,** *15*, 122–124.

Banks, J. E.; Jackson, C.; Baya, A.; Minnella, H.; Nitz, M.; Hitchcock, J.; Bruinsma, D. Forest-type preference of an Afrotropical thrush (East Coast Akalat *Sheppardia gunningi sokokensis*) in Arabuko-Sokoke Forest, Kenya. *Ostrich* **2012,** *83*, 105–108.

Bennun, L.; Dranzoa, C.; Pomeroy, D. The forest birds of Kenya and Uganda. *J. East Afr. Nat. Hist.* **1996,** *85*, 23–48.

Bennun, L.; Matiku, P.; Mulwa, R.; Mwangi, S.; Buckley, P. Monitoring important bird areas in Africa: Towards a sustainable and scaleable system. *Biodivers. Conserv.* **2005,** *14*, 2575–2590.

Bennun, L.; Njoroge, P. Important bird areas in Kenya. *Ostrich* **2000,** *71*(1–2), 164–167.

BirdLife International. *Guidelines for the Application of the IBA Criteria*. Final Version. Guidelines for the application of the IBA criteria final approved version_July2020.pdf. BirdLife International: Cambridge, UK, **2020**.

BirdLife International. IUCN red list for birds. Downloaded from http://www.birdlife.org on 10/04/2023. **2023a**.

BirdLife International. Species factsheet: *Geokichla guttata*. Downloaded from http://www.birdlife.org on 14/04/2023. **2023b**.

BirdLife International. Species factsheet: *Cisticola restrictus*. Downloaded from http://www.birdlife.org on 14/04/2023. **2023c**.

Borghesio, L.; John, J. R. M.; Mulungu, E.; Mkongewa, V.; Joho, M.; Cordeiro, N. J. Observations of threatened birds in the East Usambara Mountains, Tanzania. *Bull. Afr. Bird Club* **2008,** *15*(1), 59–70.

Borghesio, L.; Ndang'ang'a, P. K. Habitat selection and the conservation status of Fischer's turaco *Tauraco fischeri* on Unguja, Tanzania. *Oryx* **2003,** *37*(4), 444–453.

Bowie, R. C. K.; Monahan, W. B.; Fjeldså, J. Climate cycles, habitat stability, and lineage diversification in an African biodiversity hotspot. *Diversity* **2023,** *15*(3), 394; https://doi.org/10.3390/d15030394.

Britton, P. L. *Birds of East Africa*. East Africa Natural History Society: Nairobi, **1980**.

Brown, L. H.; Urban, E. K.; Newman, K. *The Birds of Africa.* Academic Press: London, **1982**. Vol. 1.

Burgess, N.; Muir, C. *Coastal Forests of Eastern Africa: Biodiversity and Conservation*. Society for Environmental Exploration/Royal Society of the Protection of Birds: UK, **1994**.

Burgess, N. D. Global importance and patterns in the distribution of Coastal Forest species. In: Burgess, N. D.; Clarke, G. P. (Eds.), *Coastal Forests of Eastern Africa*. IUCN: Cambridge, **2000,** pp. 235–248.

Burgess, N. D.; Butynski, T. M.; Cordeiro, N. J., Doggart, N.; Fjeldså, J.; Howell, K.; Kilahama, F.; Loader, S. P.; Lovett, J. C.; Mbilinyi, B.; Menegon, M.; Moyer, D. C.; Nashanda, E.; Perkin, A.; Stanley, W. T.; Stuart, S. N. The biological importance of the Eastern Arc Mountains of Tanzania and Kenya. *Biol. Conserv.* **2007,** *134*, 209–231.

Burgess, N. D.; Clarke, G. P. (Eds.). *The Coastal Forests of Eastern Africa*. IUCN: Cambridge, **2000**.

Burgess, N. D.; Clarke, G. P.; Rodgers, W. A. Coastal forests of eastern Africa: Status, endemism patterns and their potential causes. *Biol. J. Linn. Soc.* **1998,** *64*, 337–367.

Burgess, N. D.; Mlingwa, C. O. F. Evidence for altitudinal migration of forest birds between montane Eastern Arc and lowland forests in East Africa. *Ostrich* **2000,** *71*, 184–90.

Catry, P.; Mellanby, R.; Suleiman, K. A.; Salim, K. H.; Hughes, M.; McKean, M.; Anderson, N.; Constant, G.; Heany, V.; Martin, G.; Armitage, M.; Wilson, M. Habitat selection by terrestrial birds on Pemba Island (Tanzania), with particular reference to six endemic taxa. *Biol. Conserv*. **2000,** *95*(3), 259–267.

Clancey, P. A. *The Birds of Southern Mozambique*. African Bird Book Publishing, **1996,** p. 312.

Collar, N. J.; Crosby, M. J.; Stattersfield, A. *Birds to Watch 2: The World List of Threatened Birds.* BirdLife International: Cambridge, UK, **1994**.

Collar, N. J.; Stuart, S. N. *Threatened Birds of Africa and Related Islands*. International Council for Bird Preservation: Cambridge, UK, **1985**.

Colyn, R. B.; Ehlers Smith, D.; Ehlers Smith, Y.; Smit-Robinson, H.; Downs, C. T. Predicted distributions of avian specialists: A framework for conservation of endangered forests under future climates. *Divers. Distrib* **2020,** doi: 10.1111/ddi.13048.

Cooper, T. J. G. *The Effects of Land Use Changes on the Distribution of Forest Dependent Bird Species in South Africa.* MSc Thesis. Stellenbosch University, **2015**.

Davies, G. B. P.; Symes, C. T.; Chittenden, H. N.; Peek, J. R. Mangrove kingfishers (*Halcyon senegaloides*; Aves: Alcedinidae) nesting in arboreal *Nasutitermes* (Isoptera: Termitidae: Nasutitermitinae) termitaria in central Mozambique. *Ann. Ditsong Nat.l Museu. Nat. Hist*. **2012,** *2*, 146–152.

del Hoyo, J.; Collar, N. J.; Christie, D. A.; Elliott, A.; Fishpool, L. D. C.; Boesman, P.; Kirwan, G. M. *HBW and BirdLife International Illustrated Checklist of the Birds of the World, Vol. 2: Passerines*. Lynx Edicions and BirdLife International: Barcelona, Spain and Cambridge, UK, **2016**.

del Hoyo, J.; Collar, N. J.; Christie, D. A.; Elliott, A.; Fishpool, L. D. C. *HBW and BirdLife International Illustrated Checklist of the Birds of the World*. Lynx Edicions BirdLife International: Barcelona, Spain and Cambridge, UK, **2014**.

del Hoyo, J.; Elliott, A.; and Sargatal, J. (eds). *Handbook of the Birds of the World, Vol. 2: New World Vultures to Guineafowl*. Lynx Edicions: Barcelona, Spain, **1994**.

Dinesen, L.; Lehmberg, T.; Svendsen, J. O.; Hansen, L. Range extensions and other notes on some restricted-range forest birds from West Kilombero in the Udzungwa mountains, Tanzania. *Scopus* **1993,** *17*, 48–59.

Donald, P. F.; Fishpool, L. D. C.; Ajagbe, A.; Bennun, L. A.; Bunting, G.; Burfield, I. J.; Butchart, S. H. M.; Capellan, S.; Crosby, M. J.; Dias, M. P.; Diaz, D.; Evans, M. I.; Grimmett, R.; Heath, M.; Jones, V. R.; Lascelles, B. G.; Merriman, J. C.; O'brien, M.; Ramírez, I.; Waliczky, Z.; Wege, D. C. Important Bird and Biodiversity Areas (IBAs): The development and characteristics of a global inventory of key sites for biodiversity. *Bird Conserv. Int.* **2019,** *29,* 177–198.

Dowsett-Lemaire, F. Ecological and biogeographical aspects of forest bird communities in Malawi. *Scopus* **1989,** *13*, 1–80.

Eriksen, T.; Halberg, K.; Lehmberg, T.; Petersen, F. S. *A Survey of Bird life in Five Coastal Forests of South-eastern Tanzania, 1993*. Danish Ornithological Society: Copenhagen, **1994**.

Evans, T. D.; Watson, L. G.; Hipkiss, A. J.; Kiure, J.; Timmins, R. J.; Perkin, A. W. New records of Sokoke Scops Owl *Otus ireneae*, Usambara Eagle Owl *Bubo vosseleri*, and East Coast Akalat *Sheppardia gunningi* from Tanzania. *Scopus* **1994,** *18*, 40–47.

Fishpool, L. D. C.; Evans, M. I. (eds*.). Important Bird Areas in Africa and Associated Islands: Priority Sites for Conservation*. Pisces Publications and Cambridge: Newbury, UK: BirdLife International, **2001**.

Fjeldså, J.; Bowie, R. C. K. Evolutionary and ecological explanations for the elevational flexibility of several East African bird species complexes. *Front. Ecol. Evol.* **2021,** *9,* 768062.

Fjeldså, J.; Bowie, R. C. K.; Kiure, J. The forest batis, *Batis mixta*, is two species: Description of a new, narrowly distributed *Batis* species in the Eastern Arc biodiversity hotspot. *J. Ornithol.* **2006,** *147*(4), 578–590.

Fjeldså, J.; Roy, M. S.; Kiure, J. A new montane subspecies of *Sheppardia gunningi* (East-coast Akalat) from Tanzania. *Bull. Br. Ornithol. Club* **2000,** *120*, 27–33.

Gacheru, P.; Mutunga, J.; Mwinami, T. F.; Matiku, P.; Mulwa, R.; Ngoru, B.; Kariuki, L. R.; Wanjohi, H.; Mwang'ombe, J. *Kenya's Key Biodiversity Areas: Status and Trends 2021*. Nature Kenya: Nairobi, **2022**.

Greenway, P. J.; Rodgers, W. A.; Wingfield, R. J.; Mwasumbi, L. B. The vegetation of Mafia Island, Tanzania. *Kirkia* **1988,** *13*, 197–238.

HBW and BirdLife International. Handbook of the Birds of the World and BirdLife International digital checklist of the birds of the world. Version 7. **2022**. Available at: http://datazone.birdlife.org/userfiles/file/Species/Taxonomy/HBW-BirdLife_Checklist_v7_Dec22.zip.

Hipkiss, A. J.; Watson, L. G.; Evans, T. D. The Cambridge-Tanzania Rainforest Project 1992: Brief account of ornithological results and conservation proposals. *Ibis* **1994,** *136*, 107–108.

Holsten, B.; Braunlich, A.; Huxham, M. Rondo Forest Reserve, Tanzania: An ornithological note including new records of the East Coast Akalat *Sheppardia gunningi*, the Spotted Ground Thrush *Turdus fischeri*, and the Rondo Green Barbet *Stactolaema olivacea woodwardia*. *Scopus* **1991,** *14*, 125–128.

Hunter, N.; Carter, C.; Mlunga, E. Recent observations in the Udzungwa and Uluguru Mountains, Central Tanzania. *Bull. Afr. Bird Club* **1996,** *3*, 96–98.

Irwin, M. P. S. Birds. In: Timberlake, J.; Shaw, P. (Eds.), *Chirinda Forest: A Visitor's Guide*. Forestry Commission/ODA: Harare, **1994,** pp. 77–91.

IUCN. *A Global Standard for the Identification of Key Biodiversity Areas, Version 1.0*. First edition. IUCN: Gland, Switzerland, **2016**.

Jensen, F. P.; Tøttrup, A. P.; Christensen, K. D. The avifauna of coastal forests in southeast Tanzania. *Scopus* **2005,** *25*, 1–22.

Kashiagili, J.; Mbilinyi, B.; Tabor, K.; Hall, J.; Malugu, I.; Sumbi, P.; Kijazi, A.; Burgess, N. D. Tanzania is losing its biodiversity-rich coastal forests: Deforestation in Tanzania's coast forests. *The Arc Journal* **2013,** *28*, 5–7.

Keith, S.; Urban, E. K.; Fry, C. H. (Eds). *The Birds of Africa.* Academic Press: London, **1992,** Vol. IV.

Kingdon, J. *Island Africa*. Collins: London, **1990**.

Lovett, J. C.; Clarke, G. P.; Moore, R.; Morrey, G. Elevational distribution of restricted range forest tree taxa in eastern Tanzania. *Biodivers. Conserv.* **2001,** *10*, 541–550.

Lowe, A. J.; Clarke, G. P. Vegetation structure. In: Burgess, N. D.; Clarke, G. P. (Eds.), *Coastal Forests of Eastern Africa*. IUCN: Cambridge, **2000,** pp. 103–113.

MacArthur, R. H.; Wilson, E. O. An equilibrium theory of insular zoogeography. *Evol.* **1963,** *17*(4), 373–387.

Mace, G. M.; Stuart, S. N. Draft IUCN red list categories. *Species* **1994,** *21–22*, 13–24.

Mackworth-Praed, C. W.; Grant, C. H. B. *Birds of Eastern and North Eastern Africa*. Longman: London, **1960,** Vol. II.

Maclean, I. M. D.; Bray, J.; Andrews, D.; Mlawila, L.; Kitaluta, K.; Timothy, J. The status and habitat preferences of birds associated with coastal forest and grassland in Kilwa District, southern Tanzania. *Bull. Afr. Bird Club* **2012,** *19*, 144–159.

Madgwick, J. *Somalia Research Project: An Ecological Study of the Remaining Areas of Riverine Forest in the Jubba Valley, Southern Somalia*. University College London: London, **1986**.

Matiku, P. M.; Bennun, L. A.; Nemeth, E. Distribution and population size of the threatened East Coast Akalat in Arabuko-Sokoke Forest, Kenya. *Ostrich* **1999,** *71*, 282–285.

Mlingwa, C. O. F. A first record of the Stripe-cheeked Greenbul *Andropardus milanjensis* in the Pugu Hills. *Bull. Br. Ornithol. Club* **1996,** *116*, 63–64.

Mlingwa, C. O. F.; Waiyaki, E. M.; Bennun, L. A.; Burgess, N. D. Birds. In: Burgess, N. D.; Clarke, G. P. (Eds.), *Coastal Forests of Eastern Africa*. IUCN: Cambridge, **2000,** pp. 149–171.

Monadjem, A.; Virani, M. Z.; Jackson, C.; Reside, A. Rapid decline and shift in the future distribution predicted for the endangered Sokoke Scops Owl *Otus ireneae* due to climate change. *Bird Conserv. Int.* **2013,** *23*(2), 247–258.

Nahonyo, C. L.; Mwasumbi, B.; Msuya, C. A.; Masao, C. A.; Suya, T. B.; Shing'wenda, C. *Ngezi – Vumawimbi Forest Reserves Biodiversity Inventory Report*. Department of Zoology and Marine Biology, University of Dar es Salaam, **2005**.

Ndang'ang'a, P. K.; Mulwa, R.; Jackson, C. Status of the endangered Spotted Ground Thrush *Zoothera guttata fischeri* in coastal Kenyan forests. *Scopus* **2008,** *27*, 6–9.

Ndang'ang'a, P. K.; Eshiamwata, G.; Ngari, A.; Pius, E.; John, J.; Arinaitwe, J.; Mbwana, S. *Status Report for the Eastern Arc Mountains and Coastal forests of Kenya and Tanzania Region, 2008.* BirdLife International: Nairobi, Kenya, **2009**.

Ndang'ang'a, P. K.; Sande, E.; Evans, S. W.; Buckley, P.; Newbery, P.; Hoffmann, D. A.; John, J.; *International Action Plan for Spotted Ground Thrush Zoothera guttata (Threatened Birds of Africa)*. BirdLife International, Nairobi, Kenya and Royal Society for the Protection of Birds: Sandy, Bedfordshire, UK, **2005**.

Nemeth, E.; Bennun, L. Distribution, habitat selection and behaviour of the East Coast Akalat *Sheppardia gunningi*. *Bird Conserv. Int.* **2000,** *10*,115–130.

Outwater, A. H.; Locker, I. M.; Gereau, R. E. *A Garden Guide to Native Plants of Coastal East Africa*. Mkuki na Nyota Publishers Ltd: Dar es Salaam, **2019**.

Owino, A. O.; Amutete, G.; Mulwa, R. K.; Oyugi, J. O. Forest patch structures and bird species composition of a lowland riverine coastal forest in Kenya. *Trop. Conserv. Sci.* **2008,** *3*, 242–264.

Pakenham, R. H. W A new race of *Turacus*. *Bull. Br. Ornithol. Club* **1938,** *58*, 111–112.

Peake, J. F. The evolution of terrestrial faunas in the western Indian Ocean. *Philos. Trans. R. Soc B.* **1972,** *260*, 581–610.

Schürmanna, A.; Kleemanna, J.; Fürsta, C.; Teucherb, M. Assessing the relationship between land tenure issues and land cover changes around the Arabuko Sokoke Forest in Kenya. *Land Use Policy* **2020,** *95*: https://doi.org/10.1016/j.landusepol.2020.104625.

Seddon, N.; Ekstrom, J. M. M. J.; Capper, D. R.; Isherwood, I. S.; Muna, R.; Pople, R. G.; Tarimo, E.; Timothy, J. Notes on the ecology and conservation status of key bird species in Nilo and Nguu North Forest Reserves, Tanzania. *Bird Conserv. Int.* **1999,** *9*, 9–28.

Sibley, C. G.; Monroe, B. L. *Distribution and Taxonomy of Birds of the World.* Yale University Press: New Haven, **1990**.

Soi, B. C. Status of restricted-range bird species of east African coastal forests in five south coast forest of Kenya. *J. Agric. Ecol.* **2018,** *16*(3), 1–10.

Spottiswoode, C. N.; Patel, I. H.; Herrmann, E.; Timberlake, J.; Bayliss, J. Threatened bird species on two little-known mountains (Chiperone and Mabu) in northern Mozambique. *Ostrich* **2008,** *79*(1), 1–7.

Stattersfield, A. J.; Crosby, M. J.; Long, A. J.; Wege, D. C. *Endemic Bird Areas of the World: Priorities for Biodiversity Conservation.* BirdLife Conservation Series. No. 7. BirdLife International: Cambridge, **1998**.

Stevenson, T.; Fanshawe, J. *Field Guide to the Birds of East Africa: Kenya, Tanzania, Uganda, Rwanda and Burundi.* 2nd Ed. T and A D Poyser Ltd: London, **2020,** p. 286.

Stuart, S. N.; Jensen, F. P. The avifauna of the Uluguru Mountains, Tanzania. *Gerfaut* **1985,** *75*, 155–187.

Timberlake, J. *Mt Namuli—a conservation update.* Legado, Mozambique, unpublished report, **2017**.

Turner, D. A. Status and distribution of East African endemic species. *Scopus* **1977,** *1*, 2–11.

Vincent, J. The birds of northern Portuguese East Africa. Comprising a list of, and observations on, the collections made during the British Museum Expedition of, 1931–32 – Parts I-IV. *Ibis* **1933 and 1934,** *13*(3 and 4), 126–652.

Virani, M. Z.; Njoroge, P.; Gordon, I. Disconcerting trends in populations of the endangered Sokoke Scops Owl *Otus ireneae* in the Arabuko-Sokoke Forest, Kenya. *Ostrich* **2010,** *81*(2), 155–158.

Werema, C. Seasonal elevational movements of little greenbul *Andropadus virens* in the Uluguru mountains, Tanzania. *Afr. J. Ecol.* **2014,** *53*, 253–256.

White, F. The AETFAT chorological classification of Africa: History, methods and applications. *Bull. Jard. Bot. Nat. Belg. Bull. Nat. Plantentuin Belg.* **1993,** *62*, 225–281.

White, F. *The Vegetation of Africa*. UNESCO: Paris, **1983**.

APPENDIX 10.1 Geographical Distribution of the Forest-dependent Bird Species found in Coastal Forests of Eastern Africa

Common Name	Family/Scientific name	TC	FD	SO	KE	TZ	MZ	ZW	MW
	Numididae (Guineafowl)								
Eastern Crested Guineafowl	*Guttera pucherani*		F	x	x	x			
	Columbidae (Pigeons, Doves)								
Eastern Bronze-naped Pigeon	*Columba delegorguei*		FF		x	x	x	x	x
Lemon Dove	*Aplopelia larvata*		FF		x	x	x	x	x
Tambourine Dove	*Turtur tympanistria*		F	x	x	x	x	x	x
African Green-pigeon	*Treron calvus*		F		x	x	x	x	x
	Caprimulgidae (Nightjars)								
Fiery-necked Nightjar	*Caprimulgus pectoralis*		F		x	x	x	x	x
	Apodidae (Swifts)								
Mottled Spinetail	*Telacanthura ussheri*		F	*	x	x	x	x	x
Böhm's Spinetail	*Neafrapus boehmi*		F	*	x	x	x	x	x
	Cuculidae (Cuckoos)								
Whistling Yellowbill	*Ceuthmochares australis*		F	x	x	x	x	x	x
African Emerald Cuckoo	*Chrysococcyx cupreus*		F		x	x	x	x	x
Barred Long-tailed Cuckoo	*Cercococcyx montanus*		FF		x	x	x	x	x
	Rallidae (Rails, Gallinules, Coots)								
Buff-spotted Flufftail	*Sarothrura elegans*		FF		x	x	x	x	x
	Musophagidae (Turacos)								
Livingstone's Turaco	*Tauraco livingstonii*		F			x	x	x	x
Fischer's Turaco	*Tauraco fischeri*	NT	F	x	x	x			
	Strigidae (Typical Owls)								
African Barred Owlet	*Glaucidium capense*		F	x	x	x	x	x	x
Sokoke Scops-owl	*Otus ireneae*	EN	FF		x	x			
African Wood-owl	*Strix woodfordii*		F	x	x	x	x	x	x
Fraser's Eagle-owl	*Bubo poensis*		FF			x			

APPENDIX 10.1 *(Continued)*

Common Name	Family/Scientific name	TC	FD	SO	KE	TZ	MZ	ZW	MW
Pel's Fishing-owl	*Scotopelia peli*		F	x	x	x	x	x	x
	Accipitridae (Hawks, Eagles)								
Southern Banded Snake-eagle	*Circaetus fasciolatus*	NT	F	x	x	x	x	x	
Bat Hawk	*Macheiramphus alcinus*		F	x	x	x	x	x	x
Crowned Eagle	*Stephanoaetus coronatus*	NT	FF		x	x	x	x	x
Ayres' Hawk Eagle	*Hieraaetus ayresii*		F	x	x	x	x	x	x
African Goshawk	*Accipiter tachiro*		F	x	x	x	x	x	x
Black Sparrowhawk	*Accipiter melanoleucus*		F	*	x	x	x	x	x
	Trogonidae (Trogons)								
Narina Trogon	*Apaloderma narina*		F	x	x	x	x	x	x
	Bucerotidae (Hornbills)								
Silvery-cheeked Hornbill	*Bycanistes brevis*		F		x	x	x	x	x
Trumpeter Hornbill	*Bycanistes bucinator*		F		x	x	x	x	x
	Lybiidae (African barbets)								
Green Barbet	*Cryptolybia olivacea*		FF		x	x	x		x
White-eared Barbet	*Stactolaema leucotis*		F		x	x	x	x	x
Eastern Green Tinkerbird	*Pogoniulus simplex*		FF		x	x	x		x
Yellow-rumped Tinkerbird	*Pogoniulus bilineatus*		F		x	x	x	x	x
	Picidae (Woodpeckers)								
Golden-tailed Woodpecker	*Campethera abingoni*		F		x	x	x	x	x
Mombasa Woodpecker	*Campethera mombassica*		F	x	x	x			
	Psittacidae (Parrots)								
Brown-headed Parrot	*Poicephalus cryptoxanthus*		F		x	x	x	x	x
	Pittidae (Pittas)								
African Pitta	*Pitta angolensis*		FF		x	x	x	x	x

APPENDIX 10.1 *(Continued)*

Common Name	Family/Scientific name	TC	FD	SO	KE	TZ	MZ	ZW	MW
	Calyptomenidae (African and green broadbills)								
African Broadbill	*Smithornis capensis*		FF		x	x	x	x	x
	Oriolidae (Old World Orioles)								
Green-headed Oriole	*Oriolus chlorocephalus*		F		x	x	x		x
	Vangidae (Vangas and allies)								
Chestnut-fronted Helmetshrike	*Prionops scopifrons*		F	x	x	x	x	x	x
	Platysteiridae (Batises and Wattle-eyes)								
Forest Batis	*Batis mixta*		FF		x	x			
Reichenow's Batis	*Batis reichenowi*	NT	FF			x			
East Coast Batis	*Batis soror*		F		x	x	x	x	x
Black-throated Wattle-eye	*Platysteira peltata*		F	x	x	x	x	x	x
	Malaconotidae (Bush-shrikes)								
Black-backed Puffback	*Dryoscopus cubla*		F	x	x	x	x	x	x
Gorgeous Bush-shrike	*Telophorus viridis*		F	x	x	x	x	x	x
Black-fronted Bush-shrike	*Chlorophoneus nigrifrons*		FF		x	x	x	x	x
	Dicruridae (Drongos)								
Square-tailed Drongo	*Dicrurus ludwigii*		F	x	x	x	x	x	x
	Monarchidae (Monarch-flycatchers)								
Eastern Crested-flycatcher	*Trochocercus bivittatus*		FF	x	x	x	x	x	x
	Hyliotidae (Hyliotas)								
Usambara Hyliota	*Hyliota usambarae*	EN	FF			x			
	Nicatoridae (Nicators)								
Eastern Nicator	*Nicator gularis*		F	x	x	x	x	x	x
	Macrosphenidae (Crombecs and allies)								
Kretschmer's Longbill	*Macrosphenus kretschmeri*		FF		x	x	x		

APPENDIX 10.1 *(Continued)*

Common Name	Family/Scientific name	TC	FD	SO	KE	TZ	MZ	ZW	MW
	Cisticolidae (Cisticolas and allies)								
Green-capped Eremomela	*Eremomela scotops*		F		x	x	x	x	x
White-winged Apalis	*Apalis chariessa*	NT	FF		?	x	x		x
Black-headed Apalis	*Apalis melanocephala*		FF	x	x	x	x	x	x
	Pycnonotidae (Bulbuls)								
Yellow-bellied Greenbul	*Chlorocichla flaviventris*		F	x	x	x	x	x	x
Olive-headed Bulbul	*Arizelocichla striifacies*		FF		x	x	x		x
Stripe-cheeked Bulbul	*Arizelocichla milanjensis*	NT	FF				x	x	x
Little Greenbul	*Eurillas virens*		F		x	x	x		x
Lowland Tiny Greenbul	*Phyllastrephus debilis*		FF		x	x	x	x	
Terrestrial Brownbul	*Phyllastrephus terrestris*		F	x	x	x	x	x	x
Fischer's Greenbul	*Phyllastrephus fischeri*		FF	x	x	x	x		
Yellow-streaked Greenbul	*Phyllastrephus flavostriatus*		FF		x	x	x	x	x
	Scotocercidae (Bush-warblers)								
Little Yellow Flycatcher	*Erythrocercus holochlorus*		FF	x	x	x			
	Zosteropidae (White-eyes and Yuhinas)								
African Yellow White-eye	*Zosterops senegalensis*		F		x	x	x	x	x
	Pellorneidae (Ground babblers)								
Pale-breasted Illadopsis	*Illadopsis rufipennis*		FF		x	x			
	Sturnidae (Starlings)								
Black-bellied Starling	*Notopholia corusca*		F	x	x	x	x	x	
	Turdidae (Thrushes)								
Red-tailed Ant-thrush	*Neocossyphus rufus*		FF	x	x	x			
Spotted Ground-thrush	*Geokichla guttata*	VU	FF		*	x	x		x
Orange Ground-thrush	*Geokichla gurneyi*		FF		x	x	x	x	x

APPENDIX 10.1 *(Continued)*

Common Name	Family/Scientific name	TC	FD	SO	KE	TZ	MZ	ZW	MW
	Muscicapidae (Old World Flycatchers and Chats)								
Ashy Flycatcher	*Fraseria caerulescens*		F	x	x	x	x	x	x
White-starred Robin	*Pogonocichla stellata*		F		x	x	x	x	x
Swynnerton's Robin	*Swynnertonia swynnertoni*	VU	FF			x	x	x	
White-chested Alethe	*Chamaetylas fuelleborni*		FF			x	x		x
Red-capped Robin-chat	*Cossypha natalensis*		F	x	x	x	x	x	x
East Coast Akalat	*Sheppardia gunningi*	NT	FF		x	x	x		x
	Nectariniidae (Sunbirds)								
Plain-backed Sunbird	*Anthreptes reichenowi*	NT	FF		x	x	x	x	
Uluguru Violet-backed Sunbird	*Anthreptes neglectus*		F		x	x	x		
Banded Sunbird	*Anthreptes rubritorques*	VU	FF			x			
Collared Sunbird	*Hedydipna collaris*		F	x	x	x	x	x	x
Amani Sunbird	*Hedydipna pallidigaster*	EN	F		x	x			
Olive Sunbird	*Cyanomitra olivacea*		FF	x	x	x	x	x	x
	Ploceidae (Weavers)								
Clarke's Weaver	*Ploceus golandi*	EN	FF		x				
Dark-backed Weaver	*Ploceus bicolor*		F	x	x	x	x	x	x
	Estrildidae (Waxbills)								
Red-throated Twinspot	*Hypargos niveoguttatus*		F	x	x	x	x	x	x
Green Twinspot	*Mandingoa nitidula*		FF		x	x	x	x	x
	Motacillidae (Pipits and Wagtails)								
Sokoke Pipit	*Anthus sokokensis*	EN	FF		x	x			

TC = Threat category, DF = Forest dependence, SO = Somalia, KE = Kenya, TZ = Tanzania, MZ = Mozambique, ZW = Zimbabwe, MW = Malawi.

**Nonbreeding or passage (*), presence uncertain (?), breeding/resident (x). All species where TC is not indicated fall under Least Concern category*

CHAPTER 11

DISTRIBUTION AND ABUNDANCE OF MAMMALS ALONG THE BIODIVERSITY HOTSPOT OF THE COASTAL FORESTS OF EASTERN AFRICA

RICHARD D. LYAMUYA[1*], DEUSDEDITH B. FIDELIS[1], EMMANUEL RICHARD[1], ALLY J. K. NKWABI[1], and KWASLEMA MALLE HARIOHAY[2]

[1]*Tanzania Wildlife Research Institute (TAWIRI), Arusha, Tanzania*

[2]*College of African Wildlife Management, Mweka, Moshi, Tanzania*

*Corresponding author

ABSTRACT

This chapter gives an overview of the distribution and abundance of mammals along the Coastal Forests of Eastern Africa. We found that the coastal forests of Somalia, Kenya, and Tanzania supported at least 158 mammal species, including bats (58), rodents (>27), carnivores (19), primates (14), and shrews (14) belonging to 22 families and 11 orders, including the critically endangered Aders' duiker, African wild dog, African lion, Pousargues's monkey, lesser kudu, a distinctive form of giant sengi, and the vulnerable Sokoke bushy-tailed mongoose, which were widely distributed and relatively abundant. In addition, 31 species (20%) in these forests were forest specialists, and 109 species were generalists. Moreover, about 16 mammal species were found inhabiting the Zaraninge coastal forest region in Tanzania, and about 39 different mammal species were found inhabiting the northern coastal forests of Kenya in the Arabuko-Sokoke forest and the Boni-Dodori forest system. Despite of that currently, the majority of mammalian species inhabiting the Eastern African Coastal Forests are suffering mainly from their habitat loss as a result of forest fragmentation and degradation. Therefore, there is an urgent need to better conserve the habitats of these mammals to ensure their long-term survival and growth within the entire forest ecosystem.

Biodiversity Hotspot of the Coastal Forests of Eastern Africa. T. Pullaiah and Ally J. K. Nkwabi (Eds.)

DOI: 10.1201/9781998511099-11

11.1 INTRODUCTION

11.1.1 THE EASTERN AFRICA COASTAL FORESTS

The Eastern Africa Coastal Forests extend along the coastal plain of East Africa from southern Somalia to northern Mozambique (Burgess et al., 2008), are found within the so-called "Swahili regional center of endemism and Swahili-Maputaland regional transition zone in Eastern Africa" (Burgess et al., 1998). Interestingly, most of the Eastern African countries where these forests are found have set aside some area of land for their conservation. For example, according to Burgess et al. (1998), approximately 3167 km^2 of coastal forest remains 2 km^2 in Somalia, 660 km^2 in Kenya, 697 km^2 in Tanzania, 16 km^2 in Malawi, 3 km^2 in Zimbabwe, and perhaps 1790 km^2 in Mozambique, despite that most forests are small ($\leq$20 km^2), and all but 19 are under 30 km^2 in area. Moreover, over 80% of the coastal forest is located on government land, principally forest reserves; only 8.3 km^2 is found in national parks (6.2 km^2 in Kenya [Arabuko-Sokoke], 2 km^2 in Tanzania [Mafia Island] and tiny patches in Zimbabwe) (Burgess et al., 1998). Again, according to Burgess et al. (1998), coastal Eastern Africa is predominantly a plain sloping gently upward away from the sea, with variable width, interrupted in places by low hills and plateaus. While at the northern and southern limits of the coastal forest area (southern Somalia and northern Kenya, and in southern Mozambique) it extends inland over 200 km, but on the south Kenyan and northern Tanzanian coasts, and again along the northern Mozambique coast, it is less than 30 km wide (Burgess et al., 1998). This coastal plain comprises a complex sequence of mainly post-Jurassic (C135 million years old) sediments. These include both marine and freshwater sedimentary rocks (sandstones, limestones, marls, shales, mudstones, and unconsolidated sediments), which are usually of progressively younger ages the nearer they are to the present shoreline (Burgess et al., 1999; Cairncross, 2001).

11.1.2 WHY IS IT IMPORTANT TO CONSERVE THE EASTERN AFRICA COASTAL FORESTS?

The Eastern Africa Coastal Forest is currently conserved because of its large number of endemic species, high biodiversity, and concentration of rare and threatened taxa it harbors. According to Azeria et al. (2007), this uniqueness makes the coastal forests of East Africa one of the highest-priority ecosystems for conservation, both in Africa and globally. Additionally, for instance, the study by Amin et al. (2019) confirmed that the Boni-Dodori forest is of major importance to mammal conservation within the Eastern African coastal biodiversity hotspot because it contains relatively undisturbed natural habitats and also holds complete communities of predators and herbivores. Despite that these forests are currently threatened by human disturbance, through increasing fragmentation and forest degradation (Azeria et al., 2007).

11.1.3 STUDIES OF MAMMALS POPULATION IN THE EASTERN AFRICA COASTAL FORESTS

Mammalian inventories are an important source for developing conservation and management strategies (Augugliaro et al., 2019). For over a century, major efforts have been made by several institutions to develop census techniques for mammals. Different seasons have called for different inventory techniques and often for the same species (Villegas-Patraca et al., 2022). There are several advantages to carrying out mammal inventories, most of which are more rewarding to wildlife managers and conservationists (Stephenson and Stengel, 2020). This is because reliable census methods may help in understanding and devising best management practices, the ones that favor the sustainability of mammal species. However, there are important factors to consider while one is preparing to conduct mammalian inventories and select the type of census methods, which are the size of the species and its natural history (Bogoni et al., 2022a, b). In most of the mammalian inventory, several methods exist, and some have been applied and proved to be successful, including line transects, camera trapping, interviews, capture nets, and Sherman's traps (Kiwia, 2006; Andanje et al., 2015; Amin et al., 2019). Previous studies have revealed that line transects are useful in counting relatively large and conspicuous species and stand out as one of the best methods of estimating abundance (Buckland et al., 2004; Kagan, 2019), particularly in the Eastern Africa Coastal forests (Kiwia, 2006). For example, according to Kiwia (2006), the results from a line transect survey revealed that several mammalian species were sighted and recorded in the Zaraninge forest. The theory behind line transect surveys is that they can be either walked, driven, swum, or flown in a straight path while recording individuals seen and the perpendicular distance from the line transect. On the other hand, camera trapping has been a useful technique for surveying poorly documented mammalian species inhabiting the Eastern African coastal forests (Amir et al., 2019). This is because camera trapping is useful in inventorying large and medium-sized mammals, which are often elusive and therefore missed by other inventory methods (Tobler et al., 2008). According to Amin et al. (2019), for example, in Boni-Dodori forests of the coastal forests of northern Kenya, the use of the camera trap surveys revealed 37 mammal species and estimated occupancy for 31 medium-to-large terrestrial species, some of which represent range extensions. Additionally, the camera trap results showed that the critically endangered (CR) Aders' duiker was the most frequently recorded species, while the giant sengi and the vulnerable Sokoke bushy-tailed mongoose were also widely distributed and relatively abundant (Andanje et al., 2015; Amin et al., 2019). Again, other significant records of threatened species included the African wild dog, African lion, and Pousargues's monkey (Amin et al., 2019). Also, the study by Stokes et al. (2016) from the camera traps survey revealed that other species, such as the lesser kudu (*Tragelaphus imberbis*), were recorded only in the dry coastal scrub habitats of northern coastal forests, highlighting the additional importance of preserving the wider interconnected mosaic of habitats representing complete communities of herbivores and predators in the coastal habitats. The combined results overall indicate the high diversity of mammal species in the coastal region and further highlight this coastal region as a “biodiversity hotspot” (Stokes et al., 2016). Furthermore, according to Andanje et al. (2015), one Rufous sengi (National Museums of Kenya accession number NMK168236) was captured in a Sherman trap at Mararani and one four-toed sengi (NMK168261) was netted at Managai from Northern Coastal Forests of Kenya.

11.1.4 MAMMALS OF THE EASTERN AFRICA COASTAL FORESTS

A total of 960 mammalian species occur in sub-Saharan Africa (Kiwia, 2006). Previous studies have revealed the presence of several mammals inhabiting the Eastern Africa Coastal forests (Kiwia, 2006; Stokes et al., 2016; Amin et al., 2019), including African elephant *Loxodonta africana*, Lichtenstein’s hartebeest *Alcelaphus buselaphus lichtensteinii*, buffalo *Syncerus caffer*, sable antelope *Hippotragus niger*, roan antelope *H. equines*, giraffe *Giraffa camelopardalis*, lion *Panthera leo*, leopard *P. pardus*, African wild dog *Lycaon pictus*, common zebra, Iringa red colobus monkey and the Sanje crested mangabey monkey, antelopes, eland, and others (Figures 11.1 and 11.2). It was previously found that the majority of mammals occurring in the Eastern Africa Coastal Forests belong to a class of mammalia and in different families, namely Bovidae, Canidae, Cercopithecidae, Elephantidae, Equidae, Felidae, Galagidae, Giraffidae, Herpestidae, Hippopotamidae, Hyaenidae, Hystricidae, Leporidae, Macroscelididae, Manidae, Mustelidae, Orycteropodidae, Procaviidae, Rhinocerotidae, Sciuridae, Suidae, Viverridae, and orders; including the order Artiodactyla, Carnivora, Certatiodactyla, Hyracoidea, Lagomorpha, Perissodactyla, Pholidota, Primate, Proboscidea, Rodentia and Tubulidentata (Appendix 11.1).

Furthermore, it has been previously found that the coastal forests of Somalia, Kenya, and Tanzania supported several mammalian species, including bats (58), rodents (>27), carnivores (19), primates (14), and shrews (Kiwia, 2006). Additionally, Kiwia (2006) found that about 16 mammal species were inhabiting the Zaraninge coastal forest region, in Tanzania, including the Blue duiker (*Cephalophus monticola*), Harvey’s red duiker (*Cephalophus harveyi*), Suni (*Neotragus moschatus*), African buffalo, bushbuck, wild dog, blue monkey (*Cercopithecus mitis*), vervet monkey, Angolan colobus (*Colobus angolensis*), yellow baboon (*Papio hamadryas*), leopard, Rondo dwarf galago (*Galagoides rondoensis*), Zanzibar galago (*Galagoides zanzibaricus*), Aardvark (*Orycteropus afer*), common warthog (*Phacochoerus africanus*) and bush pig (*Potamochoerus porcus*).

Again, the study by Stokes et al. (2016) revealed a total of about 39 different mammal species inhabiting the northern coastal forests of Kenya in the Arabuko-Sokoke forest and the Boni-Dodori forest system (Appendix 11.2). According to Stokes et al. (2016), the northern coastal forests (Boni-Dodori region) recorded higher terrestrial mammal species richness than the Arabuko-Sokoke Forest, and the majority of medium-to-large mammal species were recorded much more frequently in the Boni-Dodori forest system. Moreover, the study by Amin et al. (2019) from Boni-Dodori forests in Kenya also recorded 37 mammalian species including the CR Aders’ duiker, African wild dog (Figure 11.2), African lion, Pousargues’s monkey, lesser kudu, distinctive form of giant sengi and the vulnerable Sokoke bushy-tailed mongoose which were widely distributed and relatively abundant.

FIGURE 11.1 Giraffes observed in Matambwe area, Nyerere National Park, Southern Tanzania (*Photo by Emmanuel Richard, 2023*).

FIGURE 11.2 African wild dogs observed in Ruaha National Park, southern Tanzania (*Photo by Emmanuel Richard, 2023*).

Also, the coastal forest of Somalia is the richest zone in terms of mammalian species, whereby the distribution map of its mammal species has been provided (Funaioli and Simonetta, 1966; Cordier et al., 2022). On the other hand, Funaioli and Simonetta (1966) produced a list of small mammals that did not provide their distribution information along the coastal forest of Somalia. However, Gedow et al. (2017) reported that 96 species of mammals occur in the coastal forest of Lag Badana Bushbushle landscape, including bats (22 species), rodents (14 species), carnivores (23 species), and primates (3 species), while herbivores had 23 species. Previous studies have also revealed that there are several common herbivorous mammals along the coastal forest areas of Somalia that have been reported and observed foraging (Funaioli and Simonetta, 1966; Okello et al., 2016; Gedow et al., 2017), including the lesser kudu *Tragelaphus imberbis*, buffalo *Syncerus caffer*, bushpig *Potamochoerus larvatus somalinenis* and desert warthog *Phacochoerus aethiopicus delamerei*, Harvey's duiker, Kirki's dik-dik *Madoqua kirkii*, Guenther's dik dik *Madoqua guentheri*, common duikers *Sylvicapra grammia* and *Cerphalopus harveyi*), Bushbuck *Tragelaphus scriptus*, common zebra, topi *Damaliscus lunatus*, and Beisa Oryx *Oryx beisa beisa* which are uncommon and can be observed in grassland and abandoned farmland in the coastal areas of Somalia.

Furthermore, the coastal forest of Somalia is renowned for having common species of carnivores, including side-striped jackal *Canis adustus*, black-backed jackal *Canis mesomelas*, bat-eared fox *Octocyon megalotis*, spotted hyaena *Crocuta crocuta*, lions *Panthera leo*, leopards *Panthera pardus*, cheetahs *Acinonyx jubatus*, African wild dog *Lycaon pictus*, and caracals *Caracal caracal*. However, the mammalian fauna of the coastal forest of Somalia faces several threats from human activities such as poaching. For example, the resident populations of the elephant, giraffe, and hunter's hartebeest have almost gone extinct because of excessive hunting. In the absence of these bigger mammal species, poaching has now focused on smaller-sized species like dik dik. Therefore, the presence of these mammal species in this region, highlight this coastal region as a 'biodiversity hotspot'.

Mammals along the Eastern Africa Coastal Forests are distributed depending on vegetation types and land use by choosing among available habitats (Bjørneraas et al., 2012). Moreover, they are distributed inside and outside protected areas, and those in outside protected areas face threats due to anthropogenic activities resulting in a decline in their population size (Fjeldså et al., 2004; Olivero et al., 2017). Some of these declines are attributed to poaching within and outside protected areas as a result mammals are found in fewer parts of the coastal forest areas outside protected areas due to deforestation driven by small-scale farming livestock grazing and bush meat hunting (Caro and Davenport, 2016).

11.1.5 MAMMALS ENDEMISM ALONG THE EASTERN AFRICA COASTAL FORESTS

The Eastern African Coastal Forests harbor several mammalian species that are endemic to the region. However, previous studies have revealed that the pattern of endemism in the Coastal Forest Mosaic is complex, reflecting the wide range of habitats and heterogeneous forest types, a high degree of turnover of local species between adjacent forest patches, and many disjunct distributions (Burgess, 2000; WWF-US, 2003). The study by Kiwia (2006) reported that only one large mammal species, Ader's duiker, is endemic to the Eastern African coastal forests, occurring only on Zanzibar Island and Arabuko-Sokoke forests in Kenya, along with the vulnerable Pemba flying fox (*Pteropus voeltzkowi*). Again, the study by Burgess and Muir (1994) reported that there are five mammalian species that are endemic to the Eastern Africa Coastal forests, including the Ader's duiker, golden-rumped elephant shrews, African woolly bats, Pemba fruit bats, and the bat *Rhinopholus deckenii*.

11.1.6 THREATS TO MAMMALS OF EASTERN AFRICA COASTAL FORESTS

There are at least the top 10 overall threats (in ranked order) to the mammalian species inhabiting the Eastern African coastal forests namely agriculture and encroachment, fire, timber extraction, pole wood cutting, population growth, charcoal production, grazing, hunting, mining, and roads (CIICIPE, 2003). The other reported threats include carving wood, salt, and tourism (Figure 11.3). However, population growth could be better considered as an ultimate factor, driving the other proximate threats (CIICIPE, 2003). Currently, the majority of mammalian species inhabiting the Eastern African coastal forests are suffering from habitat loss as a result of forest fragmentation and degradation due to increased human disturbance (Azeria et al., 2007). Moreover, agricultural encroachment due to the increased human population in the coastal area is another conservation threat to the mammalian coastal forest species (Burgess and Muir, 1994). In Tanzania, for example, the coastal forests were more extensive in the past, but due to uncontrolled anthropogenic activities, only small patches remain that serve as refuges for the forest specialists and endemic and threatened species (Kiwia, 2006). Moreover, according to Kiwia (2006), some of those forests, being too small, overexploited, and isolated, are extremely vulnerable and are unlikely to be able to conserve viable populations of primates and other large mammals. Additionally, the study by Amin et al. (2019) reported that the Boni-Dodori forest complex required urgent protection and conservation because of their current land grabs, land conversion, and felling of indigenous hardwoods associated with, and driven by, the planned development of a major seaport at Lamu and cross-country pipeline development.

Additionally, the primary threats to the mammalian fauna in the coastal forest of Somalia include illegal harvesting (poaching), charcoal production, agriculture, human-wildlife conflict, and predator poisoning due to livestock attacks, among others. As a result of these, 13 species of mammals occurring in the coastal forest of Somalia are listed in the International Union for Conservation of Nature (IUCN) Red List. Two of them are CR (black rhinoceros and hirola or hunter's hartebeest), two are endangered (African elephant *Loxodonta africana* and wild dog *Lycaon pictus*), seven are vulnerable and one near threatened like plain zebra (Appendix 11.1). The illegal hunting of wildlife species along the coastal areas of Somalia has resulted in enormous pressure on wildlife, especially during the period of insecurity and proliferation of automatic weapons. The population of wildlife (herbivores and carnivores) along the coastal forest has practically been decimated. All that is left are a few dik diks, lesser kudu, and buffalo (Gedow et al., 2017).

FIGURE 11.3 Mangrove coastal forest cleared for salt pan preparation in Lindi region, southern Tanzania (*Photo by Emmamuel Richard, 2023*).

In the past decade, for example, nearly 10,000 rhinos were killed by poachers in Africa (Ripple et al., 2015). The remaining rhino populations in coastal forests of Africa areas are steadily declining, with fewer than 30,000 animals left in 2020 from a population of 500,000 at the beginning of the 20th century (Dang Vu et al., 2022). Today, rhino horn poaching is the main threat facing rhinos, as demand has escalated in Asian countries, especially in Vietnam (Cheung et al., 2018; Olmedo et al., 2018; Dang Vu and Nielsen, 2021). The black rhino along the coastal forest of East Africa is almost extinct and even those believed to be living in Somalia, Mozambique, and Kenya coastal as well as Tanzania (Selous and Nyerere National Park) are in imminent danger of extermination due to poaching pressure. It has been believed that rhinos in the Selous ecosystem are almost extinct, the species is listed as CR by the IUCN as of 2016. Therefore, the conservation of these mammals, especially the forest specialist species, deals essentially with the conservation of their habitat, in the entire forest ecosystems (Kiwia, 2006).

11.2 CONCLUSION AND RECOMMENDATION

11.2.1 CONCLUSION

The biological importance of the Eastern Coastal Forests in harboring diverse mammal species is well recognized throughout this chapter. The chapter indicated that there have been a considerable number of mammal species and conservation efforts in this hotspot for several decades. However, there are significant gaps in our biological knowledge of this important region. There are also important opportunities for the further application of conservation science, particularly concerning mammal inventories, given the large pace at which forest fragmentation is occurring in the region. The major threats to the Eastern Coastal Forest mammals are mainly due to the large growing human population contributed mostly by immigration from mainland regions and the growing needs of impoverished local people. The prospects for the hotspots will be greatly improved if programs addressing alternative sources of income, energy, and employment are established to reduce the current over-dependence of the community living in this region on ecosystem services.

11.2.2 RECOMMENDATIONS

From this chapter, we recommend that conservation and management plans should be prioritized for the preservation of coastal forest habitats as they harbor a significant number of wild mammal species. Again, we also recommend the preservation of a diversity of habitat types, such as mangroves, and woodland, since different habitat types affect the composition, abundance, and diversity of species. Moreover, since roads and trails may often lead to development/urbanization and eventually larger disturbance intensity to mammals, therefore infrastructure development within the coastal forest habitat should take into consideration the resulting adverse impact on mammal species. In addition, more research is needed to better determine the current status of the mammalian faunal of coastal forests in Eastern Africa. Furthermore, problems associated with climate change and their potential impacts on biodiversity should be addressed in partnership with the countries having coastal forests in Eastern Africa by the adoption of a variety of mitigation measures. The measures should include controlling anthropogenic activities such as deforestation, adoption of proper land management practices (including agroforestry), changing energy technologies (e.g., the use of efficient wood stoves and biogas), and, in case of agriculture breeding crops that will be adaptable to changes of climate, that is, drought resistance crops, short-term harvested crops, etc. Additionally, the government should be involved in adopting integrated land and water management practices and augmenting collaborations between the conservation and sustainable use of biodiversity in climate change scenarios.

KEYWORDS

- **biodiversity hotspot**
- **charismatic mammals**
- **coastal forest**
- **conservation status**
- **endemism**

REFERENCES

Amin, R.; Wacher, T.; Bowkett, A. E.; Ogwoka, B.; Morris, M.; Agwanda, B. R. Africa's forgotten forests: The conservation value of Kenya's northern coastal forests for large mammals. *J. East Afr. Nat. Hist.* **2019,** *107*(2), 41–61.

Andanje, S.; Davey, K.; Ogwoka, B.; Agwanda, B.; Ali, A.; Bruce, T.;…; Amin, R. *Mammal Diversity Surveys in the Coastal Forests: Kenya.* The Zoological Society of London: Regents Park, London, **2015**.

Augugliaro, C.; Paniccia, C.; Janchivlamdan, C.; Monti, I. E.; Boldbaatar, T.; Munkhtsog, B. J. C. L. Mammal inventory in the Mongolian Gobi, with the southeasternmost documented record of the Snow Leopard, *Panthera uncia* (Schreber, 1775), in the country. *Checklist* **2019,** *15*(4), 565–578.

Azeria, E. T.; Sanmartín, I.; Ås, S.; Carlson, A.; Burgess, N. Biogeographic patterns of the East African coastal forest vertebrate fauna. *Biodiv. Conserv.* **2007,** *16*(4), 883–912. https://doi.org/10.1007/s10531-006-9022-0

Bjørneraas, K.; Herfindal, I.; Solberg, E. J.; Sæther, B.-E.; van Moorter, B.; Rolandsen, C. M. Habitat quality influences population distribution, individual space use and functional responses in habitat selection by a large herbivore. *Oecologia* **2012,** *168*, 231–243. https://doi.org/10.1007/s00442-011-2072-3

Bogoni, J. A.; Carvalho-Rocha, V.; Silva, P. G. D. Spatial and land-use determinants of bat species richness, functional diversity, and site uniqueness throughout the largest Tropical country, Brazil. *Mamm. Rev.* **2022a,** *52*(2), 267–283. https://doi.org/10.1111/mam.12279

Bogoni, J. A.; Peres, C. A.; Ferraz, K. M. P. M. B. Medium- to large-bodied mammal surveys across the Neotropics are heavily biased against the most faunally intact assemblages. *Mamm. Rev.* **2022b,** *52*(2), 221–235. https://doi.org/10.1111/mam.12274

Buckland, S. T.; Anderson, D. R.; Burnham, K. P.; Laake, J. L.; Borchers, D. L.; Thomas, L. *Advanced Distance Sampling; Estimating Abundance of Biological Populations*. Oxford University Press: Oxford, **2004**.

Burgess, N. D.; Clarke, G. P.; Rodgers, W. A. Coastal forests of eastern Africa: Status, endemism patterns and their potential causes. *Biol. J. Linn. Soc.* **1998,** *64*(3), 337–367.

Burgess, N. D.; Clarke, G. P.; Rodgers, W. A. Coastal forests of eastern Africa: Status, endemism patterns and their potential causes. *Biol. J. Linn. Soc.* **2008,** *64*(3), 337–367. https://doi.org/10.1111/j.1095-8312.1998.tb00337.x

Burgess, N. D.; Muir, C. Coastal forests of Eastern Africa: Biodiversity and conservation. In: *Society for Environmental Exploration*. Royal Society for the Protection of Birds: UK, **1994,** p. 112.

Burgess, N. D.; Ponder, K. L.; Goddard, J. Surface and leaf-litter arthropods in the coastal forests of Tanzania. *Afr. J. Ecol.* **1999,** *37*(3), 355–365.

Burgess, N. Global importance and patterns in the distribution of coastal forest species. *Coastal Forests of Eastern Africa* **2000,** pp. 235–248.

Cairncross, B. An overview of the Permian (Karoo) coal deposits of southern Africa. *J. Afr. Earth Sci.* **2001,** *33*(3–4), 529–562.

Caro, T.; Davenport, T. R. Wildlife and wildlife management in Tanzania. *Conserv. Biol.* **2016,** *30*(4), 716–723. https://doi.org/10.1111/cobi.1265

Cheung, H.; Mazerolle, L.; Possingham, H. P.; Biggs, D. Medicinal use and legalized trade of rhinoceros horn from the perspective of traditional Chinese medicine practitioners in Hong Kong. *Trop. Conserv. Sci.* **2018,** *11*, 1–8. https://doi.org/10.1177/1940082918787428

CIICIPE. *Eastern Arc Mountains & coastal forests of Tanzania & Kenya*. **2003**.

Cordier, C. P.; Smith, D. A. E.; Smith, Y. E.; Downs, C. T. Camera trap research in Africa: A systematic review to show trends in wildlife monitoring and its value as a research tool. *Global Ecol. Conserv.* **2022,** *40*, e02326.

Dang Vu, H. N.; Nielsen, M. R. Evidence or delusion: A critique of contemporary rhino horn demand reduction strategies. *Hum. Dimens. Wildl.* **2021,** *26*(4), 390–400. https://doi.org/10.1080/10871209.2020.1818896

Dang Vu, H. N.; Nielsen, M. R.; Jacobsen, J. B. Conserving rhinos by legal trade: Insights from a choice experiment with rhino horn consumers. *Ecol. Econ.* **2022,** *193*, 1–9. https://doi.org/10.1016/j.ecolecon.2021.10728

Fjeldså, J.; Burgess, N. D.; Blyth, S.; De Klerk, H. M. Where are the major gaps in the reserve network for Africa's mammals? *Oryx* **2004,** *38*(1), 17–25.

Funaioli, U.; Simonetta, A. M. The mammalian fauna of the Somàli republic: Status ànd conservàtion problems: Ricerche sulla Fauna della Somalia promosse dall'Istituto di Zoologia e dal Museo Zoologico dell'Università di Firenze: XXVII. *Monitore Zoologico Italiano. Supplemento* **1966,** *1*(1), 285–347.

Gedow, A.; De Leeuw, J.; Koech, G. *Assessment of the biodiversity in terrestrial and marine landscapes of the proposed Lag Badana National Park and surrounding areas,* Jubaland, Somalia. Working paper. **2017,** https://doi.org/10.13140/RG.2.2.34736.66560

Kagan, S. *How to Count Animals, More Or Less*. Oxford University Press, **2019**.

Kiwia, H. Species richness and abundance of large mammals in Zaraninge forest, Coast region, Tanzania. *Tanzania J. Sci.* **2006,** *32*(2), 105–116.

Okello, M. M.; Kiringe, J. W.; Muruthi, P.; Kenana, L.; Maliti, H.; Sitati, N. W.; Ndambuki, S. Post drought population status and trend of specialized browsers in the mid Kenya-Tanzania Borderland. *Nat. Resour.* **2016,** *7*(10), 580–610. http://dx.doi.org/10.4236/nr.2016.710048

Olivero, J.; Fa, J. E.; Real, R.; Farfán, M. Á.; Márquez, A. L.; Vargas, J. M.;…; Nasi, R. Mammalian biogeography and the Ebola virus in Africa. *Mamm. Rev.* **2017,** *47*(1), 24–37.

Olmedo, A.; Sharif, V.; Milner-Gulland, E. Evaluating the design of behavior change interventions: A case study of rhino horn in Vietnam. *Conserv. Lett.* **2018,** *11*(1), e12365. https://doi.org/10.1111/conl.12365

Ripple, W. J.; Newsome, T. M.; Wolf, C.; Dirzo, R.; Everatt, K. T.; Galetti, M.;…; Lindsey, P. A. Collapse of the world's largest herbivores. *Sci. Adv.* **2015,** *1*(4), e1400103. https://doi.org/10.1126/sciadv.1400103

Stephenson, P. J.; Stengel, C. An inventory of biodiversity data sources for conservation monitoring. *PLoS One* **2020,** *15*(12), e0242923. https://doi.org/10.1371/journal.pone.0242923

Stokes, H.; Ogwoka, B.; Bett, J.; Wacher, T.; Amin, R. Mammal diversity survey in the northern coastal forests of Kenya: Arabuko-sokoke forest and the boni-dodori forest system. *Unpublished report to the Zoological Society of London,* London. https://www. zsl. org/sites/default/files/media/2017-04/Boni-Dodori% 20Bird% 20Survey% 20Report **2016,** *20*(282015), 29.

Tobler, M. W.; Carrillo-Percastegui, S. E.; Leite Pitman, R.; Mares, R.; Powell, G. An evaluation of camera traps for inventorying large- and medium-sized terrestrial rainforest mammals. *Anim. Conserv.* **2008,** *11*(3), 169–178. https://doi.org/10.1111/j.1469-1795.2008.00169.x

Villegas-Patraca, R.; Aguilar-López, J. L.; Hernández-Hernández, J. C.; Muñoz-Jiménez, O. Diversity and conservation of terrestrial vertebrates (birds, mammals, and reptiles) of Sierra Cucapá, Mexicali, Baja California, Mexico. *ZooKeys* **2022,** *1088*, 17. https://doi.org/10.3897/zookeys.1088.76134

WWF-US. *Ecoregional Reports: Northern Zanzibar-Inhambane Coastal Forest Mosaic*. **2003**.

APPENDIX 11.1 Mammals' Species List Recorded Along the Eastern Africa Coastal Forest

Sr. No.	Common Species Name	Scientific Name	Order	Family	IUCN Status
1	Aardvark	*Orycteropus afer*	Tubulidentata	Orycteropodidae	LC
2	Impala	*Aepyceros melampus*	Artiodactyla	Bovidae	LC
3	Lichtenstein's hartebeest	*Alcelaphus buselaphus*	Artiodactyla	Bovidae	LC
4	Black-backed jackal	*Canis mesomelas*	Carnivora	Canidae	LC
5	Blue duiker	*Cephalophus monticola*	Artiodactyla	Bovidae	LC
6	Abbott's duiker	*Cephalophus spadix*	Artiodactyla	Bovidae	VU
7	Red duiker	*Cepholophus natalensis*	Artiodactyla	Bovidae	LC
8	Sanje mangabey monkey	Cercocebus sanjei	Primates	Cercopithecidae	LC
9	Vervet monkey	*Cercopithecus aethiops*	Primates	Cercopithecidae	LC
10	Blue monkey	*Cercopithecus mitis*	Primates	Cercopithecidae	LC
11	Civet (African civet)	*Civettictis civetta*	Carnivora	Viverridae	LC
12	Black and white colobous monkey	*Colobus guereza*	Primates	Cercopithecidae	LC
13	Wildbeest (common)	*Connochaetes taurinus*	Artiodactyla	Bovidae	LC
14	Nyasa wildbeest	*Connochaetes taurinus johnstoni*	Artiodactyla	Bovidae	LC
15	Tree hyrax	*Dendrohyrax arboreus*	Hyracoidea	Procaviidae	LC
16	Black rhinoceros*	*Diceros bicornis*	Perissodactyla	Rhinocerotidae	CR
17	Burchell's zebra	*Equus burchellii*	Perissodactyla	Equidae	NT
18	Serval cat	Felis *serval*	Carnivora	Felidae	LC
19	Wild cat	*Felis silvestris*	Carnivora	Felidae	LC
20	Rungwe dwarf galago	*Galagoides* sp.	primates	Galagidae	
21	Small spotted genet	*Genetta genetta*	Carnivora	Viverridae	LC
22	Large-spotted genet	*Genetta tigrina*	Carnivora	Viverridae	LC
23	Giraffe	*Giraffa camelopardalis*	Artiodactyla	Giraffidae	VU
24	Bush hyrax	*Heterohyrax brucei*	Hyracoidea	Procaviidae	LC
25	Hippopotamus	*Hippopotamus amphibious*	Artiodactyla	Hippopotamidae	VU
26	Roan antelope	*Hippotragus equinus*	Artiodactyla	Bovidae	LC
27	Sable antelope	*Hippotragus niger*	Artiodactyla	Bovidae	LC

APPENDIX 11.1 *(Continued)*

Sr. No.	Common Species Name	Scientific Name	Order	Family	IUCN Status
28	East African porcupine	*Hystrix africaesustrails*	Rodentia	Hystricidae	LC
29	Common waterbuck	*Kobus ellipsiprymnus*	Artiodactyla	Bovidae	LC
30	Waterbuck	*Kobus ellipsiprymnus*	Artiodactyla	Bovidae	LC
31	Puku	*Kobus vardoni*	Artiodactyla	Bovidae	NT
32	Hare	*Lepuss saxatilis*	Lagomorpha	Leporidae	LC
33	African elephant	*Loxodonta africana*	Proboscidea	Elephantidae	EN
34	Wild dog	*Lycaon pictus*	Carnivora	Canidae	CR
35	Kirk's dikdik	*Madoqua kirkii*	Artiodactyla	Bovidae	LC
36	Pangolin	*Manis temmincki*	Pholidota	Manidae	CR
37	Ratel (honey-badger)	*Mellivora capensis*	Carnivora	Mustelidae	LC
38	Suni	*Neotragus moschatus*	Artiodactyla	Bovidae	LC
39	Bat-eared fox	*Octocyon megalotis*	Carnivora	Canidae	LC
40	Klipspringer	*Oreotragus oreotragus*	Artiodactyla	Bovidae	LC
41	Oribi	*Ourebia ourebia*	Artiodactyla	Bovidae	LC
42	Lion	*Panthera leo*	Carnivora	Felidae	VU
43	Leopard	*Panthera pardus*	Carnivora	Felidae	VU
44	Yellow baboon	*Papio cynocephalus*	Primates	Cercopithecidae	LC
45	Common warthog	*Phacochoerus africanus*	Perissodactyla	Suidae	LC
46	Spotted hyaena	*Crocuta crocuta*	Carnivora	Hyaenidae	LC
47	Bush pig	*Potamochoerus larvatus*	Artiodactyla	Suidae	LC
48	Sharpe's grysbok	*Raphicerus sharpie*	Artiodactyla	Suidae	LC
49	Southern reedbuck	*Redunca arundinum*	Artiodactyla	Suidae	LC
50	Mountain reedbuck	*Redunca fulvorufula*	Artiodactyla	Suidae	LC
51	Bohor reedback	*Redunca redunca*	Artiodactyla	Bovidae	LC
52	Kipunji monkey	*Rungwecebus kipunji*	Primates	Cercopithecidae	CR
53	Common duiker	*Sylvicapra grimmia*	Artiodactyla	Bovidae	LC
54	African buffalo	*Syncerus caffer*	Artiodactyla	Bovidae	NT

APPENDIX 11.1 *(Continued)*

Sr. No.	Common Species Name	Scientific Name	Order	Family	IUCN Status
55	Eland	*Taurotragus oryx*	Artiodactyla	Bovidae	LC
56	Lesser kudu	Tragelaphus imberbis	Artiodactyla	Bovidae	NT
57	Bushbuck	*Tragelaphus scriptus*	Artiodactyla	Bovidae	LC
58	Greater kudu	*Tragelaphus strepsiceros*	Artiodactyla	Bovidae	LC
59	Giant elephant shrew	*Rhynchocyon cimei hendersoni*	Macroscelidea	Macroscelididae	LC
60	Tanganyika mountain squirrel	*Paraxerus lucifer lucifer*	Rodentia	Sciuridae	LC
61	Zanzibar galago	*Galagoides zanzibaricus*	Primates	Galagidae	LC
62	Rondo galago	*Galagoides rondoensis*	Primates	Galagidae	EN
63	Common warthog	*Phacochoerus africanus*	Artiodactyla	Suidae	LC
64	Red bush squirrel	*Paraxerus palliatus*	Rodentia	Sciuridae	LC
65	Desert warthog	*Phacochoerus aethiopicus*	Artiodactyla	Suidae	LC
66	Carcal	*Caracal caracal*	Carnivora	Felidae	LC
67	Aders' duiker	Cephalophus adersi	Artiodactyla	Bovidae	CR
68	Sykes's monkey	*Cercopithecus mitis*	primates	Cercopithecidae	LC
69	Harvey's duiker	*Cephalophus harveyi*	Artiodactyla	Bovidae	LC
70	Central African large-spotted genet	*Genetta maculata*	Carnivora	Viverridae	LC
71	Common dwarf mongoose	*Helogale parvula*	Carnivora	Herpestidae	LC
72	Marsh mongoose	*Atilax paludinosus*	Carnivora	Herpestidae	LC
73	Honey badger	*Mellivora capensis*	Carnivora	Mustelidae	LC
74	Slender mongoose	*Herpestes sanguineus*	Carnivora	Herpestidae	LC
75	White-tailed mongoose	*Ichneumia albicauda*	Carnivora	Herpestidae	LC
76	Golden-rumped sengi or elephant-shrews	*Rhynchocyon chrysopygus*	Macroscelidea	Macroscelididae	EN
77	Boni giant sengi or or elephant-shrews	*Rhynchocyon* sp.	Macroscelidea	Macroscelididae	LC

CR = critically endangered; EN = endangered; LC = least concern; NT = near threatened; VU = vulnerable.

**Indicates forest specialist species.*

APPENDIX 11.2 Showing Mammals' Species List Recorded in the Northern Coastal Forests of Kenya in Arabuko-Sokoke Forest and the Boni-Dodori Forest System

Species Category	Species Name	Scientific Name	Trophic Level
Ungulates	Aders' duiker	*Cephalophus adersi*	Herbiviores
	Harvey's duiker	*Cephalophus harveyi*	
	Blue duiker	*Philantomba monticola*	
	Common duiker	*Sylvicapra grimmia*	
	Suni	*Nesotragus moschatus*	
	Kirk's dik-dik	*Madoqua kirkii*	
	Bushbuck	*Tragelaphus scriptus*	
	Lesser kudu	*Tragelaphus imberbis*	
	Waterbuck	*Kobus ellipsiprymnus*	
	African buffalo	*Syncerus caffer*	
	Common hippopotamus	*Hippopotamus amphibius*	
	Bushpig	*Potamochoerus larvatus*	
	Common warthog	*Phacochoerus africanus*	
	Desert warthog	*Phacochoerus aethiopicus*	
Carnivores	African wild dog	*Lycaon pictus*	Carnivores
	Caracal	*Caracal caracal*	
	African lion	*Panthera leo*	
	Leopard	*Panthera pardus*	
	African civet	*Civettictis civetta*	
	Central African large-spotted genet	*Genetta maculata*	
	Spotted hyaena	*Crocuta crocuta*	
	Honey badger	*Mellivora capensis*	
	Common dwarf mongoose	*Helogale parvula*	
	Marsh mongoose	*Atilax paludinosus*	
	Slender mongoose	*Herpestes sanguineus*	
	Sokoke bushy-tailed mongoose	*Bdeogale omnivora*	
	White-tailed mongoose	*Ichneumia albicauda*	
Afrotheria	African elephant	*Loxodonta africana*	Herbivores
	Aardvark Four-toed sengi	*Orycteropus afer*	
	Four toed sengi	*Petrodomus tetradactylus*	
	Giant sengi	*Rhynchocyon* sp.	Insectivores
Primates	Sykes's monkey	*Cercopithecus mitis*	Omnivores
	Vervet monkey	*Chlorocebus pygerythrus*	
	Yellow baboon	*Papio cynocephalus*	
	Unidentified Galago spp.		

APPENDIX 11.2 *(Continued)*

Species Category	Species Name	Scientific Name	Trophic Level
Rodents	Porcupine	*Hystrix* sp.	Omnivores
	Northern giant pouched rat Red bush squirrel	*Cricetomys gambianus*	
	Lagomorphs	*Paraxerus palliatus*	
Lagomorph	Hare	*Lepus* sp.	Herbivores

Source: Andanje et al., 2015.

CHAPTER 12

SMALL MAMMAL DISTRIBUTION AND DIVERSITY ALONG THE COASTAL FORESTS OF EASTERN AFRICA

NSAJIGWA EMMANUEL MBIJE[1*], LEVINA Z. MCHARO[2], DAVID J. KAVANA[3], TECLAMAGENI S. MAYEJI[4], and MESHACK HAULE[4]

[1]*Sokoine Umiversity of Agriculture, College of Forestry, Wildlife and Tourism, Department of Wildlife.Management, Morogoro Tanzania*

[2]*Protect Nature, Save Farmlands Company Limited, Arusha, Tanzania*

[3]*Department of Natural Resources Management and Conservation, Sokoine University of Agriculture, Mizengo Pinda Campus College, Mpanda, Katavi, Tanzania.*

[4]*Sokoine University of Agriculture, Institute of Pest Management, Morogoro, Tanzania*

*Corresponding author

ABSTRACT

The East African coast is a renowned hub with a high diversity of small mammals, an important group of animals that contribute to the region's biodiversity. Among these, bats, rodents, carnivores, and shrews form a very diverse group. The occurrence and diversity of these mammals are certainly influenced by various factors, including vegetation type, altitude, climatic conditions, disease prevalence, seasonal food and water availability, human disturbance, the presence of predators, and life history. Interestingly, the mosaic of coastal forests in East Africa is home to more than 90 species of small mammals that are found in different habitat types. Among these microhabitats, the lowland forest habitat stands out as exceptionally rich in biodiversity with approximately 52 species, including small mammals that are endemic to this area. Moreover, the adjacent nonforested areas within the Eastern African coast support about 45 species that are exclusive to this unique biome. The occurrence of humid forests versus dry woodland habitats, geological history, ecological barriers that limit connectivity, and past ecological perturbation such as volcanism may explain this. Based on this heterogeneity, an understanding of the characteristics of small mammal species within coastal forest ecosystems of East Africa is a key start for conservation and management. Intensive field surveys and associated morphological and

Biodiversity Hotspot of the Coastal Forests of Eastern Africa. T. Pullaiah and Ally J. K. Nkwabi (Eds.)

DOI: 10.1201/9781998511099-12

molecular genetics analyses are strongly advised to document the small mammals of the coastal forests of East Africa for conservation purposes.

12.1 INTRODUCTION

Small mammals in the coastal forest of Eastern Africa represent a diverse and ecologically important group of animals that contribute to the region's biodiversity (Burgess et al., 1998). They comprise different animal groups, including flying mammals such as Chiroptera, represented by bats (Barnett, 1992; Burgess et al., 1998). The Order Chiroptera comprises one of the most diverse and widespread groups of small mammals in the East African coastal forests, with numerous species belonging to the family Pteropodidae (fruit bats) and Vespertilionidae (insect-eating bats) known to inhabit various ecosystems such as forests, caves, and urban areas (Kingdon, 2015). The small mammal group also includes other nonflying mammals of the Order Soricomorpha represented by shrews, Erinaceomorpha represented by hedgehogs, and the Order Macroscelidea which is represented by elephant shrews or sengis (Barnett, 1992; Kingdon, 2015), the Order Soricomorpha includes families such as Soricidae, represented by species like the African giant shrew of the genus *Crocidura* known to inhabit a range of habitats including lowland coastal forests (Burgess et al., 1998; Kiwia, 2009; Sabuni et al., 2015). The Order Macroscelidea comprises elephant shrews (with several genera including *Elephantulus, Rhynchocyon,* and *Petrodromus*), the group members are distributed in the coastal forests of Tanzania, Kenya, and Mozambique. Kingdon (2015) reported that the chequered elephant shrew (*Rhynchocyon cirnei*) observed to occur in the southern coastal part of Tanzania in the Lindi and Mtwara region as well as on the coast of Mozambique. Also, the Zanj elephant shrew (*R. petersi*) inhabits the coastal forest of Rufiji River, Mafia, and Zanzibar in Tanzania. In Kenya, coastal forest, the species accommodates the forest of Mombasa. The golden-rumped elephant shrew (*R. chrysopygus*) is vulnerable (VU) and observed to be confined to very small areas of the Kenya coastal thickets, while the species is facing conservation challenges leading to serious concern for its survival (Kingdon, 2015, Amin et al., 2018).

Other nonflying small mammals include mongoose (Carnivora), hedgehogs (Erinaceomorpha), hyrax (Hyracoidea), and hares (Lagomorpha) (Kingdon, 2015). The Order Rodentia is another group of nonflying small mammals, is the most diverse and widely distributed throughout East Africa, including coastal forests, and inhabits a variety of habitats, including grasslands, savannas, and agricultural areas (Kiwia, 2009; Kingdon, 2015; Sabuni et al., 2015). Most of these animals are well adapted to nocturnal activity with few rodents demonstrating diurnal activities (Happold, 2013). For example, grass rats (*Arvicanthis* spp.) were trapped by Nkwabi et al. (2018) foraging in the daytime in the five Wildlife Management Areas of southern Tanzania bordering Nyerere National Park and Niassa corridor. Due to their cryptic elusive nature, the group has often been disregarded when conservation strategies are initiated (Avenant and Cavallini, 2007).

Notably, the mosaic of coastal forests in East Africa is home to more than 90 species of small mammals that are found in different microhabitats of this region (Burgess et al., 1998; Habel et al., 2017; Koch and Kaplan, 2022). Among these microhabitats, the lowland forest habitat stands out as exceptionally rich in biodiversity with approximately 52 vertebrate mammal species, including small mammals that are endemic to this area (Burgess et al., 2017). Moreover, the adjacent nonforested areas along the eastern African coast also play a vital role, as they support a minimum of 45 vertebrate mammal species that are exclusive to this unique biome (Amin et al., 2018). Many of the region's endemic species inhabit small and limited geographical areas, often restricted to a single location (Kiwia, 2009; Sabuni et al., 2015). This, coupled with the ongoing reduction of coastal forest habitat, poses a severe threat to many of these species, increasing their risk of extinction (Habel et al., 2017). The ongoing research in coastal forests continues to unveil new species of small mammals, for instance, in the last decade, scientists have documented the discovery of four mammal species (Burgess et al., 2017), this discovery is a milestone achievement in conservation sciences and it figures the position of the Eastern African coastal ecoregion in biodiversity hotspot contribution. Information on the species presence and diversity of small mammals in the coastal forests of Eastern Africa has been reported since the 20th century (Rodgers et al., 1982; Burgess et al., 1992; Burgess and Muir, 1994; Andrews, 1998; Cockle et al., 1998). At present, there are several other studies exist in the area (Bloesch and Frank-Klötzli, 2004; Kiwia, 2009; Azeria et al., 2007; Verheyen et al., 2007; Kock and Stanley, 2009; Cheke, 2011; Denys et al., 2011; Sabuni et al., 2015; Russell et al., 2016; Amin et al., 2018). These studies have enriched our knowledge of understanding the status, composition, and species diversity of small mammals on a large scale, and some of these have yielded the discovery of new species, for example, Verheyen et al. (2007) and Denys et al. (2011).

12.2 DIVERSITY OF SMALL MAMMALS ALONG THE COASTAL FOREST AREAS

In this chapter, small mammals refer to "mammal species weighing from a maximum of 5 kg to 0.4 g when adult" (Hafner, 2007; Merritt, 2010). Being part of the African continent the East African coastal ecoregion supports more than 99 species of small mammal fauna (Burgess et al., 1998). These include bats (58), rodents (27), and 14 species of shrews (Burgess et al., 1998), in which their distribution is not uniform across all habitats (Burgess and Clarke, 2000; Andanje, et al., 2010; Amin et al., 2018). The occurrence and diversity are certainly influenced by various factors, including vegetation type, altitude, climatic conditions, disease prevalence, seasonal food and water availability, human disturbance, the presence of predators, and life history strategies (Clarke, 1998; Olson et al., 2001; Habel et al., 2017). Among the groups, small mammals such as bats, rodents, carnivores, and shrews are a more diverse group and they occupy different habitat types (Burgess et al., 1998; Kiwia, 2009; Sabuni et al., 2015; Amin et al., 2018). The Eastern African coastal forests have been shown to consist of distinct levels of endemism that make them a globally significant hub of endemism with a significant number of endemic species of small mammals (Burgess and Muir, 1994; Kiwia, 2009; Kock and Stanley, 2009; Sabuni et al., 2015; Amin et al., 2018). Within this remarkable biodiversity hotspot, the diverse assemblage of small mammals plays a pivotal role in shaping the ecological dynamics and contributing to the region's uniqueness (Kock and Stanley, 2009). The presence of a diverse group of small mammals' interactions with other biota highlights the complexity of these coastal forests in this region (Azeria et al., 2007). With its unique diversity, the ecoregion forms part of the Zanzibar-Inhambane, Somali-Maasai, and Zambezian regional centres of endemism of species elements, where the vegetation types are classified into natural and modified habitats based on the criteria within the International Finance Corporation performance standard 6 (Clarke, 1998). It extends along the eastern coastline of East Africa, covering four countries, including Somalia, Kenya, Tanzania, and Mozambique (Figure 12.1), forming a fragmented link with some coastal forest patches of Malawi and Zimbabwe (Clarke, 1998; Salehe and Stirton, 2006). The aim of this chapter is to provide an overview of the species composition and diversity of small mammals within Eastern African coastal forests while emphasizing the geographical distribution of these small mammals within this region.

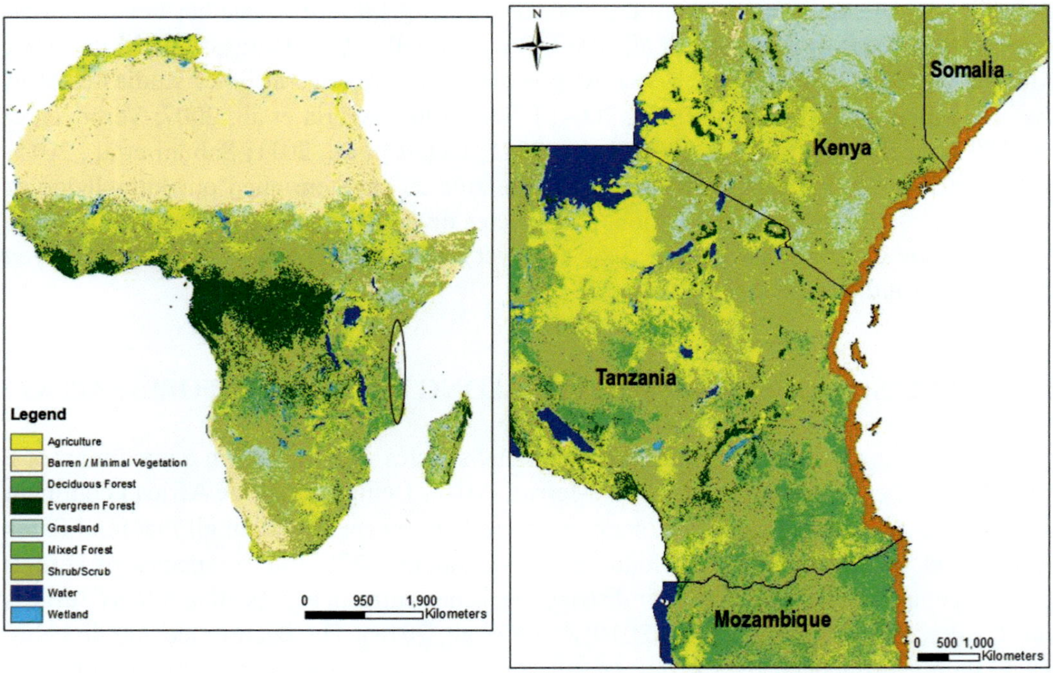

FIGURE 12.1 The coastal forests of Eastern Africa (in orange) from the right-hand map, and the left-hand map represent the major biomes of Africa (*Map by David J. Kavana SUA, Department of Natural Resources Management and Conservation*).

12.2.1 LITERATURE REVIEW

The information obtained in this chapter was also gathered through systematic literature searching and review, as shown in the desk review of existing secondary information and literature through the coastal ecoregion on small mammals (Figure 12.2).

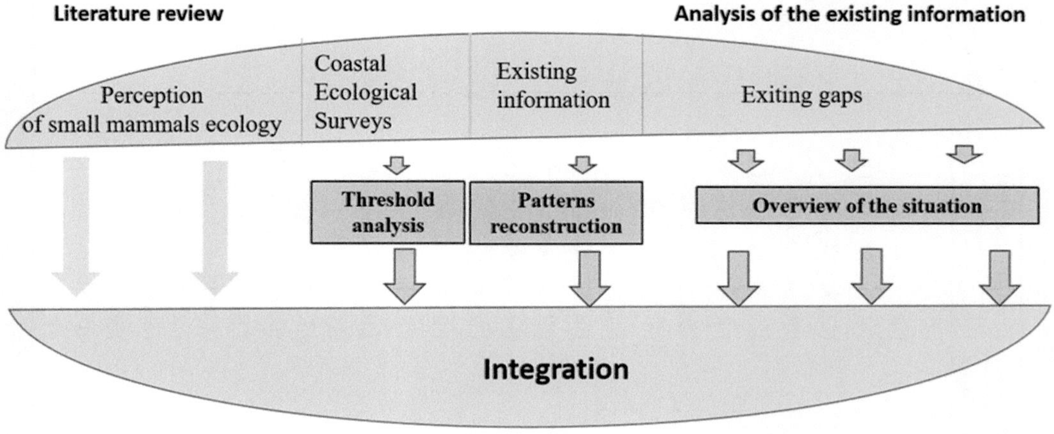

FIGURE 12.2 Flowchart of the methodological approach used: first, a literature review was conducted and analysis of available was done from the collected information. Second, various analyses were applied to rodents' ecological surveys.

12.3 SMALL MAMMALS' DISTRIBUTION PATTERNS ALONG THE COASTAL AREAS

The distribution pattern of small mammals along the coastal areas of East Africa is not uniform and is influenced by a complex interplay of ecological factors, including habitat types, elevation, climate, and human activities (Kiwia, 2009; Kock and Stanley, 2009; Amin et al., 2018). This distribution pattern provides species-habitat interaction and is shaped by the need for small mammals to find suitable niches and resources within the diverse coastal ecosystems (Sabuni et al., 2015). As earlier mentioned, different groups of small mammals inhabit the landscapes of East African coastal forests, and they include the following orders:

12.3.1 CHIROPTERA

The Order Chiroptera, commonly represented by bats, is a most diverse and unique group of mammals in East African coastal forests characterized by their ability to fly (Barnett, 1992). The order comprises several families known to inhabit the coastal forest (Burgess et al., 1998). Different studies, such as Burgess et al. (1998), Kock and Stanley (2009), and Cheke (2011), conducted along the coast of East Africa presented approximately eight families of Order Chiroptera, including Pteropodidae, Nycteridae, Megadermatidae, Rhinolophidae, Hipposideridae, Vespertilionidae, Emballonuridae, and Molossidae (Table 12.1). Family Vespertilionidae is the most species-rich family, with 13 species known to inhabit the coastal areas of East Africa (Burgess et al., 1998). Among the 13 species nine species, including *Neoromicia capensis*, *N. rendalli, N. somalica,*

TABLE 12.1 A Species Checklist of Bats Recorded From the East African Coastal Forests; The Checklist Includes Small Mammals Identified to Species Level

Order	Family	Species	Possible Geographic Range in EACR				Primary Habitat	IUCN Category
			Somalia	Kenya	Tanzania	Mozambique		
Chiroptera	Pteropodidae	*Pteropus seychellensis comorensis* (Nicoll, 1908)			*		FR, AR/TR	LC
		Pteropus voeltzkowi (Matschie, 1909)			*		FR	VU
		Epomophorus wahlbergi (Sundevall, 1846)	*	*	*	*	FR, Sn, SL, AR/TR	LC
		Epomophorus labiatus (Temminck, 1837)		*	*		FR, Sn, SL, GL	LC
		Eidolon helvum (Kerr, 1792)		*	*	?	FR, Sn, AR/TR	NT
		Rousettus aegyptiacus (Geoffroy, 1810)		*	*	?	FR, Sn, SL, GL, Cv	LC
		Lissonycteris angolensis (Bocage, 1898)		*	*	*	FR, Sn, SL, GL, Ro, Cv	LC
		Myonycteris relicta (Bergmans, 1980)		*	*		FR, Sn	LC
	Nycteridae	*Nycteris hispida* (Schreber, 1775)	*	*	*	*	FR, Sn	LC
		Nycteris aurita (Andersen, 1912)	*	*	*		Sn, De	LC
		Nycteris grandis (Peters, 1865)		*	*	*	FR, Sn, AR/TR	LC
		Nycteris macrotis (Dobson, 1876)	*	*	*	?	FR, Sn, AR/TR	LC
		Nycteris thebaica (Geoffroy, 1818)	*	*	*	*	FR, Sn, SL, GL, Cv, De	LC
	Megadermatidae	*Lavia frons* (Geoffroy, 1810)	*	*	*		FR, Sn, SL	LC
		Cardioderma cor (Peters, 1872)	*	*	*		FR, Sn, SL, AR/TR	LC
	Rhinolophidae	*Rhinolophus landeri* (Martin, 1838)	*	*	*	*	FR, Sn, SL, Cv	LC
		Rhinolophus hildebrandtii (Peters, 1878)		*	*	*	FR, Sn, Cv and Sb (nonaquatic)	LC
		Rhinolophus deckenii (Peters, 1868)		*	*		FR, Cv and Sb (nonaquatic), AR/TR	LC
	Hipposideridae	*Macronycteris commersoni* (Geoffroy, 1813)	?	?	?	?	FR, Cv and Sb (nonaquatic), AR/TR	NT
		Hipposideros ruber (Noack, 1893)		*	*	*	FR, Sn, Cv and Sb (nonaquatic)	LC
		Hipposideros caffer (Sundevall, 1846)				*	FR, Sn, SL, WL, Cv, and Sb (nonaquatic)	LC
		Hipposideros cyclops (Temminck, 1853)		*	*		FR, Sn	LC

TABLE 12.1 *(Continued)*

Order	Family	Species	Possible Geographic Range in EACR				Primary Habitat	IUCN Category
			Somalia	Kenya	Tanzania	Mozambique		
		Triaenops persicus (Dobson, 1871)		*	*	*	Sn, SL, Cv, and Sb (nonaquatic)	LC
		Cloeotis percivali (Thomas, 1901)				?	FR, Sn, Cv, and Sb (nonaquatic)	LC
	Vespertilionidae	*Myotis tricolor* (Temminck, 1832)		*	*	*	FR, Sn, SL	LC
		Neoromicia capensis (A. Smith, 1829)	*	*	*	*	FR, Sn, SL, GL, AR/TR	LC
		Neoromicia rendalli (Thomas, 1889)	*	*	*	*	Sn, SL, AR/TR	LC
		Neoromicia somalica (Thomas, 1901)	*	*	*		FR, Sn	LC
		Neoromicia helios (Heller, 1912)	*	*	?		?	DD
		Scotoecus hirundo (de Winton, 1899)	*	*	*	*	FR, Sn	LC
		Scotophilus viridis (Peters, 1852)	?	*	*	*	Sn, AR/TR	LC
		Scotophilus dinganii (A. Smith, 1833)	*	*	*		Sn, AR/TR	LC
		Scotophilus nigrita (Schreber, 1774)		*	*	?	FR, Sn	LC
		Glauconycteris variegata (Tomes, 1861)	*	*	*	*	FR, Sn, SL, AR/TR	LC
		Glauconycteris argentata (Dobson, 1875)		*	*		FR, Sn	LC
		Kerivoula africana (Dobson, 1878)			*		FR	EN
		Miniopterus minor (Peters, 1867)	?	*	*	?	FR?	DD
	Emballonuridae	*Coleura afra* (Peters, 1852)	*	*	*		FR, Sn, SL, Cv, and Sb (nonaquatic), De	lLC
		Taphozous hildegardeae (Thomas, 1909)		*	*		FR, Cv, and Sb (nonaquatic), AR/TR	EN
	Molossidae	*Tadarida brachyptera* (Peters,1852)		*	*	*	FR	LC
		Otomops martiensseni (Matschie, 1897)		*	*	?	FR, Sn, AR/TR	NT

*****Indicates species presence.*

?*Indicates species presence is not well known.*

AR = *artificial;* **Cv** = *caves;* **DD** = *data deficient;* **De** = *desert;* **EACR** = *east African coastal region;* **EN** = *endangered;* **FR** = *forest;* **GL** = *grassland;* **LC** = *least concern;* **MC** = *marine coastal;* **NT** = *near threatened;* **Ro** = *rocky areas;* **Sb** = *subterranean habitats;* **SL** = *shrubland;* **Sn** = *savanna;* **TR** = *terrestrial;* **VU** = *vulnerable;* **WL** = *wetlands.*

Scotophilus viridis, S. hirundo, S. dinganii, S. nigrita, Glauconycteris variegata, G. argentata are least concern (LC); two species, including *N. helios* and *Miniopterus minor*, are data deficient (DD) and one endangered (EN) *Kerivoula africana* (Baillie et al., 1996). In the family of Pteropodidae, at least eight species are known from the coastal forests of East Africa, three species are known to occur from Tanzania, including LC *Pteropus seychellensis comorensis* and VU *Pteropus voeltzkowi* and near threatened (NT) *Eidolon helvum* (Burgess et al., 1998). The mentioned species are LC, VU, and NT, respectively, and they inhabit forested habitats (Baillie et al., 1996). Other bat species of the family Pteropodidae include *Epomophorus labiatus, Rousettus aegyptiacus, Lissonycteris angolensis,* and *Myonycteris relicta*, they are all LC species and are known to inhabit different habitats of Tanzania and Kenya's coastal areas, including the forest, savanna, grassland, shrubland, and artificial habitats (Burgess et al., 1998; Kock and Stanley, 2009). Other families comprise a varying number of species, for instance, Hipposideridae comprises six species of the genus *Macronycteri*, *Hipposideros*, *Triaenops,* and *Cloeotis*; Nycteridae comprises five species of the genus *Nycteris*; Rhinolophidae comprises three species of the genus *Rhinolophus*; Emballonuridae consists of two species of the genus *Coleura* and *Taphozous;* Megadermatidae also comprises two species in the genera *Lavia* and *Cardioderma;* and Molossidae comprises two species in the genera, *Tadarida* and *Otomops* (Kock and Stanley, 2009).

12.3.2 RODENTIA

The Order Rodentia, commonly represented by rats, mice, and squirrels, is a highly diverse and widespread group of mammals characterized by their continuously growing incisor teeth (Merritt, 2010). In East African coastal forests, several rodent families are known to inhabit the ecoregion. Different studies such as Rodgers et al. (1982), Burgess et al. (1998), Kiwia (2009), Azeria et al. (2007), Mulungu et al. (2008), Kock and Stanley (2009), Amin et al. (2018), and Sabuni et al. (2015) conducted in East African coastal forests have reported five families of rodents, including Muridae, Nesomyidae, Gliridae, Sciuridae, and Hystricidae (Table 12.2). The families comprise a varying number of species, for instance, Gliridae comprises two species of the genus *Graphiurus* (Figure 12.3a); Sciuridae comprises two species of the genus *Paraxerus,* and Hystricidae comprises one species *Hystrix cristata* (crested porcupine), this species is found along forest margins, savannas, woodlands, steppes, and uplands of coastal areas of East Africa. These rodent species are known to inhabit different habitats of the East African coastal region, with some, such as *Beamys hindei*, known to be forest specialists in the coastal area of Tanzania and Kenya. Family Muridae is the most species-rich, comprising seven species, including *Mastomys natalensis, Rattus rattus* (Figure 12.3b), *Grammomys dolichurus, G. caniceps, G. cometes, G. dryas, Mus minutoides,* and *Pelomys fallax* which are known to inhabit forest, savanna, bushland, grassland and artificial habitats of East African coast. The second large family is Nesomyidae, comprising three species including *Beamys hindei, Cricetomys gambianus,* and *C. ansorgei.*

TABLE 12.2 A Species Checklist of Rodents Recorded From the East African Coastal Forests; the Checklist Includes Small Mammals Identified to Species Level

Order	Family	Species	Possible Geographic Range in EACR				Primary Habitat	IUCN Category
			Somalia	Kenya	Tanzania	Mozambique		
Rodentia	Nesomyidae	*Beamys hindei* (Thomas, 1909)		*	*	*	FR	LC
		Cricetomys gambianus (Waterhouse, 1840)		*	?		FR, Sn, AR/TR	LC
		Cricetomys ansorgei (Thomas, 1904)		*	*	*	FR, Sn, AR/TR	LC
	Muridae	*Mastomys natalensis* (Smith, 1834)	*	*	*	*	Sn, SL, TR, AR	LC
		Rattus rattus (Linnaeus, 1758)	?	*	*	*	FR, SL, GL, WL, AR/TR	LC
		Grammomys dolichurus (Smuts, 1832)			*	*	FR, SL, GL, AR/TR	LC
		Grammomys caniceps (Hutterer and Dieterlen, 1984)	*	*			SL?	DD
		Grammomys cometes (Thomas and Wroughton, 1908)				*	FR, SL	LC
		Mus minutoides (Smith, 1834)		?	?	*	FR, Sn, SL, GL, AR/TR	LC
		Pelomys fallax (Peters, 1852)			*	*	Sn, AR/TR	LC
	Gliridae	*Graphiurus murinus* (Desmarest, 1822)			*		FR, Sn, SL, GL, WL, AR/TR	LC
		Graphiurus kelleni (Reuvens, 1890)		*	*		FR, Sn	LC
	Sciuridae	*Paraxerus palliatus* (Peters, 1852)	*	*	*	*	FR, Sn, AR/TR	LC
		Paraxerus flavovittis (Peters, 1852)		*	*	*	Sn, AR/TR	LC
	Hystricidae	*Hystrix cristata* (Linnaeus, 1758)		*	*		FR, Sn, SL, GL, Ro, AR/TR	LC

**Indicates species presence.*

?*Indicates species presence is not well known.*

AR = *artificial;* **Cv** = *caves;* **DD** = *data deficient;* **De** = *desert;* **EACR** = *east African coastal region;* **EN** = *endangered;* **FR** = *forest;* **GL** = *grassland;* **LC** = *least concern;* **MC** = *marine coastal;* **NT** = *near threatened;* **Ro** = *rocky areas;* **Sb** = *subterranean habitats;* **SL** = *shrubland;* **Sn** = *savanna;* **TR** = *terrestrial;* **VU** = *vulnerable;* **WL** = *wetlands.*

FIGURE 12.3 (a) *Graphiurus* sp. photo from miombo woodland of Selous (today Nyerere National Park)-Niassa corridor (*Photo by Dr. Ally J. K. Nkwabi, April 2014*), (b) *Rattus rattus* from Mafia marine park (*Photo by Teclamageni Mayeji, September, 2021*).

12.3.3 CARNIVORA

Order Carnivora is represented by the mammals that largely feed on meat (Table 12.3). This order is a representation of the specialized feeding behavior of some small mammals (Kingdon, 2015). Studies conducted by Burgess et al. (1998), Kock and Stanley (2009), and Amin et al. (2018) reported five families of the Order Carnivora, including Herpestidae, Viverridae, Canidae, Felidae and Nandiniidae. Family Herpestidae is the most species-rich, with six species of small mammals, including *Atilax paludinosus, Ichneumia albicauda, Bdeogale omnivora, B. crassicauda, Herpestes sanguineus,* and *Helogale parvula,* known to inhabit different habitats of East African coast. Family Viverridae comprises two species, *Genetta maculata* (large-spotted genet) and *Civettictis civetta* (African civet) which are widely distributed and common along the coastal woodlands and secondary forests of Somalia, Kenya, Tanzania, and Mozambique; Canidae comprises one species, the *Otocyon megalotis,* commonly known as bat-eared fox, is the only living species of the genus *Otocyon*. It is widely distributed and common along the coastal forests of Somalia, Kenya and Tanzania whereas the Family Felidae comprises one species of the genus *Felis*; example *Felis sylvestris* (wild cat) can be observed along the coastal forest of Somalia, Kenya, Tanzania, and Mozambique. *F. sylvestris* is a widespread and common species. This species can be affected and suspected of hybridizing with domestic cats and can be predated with dogs when in areas nearby human-inhabited areas. Family Nandiniidae comprises one species, *Nandinia binotata* (African palm civet); this species occurs in coastal forest of Kenya.

12.3.4 SORICOMORPHA AND MACROSCELIDEA

The Order Soricomorpha is represented by shrews, a very ancient group with recognizable fossil ancestors dating back more than 40 million years in Eurasia, but the first shrew fossils

TABLE 12.3 A Species Checklist of Other Groups of Small Mammals Recorded From the East African Coastal Forests; the Checklist Includes Small Mammals Identified to Species Level

Order	Family	Species	Possible Geographic Range in EACR				Primary Habitat	IUCN Category
			Somalia	Kenya	Tanzania	Mozambique		
Soricomorpha	Soricidae	*Crocidura fuscomurina* (Heuglin, 1865)		*	*	*	Sn, GL, De, AR/TR	LC
		Crocidura hildegardeae (Thomas, 1904)		*	*		FR, AR/TR	LC
		Crocidura jacksoni (Thomas, 1904)		*	*		FR, AR/TR	LC
		Crocidura parvipes (Osgood, 1910)		*	*		FR, Sn, SL	LC
Primates	Galagidae	*Paragalago rondoensis* (Honess in Kingdon, 1997)			*		FR	EN
		Paragalago zanzibaricus (Matschie, 1893)			*		FR	NR
		Paragalago cocos (Heller, 1912)		*	*		FR, AR/TR	LC
		Otolemur crassicaudatus (Geoffroy Saint-Hilaire, 1812)		*	*	*	FR, Sn, SL, AR/TR	LC
		Otolemur garnettii ssp. *garnettii* (Ogilby, 1838)			*		FR, SL, AR/TR	VU
		Otolemur garnettii ssp. *lasiotis* (Peters, 1876)	*	*	*		FR	LC
Hyracoidea	Procaviidae	*Dendrohyrax validus* (True, 1890)		*	*		FR	NT
		Dendrohyrax arboreus (Smith, 1827)		*	*	*	FR, Sn, Ro	LC
Macroscelidea	Macroscelididae	*Rhynchocyon chrysopygus* (Günther, 1881)		*			FR, SL	EN
		Rhynchocyon petersi (Bocage, 1880)		*	*		FR, SL, AR/TR	LC
		Elephantulus rufescens (Peters, 1878)	*	*	*		Sn, GL	LC
		Petrodromus tetradactylus (Peters, 1846)		*	*	*	FR, Sn, SL	LC
Carnivora	Herpestidae	*Ichneumia albicauda* (Cuvier, 1829)	*	*	*	*	FR, Sn, SL, GL, WL, AR/TR	LC
		Bdeogale omnivora (Heller, 1913)		*	?		FR	VU
		Bdeogale crassicauda (Peters, 1852)		*	?	*	FR, Sn, SL	LC
		Herpestes sanguineus (Rüppell, 1835)	*	*	*	*	FR, Sn, SL, GL, AR/TR	LC
		Atilax paludinosus (G.[Baron] Cuvier, 1829)		*	*	*	FR, GL, WL, MC	LC

TABLE 12.3 *(Continued)*

Order	Family	Species	Possible Geographic Range in EACR				Primary Habitat	IUCN Category
			Somalia	Kenya	Tanzania	Mozambique		
		Helogale parvula (Sundevall, 1847)	*	*	*	*	Sn, GL	LC
	Viverridae	*Genetta maculata* (Gray, 1830)	*	*	*	*	FR, Sn, SL, AR/TR	LC
		Civettictis civetta (Schreber, 1776)	*	*	*	*	FR, Sn, SL, WL, AR/TR	LC
	Canidae	*Otocyon megalotis* (Desmarest, 1822)	*	*	*		Sn, SL, GL	LC
	Felidae	*Felis lybica* (Forster, 1780)	*	*	*	*	FR, Sn, SL, GL, De, AR/TR	LC
	Nandiniidae	*Nandinia binotata* (Gray, 1830)		*	*	*	FR, Sn, SL, AR/TR	LC

*Indicates species presence.

?*Indicates species presence is not well known.*

AR = *artificial;* **Cv** = *caves;* **DD** = *data deficient;* **De** = *desert;* **EACR** = *east African coastal region;* **EN** = *endangered;* **FR** = *forest;* **GL** = *grassland;* **LC** = *least concern;* **MC** = *marine coastal;* **NT** = *near threatened;* **Ro** = *rocky areas;* **Sb** = *subterranean habitats;* **SL** = *shrubland;* **Sn** = *savanna;* **TR** = *terrestrial;* **VU** = *vulnerable;* **WL** = *wetlands.*

in Africa are from between 18 and 15 mya (Kingdon, 2015). In the East African coast, one family Soricidae is known with several species, including the *Crocidura fuscomurina, C. hildegardeae, C. jacksonii,* and *C. parvipes* which are known to inhabit forest, savanna, bushland, grassland, and artificial habitats of the coastal forest of East Africa (Happold, 2013; Kingdon, 2015). The Order Macroscelidea, a sister group of Soricomorpha, is represented by elephant shrews (sengis), which are mouse- or rat-sized animals with large eyes and ears, slender limbs and a long, bare tail (Kingdon, 2015). Sengis have a tubular snout protruding from a strongly tapered skull. In Africa the Order is represented by two extant families (Macroscelidinae and Rhynchocyoninae). Family Macroscelidinae comprises three genera which are *Petrodromus, Macroscelides,* and *Elephantulus.* Family Rhynchocyoninae comprises a single genus *Rhynchocyon.* The elephant shrews have adapted to various microhabitats ranging from rock caves, underground burrows, forested, bushy, or scrubby habitats and feed on a variety of invertebrates such as insects (Butler, 1995; Douady and Douzery, 2009). In the East African coast, some species of elephant shrews are inhabiting the region and they are endemic to this region, for example, the *Rhynchocyon chrysopygus* (Golden-rumped sengi) is confined to two very small areas of the Kenya coast, where it lives in dry evergreen thickets (Kock and Stanley, 2009; Kingdon, 2015; Amin et al., 2018). All species of elephant shrews minimize exposing themselves to predation by remaining under vegetation and rocks and only crossing open ground at high speed. Some species, such as *Elephantulus rufescens,* maintain conspicuous paths that are daily cleared of litter by scuffling with the legs and may sometimes make use of vacant rodent holes (Kingdon, 2015).

12.3.5 PRIMATES

The order Primates comprises an intelligent group of mammals, the majority being adapted to arboreal life, like Galagos also known as bush babies (Kingdon, 2015). On the East African coast, the order Primates comprises one family of Galagidae, which is dominated by small mammals with several species (Table 12.3), including *Paragalago cocos*, which is endemic to Kenya and Tanzania, inhabiting the coastal forests in the north coast of Kenya to the northern coast of Tanzania, particularly the Mgambo Forest Reserve and Kilulu Hill Forest Reserve (Butynski et al., 2006). Another endemic species is the *P. zanzibaricus*, which is a forest specialist and endemic to Unguja (Zanzibar) Island, Tanzania, with small populations in the north (Kiwenga Forest Reserve), west (Masingini Forest Reserve), and on Uzi Island off the southwest coast (Honess et al., 2013). Another endemic species is the EN *P. rondoensis*, which is endemic to several Tanzania forest patches, including the Zaraninge forest, Pande Game Reserve, Pugu/Kazimzumbwe, Ruvu South, Noto forest reserves, and Rondo Nature Reserve (Burgess and Clarke, 2000). Other species are known to inhabit different habitats including forest, savanna, bushland, grassland, and artificial habitats of the East African coast.

12.3.6 HYRACOIDEA

Hyracoidea is an order of small mammals that includes hyraxes, which resemble rodents or rabbits, but they are not closely related to either (Happold, 2013; Kingdon, 2015). The Order comprises one family Procavidae (tree hyraxes), with two species (Table 12.3), including the NT *Dendrohyrax validus*, which is confined to the coastal forests of Tanzania, southern Kenya, and offshore islands, including Pemba, Zanzibar, and Tumbatu, the species can also be observed in montane forests of Mt. Kilimanjaro and Mt. Meru and the Eastern Arc Mountains (Milner and Gaylard, 2013). The other species is the *D. arboreus*, which has a wide geographic range in the East African coast, it is known to inhabit several forest patches and rock cliffs of Kenya and Tanzania (Milner and Gaylard, 2013).

12.4 ENDEMISM OF SMALL MAMMALS IN THE COASTAL FOREST OF EASTERN AFRICA

The East African coastal forests are one of the world's primary endemism centers, supporting species that are unique to our planet (Burgess and Clarke, 2000). These forests are among the ecosystems with the greatest conservation priority due to the abundance of endemic and near-endemic species (Burgess et al., 1998). The biogeographic investigations of the biota of coastal forests have increased during the past 10 years (Burgess et al., 2017). By simply recording the number of endemic species found in coastal forests, the majority of studies, such as Rodgers et al. (1982), Burgess et al. (1998, 2017), Kiwia (2009), Azeria et al. (2007), Kock and Stanley (2009), Sabuni et al. (2015), and Amin et al. (2018) have examined the distributional patterns of different endemic species. It is important to note that the Tanzania and Kenya coastal regions have greater levels of endemism of small mammals than other ecoregions like Mozambique, Zimbabwe, and Somaliland. This is likely related to species richness (Burgess et al., 1998; Amin et al., 2018). Out of the approximately 30 endemic or near-endemic species found in the coastal forests of East Africa, 25 are small mammals, comprising 11 bats, 10 shrews, and 4 rodents (Rescigno and Maletta, 2011; Rocliffe and Udelhoven, 2011). The coastal forests are home to a disproportionate number of these unique species. For instance, species like the Decken's horseshoe bat (*Rhinolophus deckenii*), Commerson's roundleaf bat (*Hipposideros commersoni*), African Woolly Bat (*Kerivoula Africana*), Pemba Fruit fox (*Pteropus voeltzkowi*), and Golden-rumped Elephant Shrew (*Rhynchocyon chrysopygus*) are restricted to the eastern African coastal forests (FitzGibbon and Rathbun, 2015; Entwistle and Juma, 2016; Cooper-Bohannon et al., 2020; Shapiro et al., 2020). The other two new species of bat described in the coastal forest of Tanzania include Bakari's free-tailed bat (*Mops bakarii*) and Maendeleo horseshoe bat (*R. maendeleo*) (Kock et al., 2000; Shapiro, 2017). The Bakari's free-tailed bat (*Mops bakarii*) is a new species of mastiff bat that was described in 2008; this species is endemic to Pemba Island, particularly in the Ngenzi forest reserve. This species is listed as DD in the International Union for Conservation of Nature (IUCN) Red List, due to only one known locality and lack of further data (Shapiro, 2017). While Maendeleo horseshoe bat (*R. maendeleo*) is also a recently discovered species of bat in the family Rhinolophidae. It inhabits caves of the Coastal Lowland forests of Tanzania, particularly in Amboni Cave Forest, Mkulumuzi River Gorge, and Mazumbai Forest Reserve in the Usambara Mountains (Kock et al., 2000; Shapiro, 2017). An additional group of small mammals is restricted to the Coastal Forests and Eastern Arc Mountains, particularly the lowlands. This contains the Black and rufous elephant shrew (*Rhynchocyn petersi*) and two species of bats, The East African little-collared fruit bat also known as Bergman's collared fruit bat (*Myonycteris relicta*) and Short-winged free-tailed bat (*Tardarida brachyptera*) (Taylor, 2016). The other species that was studied by Sabuni et al. (2015), the lesser pouched rat (*Beamys hindei*), is a rare species found in the forests of southern Tanzania and Malawi, as well as a few coastal forests in Kenya and Tanzania, notably, it was believed to be among the rarest rodents in Africa until recently, but it has now been shown to be more widely dispersed than previously believed and is frequently prevalent at the locations where it does exist (Fitzgibbon et al., 1995; Sabuni et al., 2015).

Furthermore, two specimens, one of bat and one of rodent are among the other new small mammal species found in coastal forests that have not yet been fully characterized and classified. Most endemic species found in coastal forests have restricted distribution ranges; they are frequently typified by fragmented and disjunction patterns or single-site endemism (Kiwia, 2009; Sabuni et al., 2015; Amin et al., 2018). These species are thought to be remnants of the old Pan-African rainforest, which split into two distinct regions during the Miocene epoch, rather than the product of more recent evolutionary processes (Clarke, 1998). According to Sabuni et al. (2015) and Amin et al. (2018), in this context, coastal forests are seen as a declining sanctuary where endemic species are gradually becoming increasingly extinct. It is significant to remember that the Eastern Arc and Coastal forests are considered to be mainland Africa's second-most significant endemism hotspot.

In Mafia island, the Seychelles fruit bat (*Pteropus seychellensis*) occurs in large colony, the island represents the only known population in Tanzania mainland while the other population is distributed in Comoros and Seychelles islands (Table 12.4). These bats are limited to a few Indian Ocean islands worldwide (Bergmans et al., 2017). Other endemic species may be found in Kenya's coastal forests, mostly in the Ngangao and Chawia habitats on the Taita hill. One such species is *Praomys taitae*, which is conspecific with *P. delectorum* (Bytebier, 2001). The other species restricted in the coastal forest of Kenya, particularly in 'Galole Road, about 8.5 km north of Garsen, a coastal province in Kenya. The other species is the wattled or butterfly bat (*Glauconycteris kenyacola*) which is a newly described species of vesper bat and a member of the family Vespartilionidae, which is classified as DD according to the IUCN Red List, as it is known only from the holo-type and might be threatened by agriculture and logging (Wilson and Mittermeier, 2019). Research is necessary to find out if these significant groups are endemic or near-endemic to Eastern African coastal forests.

The majority of endemic species within coastal forests exhibit limited distribution ranges, often characterized by single-site endemism or fragmented and disjunctive patterns (Kiwia, 2009; Sabuni et al., 2015; Amin et al., 2018). Rather than being the outcome of recent evolutionary processes, these species are more regarded as relics from the division of the ancient Pan-African rainforest into two separate parts during the Miocene era (Clarke, 1998). In this context, coastal forests are perceived as a diminishing sanctuary, where endemic species are progressively evolving into more relics (Sabuni et al., 2015; Amin et al., 2018). The occurrence of humid forests versus drier woodland habitats, geological history, ecological barriers that limit connectivity, and past ecological perturbation such as volcanism may explain this.

TABLE 12.4 Endemic Small Mammals of the East African Coastal Forests

Species	Summary Distribution	IUCN status
Pemba flying fox *Pteropus voeltzkowi*	Pemba, Zanzibar	VU
Howell's forest shrew *Sylvisorex howelli*	Uluguru, Nguru, Nguu, Ukaguru, and Usambaras	LC
Xanthippe's shrew (*Crocidura xantippe*)	Voi, Tsavo, Nyiru and Usambaras Mountains	LC
Fiery spiny mouse *Acomys ignites*	Usambara Mountains	LC
Usambara shrew (*Crocidura usambarae*)	Usambara Mountains	VU
Myosorex sp	Uluguru Mountains	EN
Tanzania white-toothed shrew (*Crocidura tansaniana*)	Usambara Mountains	EN
Decken's horseshoe bat (*Rhinolophus deckenii*)	Usambara Mountains	NR
Rhinolophus sp. *Nov*.,		EN
Commerson's roundleaf bat (*Hipposideros commersoni*)		EN
African woolly bat (*Kerivoula Africana*)	Genda Genda, and the Tong'omba coastal forests	EN
Golden-rumped elephant shrew (*Rhynchocyon chrysopygus*)	Arabuko-Sokoke Forest	VU
Bakari's free-tailed bat (*Mops bakarii*)	Ngenzi forest reserve	DD
Maendeleo horseshoe bat (*Rhinolophus maendeleo*)	Usambara Mountains	DD
Black and rufous elephant shrew (*Rhynchocyn petersi*)	Eastern Arc Mountains/Coastal forests	VU
Bergman's collared fruit bat (*Myonycteris relicta*)	Shimba Hills, Usambara, Zanzibar-Inhambane and Nguru Mountains	LC
Short-winged free-tailed bat (*Tardarida brachyptera*)		
Seychelles fruit bats (*Pteropus seychellensis*)	Mafia Island	NR
Praomys taitae (Heller, 1912)	Taita hills	EN
Kenya wattled or butterfly bat (*Glauconycteris kenyacola*)	Galole Road, Coast Province	DD
The golden-rumped elephant shrew, *Rhynchocyon chrysopygus*	Mombasa, Kilifi and Malindi	EN
Lesser pouched rat *Beamys hindei*	Various forest patches of the East African coast	LC

CR = critically endangered; DD = data deficient; EN = endangered; LC = least concern; NR = near threatened; VU = vulnerable.

During recent decades, several studies of small mammals have been conducted in various habitats on the Eastern African coast. The studies include Cockle et al. (1998), Burgess et al. (1998, 2017), Kiwia (2009), Azeria et al. (2007), Kock and Stanley (2009), Amin et al. (2018), Stanley et al. (2011), and Sabuni et al. (2015). The findings from these studies provided a compilation of information including a checklist of identified endemic small mammal species inhabiting the Eastern African Coastal ecoregion (see Table 12.4).

12.5 THREATS TO COASTAL FOREST BIODIVERSITY

The Eastern African coastal forests have undergone significant transformation due to agricultural conversion and urbanization, resulting in a loss of a substantial portion of the original forest cover (Salehe and Stirton, 2006; Gereau et al., 2016; Ntukey et al., 2022). Currently, only around 10% of the undisturbed vegetation remains, this remaining habitat is fragmented into over 400 patches of lowland forests, collectively spanning an area of approximately 6250 km^2 (Riggio et al., 2019). The detailed threats to the East African Coastal Forest biodiversity are detailed below.

12.5.1 SHIFTING CULTIVATION

Over millennia of human habitation, a large portion of the ecoregion's natural landscapes has been transformed into agricultural land (Ntukey et al., 2022). Following intensive clearing of forests for agriculture, only small patches of dense, closed-canopy lowland forest persist within the Eastern Africa coastal region (Salehe and Stirton, 2006; Kibet, 2011). These areas, encompassing a minimum of about 6200 km^2, constitute roughly 2% of the ecoregion's total area (Gereau et al., 2016; Riggio et al., 2019). These remaining coastal forests are primarily distributed in fragmented patches of modest size (Sabuni et al., 2015; Ntukey et al., 2022). Due to the limited fertility of the soil, the predominant agricultural practices focus on short-term shifting cultivation primarily for food crops like cassava, maize, and bananas (Kiwia, 2009; Sabuni et al., 2015). As the human population continues to grow, the annual demand for more agricultural land also increases which leads to encroachment of coastal forests to sustain human needs (Habel et al., 2017). Additionally, the expansion of commercial agriculture, including the establishment of plantations for crops such as coconuts, sisal, cloves, cardamom, and cashew nuts, has contributed to the depletion of lowland coastal forests and other natural habitats (Habel et al., 2017; Ntukey et al., 2022). These human-driven changes pose threats to the survival of the entire ecosystem, including its biota, such as the small mammals, which form part of the ecosystem, and they are renowned as the powerhouse of this fragmented ecosystem (Kiwia, 2009; Sabuni et al., 2015).

12.5.2 HUMAN POPULATION AND URBANIZATION

Stretching across the East African coast, more than 20 million people are living in the coastal regions of Somalia, Kenya, Tanzania, and Mozambique (Burgess et al., 2017).

The regions include large and rapidly growing population cities like Mombasa, Dar es Salaam, Zanzibar, and Beira (Salehe and Stirton, 2006). Following the growing population in these coastal regions, human settlements are expanding; the dependence on goods and raw material from coastal forests increases, and agriculture expands forcing the more proximal communities, particularly the rural communities to increase reliance on natural resources to sustain livelihood (Burgess et al., 1998, 2017; Gereau et al., 2016; Ntukey et al., 2022). Notably, in the rural coastal areas, local communities have adopted livelihood strategies of small-scale agriculture, fishing, tree cultivation, and the utilization of forest resources (Rocliffe and Udelhoven, 2011). Among these strategies, subsistence farming of essential crops like cassava and maize takes precedence as the principal economic activity (Salehe and Stirton, 2006; Burgess et al., 2017). Meanwhile, the coastal forests and woodlands serve as indispensable sources, providing a wide spectrum of both timber and nontimber products that cater to local needs and foster income generation, on the other side of the coin, the face of coastal forests is being degraded due to overexploitation of forest resources leading to biodiversity loss (Pollard, 2012). The rapid decline in forest areas with the most dramatic forest losses occurring close to the populated regions is considered the main threat to the mammals of East Africa including small mammals (Azeria et al., 2007; World Bank, 2008; Andanje et al., 2010; Tabor et al., 2010; Cheke, 2011; Habel et al., 2017). For example, outside of Dar es Salaam, the Pugu/Kazimzumbwi, Ruvu South and Masanganya Forests Reserves have been facing extensive forest loss where regeneration has been impaired in the last two decades (Ntukey et al., 2022). However, following the establishment and gazettement of protected areas, most coastal forest fragments are now found within government forest reserves which create hope for the conservation of eastern African coastal forests and their biodiversity (Burgess et al., 1998; World Bank, 2008).

12.5.3 UNPRESCRIBED BURNING AND ILLEGAL LOGGING

Charcoal production through the burning of woody plants has been a significant driver of habitat loss, particularly in the proximity to coastal towns and along major roadways in eastern African coastal regions (Salehe and Stirton, 2006; Gereau et al., 2016). Forests located near tourist destinations, such as the Arabuko-Sokoke Forest near Malindi and Watamu in Kenya, are faced with heightened pressure due to the soaring demand for wood carving and timber, largely fueled by the construction of hotels, private residences, and tourist attractions (Tabor et al., 2010; Gereau et al., 2016). Furthermore, uncontrolled burning practices, often employed to clear farmland, drive animals for hunting, collect honey, and reduce the presence of pests, pose a persistent threat to lowland coastal forests and thicket patches (Gereau et al., 2016). This kind of burning frequently results in the replacement of rare and endemic coastal forest species with more common, fire-adapted species (Kibet, 2011). Adding to these challenges is the issue of illegal logging, typically involving pit-sawing techniques, which has become a prevalent concern in nearly all remaining coastal forests, particularly in northern Mozambique and southern Tanzania

(Salehe and Stirton, 2006; Gereau et al., 2016). This illicit activity threatens the few timber trees that still endure in these regions (Ngumbau et al., 2020). Also, with the loss of these forests which form a large portion of habitats for animals like small mammals, the ecoregion is likely to lose part if not whole of its endemic and cosmopolitan species of mammals and other taxa such as birds, reptiles, amphibians and invertebrates (Azeria et al., 2007; Burgess et al., 2017; Sabuni et al., 2015).

12.5.4 MINING

Alongside its significant biological resources, the Eastern African coast has mineral reserves (Salehe and Stirton, 2006). These mineral resources encompass natural gas, gemstones, iron, titanium, and limestone (Rescigno and Maletta, 2011). However, the extraction of these minerals can give rise to destructive mining practices that endanger a large part of the natural habitat (Ntukey et al., 2022). Coastal areas, in particular, bear the impact of these activities, for instance, the mining of titanium ore found in the coastal sands leads to the destruction of native vegetation (Gereau et al., 2016). Further mining activities include the extraction of iron and manganese, conducted on a relatively small scale within the Kwale Kaya forests of coastal Kenya (Gereau et al., 2016; Burgess et al., 2017). While these mineral resources hold economic potential, the challenge lies in balancing their extraction with the conservation of the region's diverse natural habitats, whereas an imbalance of these high-income-generating activities with the conservation of biodiversity in Eastern African coastal forests poses threats to the survival of flora and fauna, including the small mammal group which has been less considered when conservation and development initiatives are planned for implementation (Pollard, 2012; Gereau et al., 2016).

12.6 CONSERVATION INITIATIVES OF THE EAST AFRICAN COASTAL FORESTS BIODIVERSITY

Despite the fact that the ecoregion has lost a large part of its natural habitat, leaving the remnants of numerous forest patches, biodiversity conservation is still alive in this region, where the presence of unique habitats and endemic species of flora and fauna form an amalgamation of a mosaic ecosystem, which serves among the biodiversity hotspots of the world (Francis et al., 2002; Rocliffe and Udelhoven, 2011; Holly et al., 2022). The majority of these forest patches are now under formal protected status and managed by the government or local community authorities, particularly in Kenya and Tanzania which promotes habitat conservation and restoration (Burgess et al., 2017; Holly et al., 2022). Among the positive initiatives for the conservation of the remaining coastal forests and their biodiversity is the establishment of protected areas such as the Tanga Coelacanth Marine Park, Dar es Salaam Marine Reserve, Jozani-Chwaka Bay National Park, Ngezi-Vumawimbi Nature Forest Reserve on Pemba, Mafia Island Marine Park, Ruvuma and Mtwara Marine

Park, Saadani National Park, Pugu Forest, and Kazimzumbwi forest reserves of Tanzania play a crucial role in both habitat and species conservation (Francis et al., 2002; Riggio et al., 2019). Relatively, the establishment of other protected areas in Kenya, such as the Arabuko Sokoke Forest Reserve, Shimba Hills and Boni National Reserves of Kenya, and the Bazaruto Archipelago and Quirimbas National Parks of Mozambique, also play a significant role in habitat and species conservation in these forests (Salehe and Stirton, 2006; Riggio et al., 2019). The establishment and gazettement of these protected areas can also be foreseen in the following context.

12.6.1 HABITAT CONSERVATION

Protected areas are designated to safeguard and conserve critical natural habitats (Riggio et al., 2019). Each of these areas represents a unique and ecologically significant ecosystem, including marine environments, coastal forests, and terrestrial landscapes (Ntukey et al., 2022). Through the establishment of forest reserves, marine parks, and national parks, these regions are shielded from destructive practices such as overfishing, logging, and habitat conversion for agriculture or development, which creates a good environment for the survival of fauna such as the rodents, bats, shrews, and other small mammal group (Francis et al., 2002; Rescigno and Maletta, 2011; Holly et al., 2022).

12.6.2 BIODIVERSITY PROTECTION

The protection of habitats within these areas directly contributes to the conservation of various species, including the small mammals (Rescigno and Maletta, 2011). Coastal and marine protected areas like Tanga Coelacanth and Mafia Island Marine Parks are vital for safeguarding diverse marine life, including coral reefs, fish, and EN species of flora and fauna (Riggio et al., 2019; Holly et al., 2022). On the other hand, terrestrial protected areas like Pande Game Reserve in Dar es Salaam, Arabuko Sokoke, and Pugu/Kazimzumbwi Forest Reserves, Shimba Hills and Boni National Reserves, the Bazaruto Archipelago, Saadani and Quirimbas National Parks protect a variety of wildlife, from large mammals to small mammals and other groups such as birds, amphibians, reptiles and insects (Riggio et al., 2019).

12.6.3 ENDANGERED SPECIES CONSERVATION

Protected areas are crucial for the conservation of EN and endemic species (Rescigno and Maletta, 2011). For instance, Arabuko Sokoke Forest Reserve and the Jozani Chwaka National Park are home to various rare and EN species, including the Pemba flying fox and the golden ramped sengi (Amin et al., 2018; Ngumbau et al., 2020). By providing a secure habitat, these areas offer these species a chance for survival, reproduction, and recovery (Riggio et al., 2019).

12.6.4 RESEARCH AND EDUCATION

The protected areas also serve as living laboratories for researchers and educational centers for local communities and visitors (Ndang'ang'a et al., 2008). They offer opportunities to study the natural world, understand ecological processes, and raise awareness about the importance of conservation. On the other hand, they provide room for scientific research that promotes species conservation and habitat restoration (Kiwia, 2009; Azeria et al., 2007; Tabor et al., 2010; Sabuni et al., 2015; Burgess et al., 2017; Ntukey et al., 2022).

12.6.5 COMMUNITY-BASED CONSERVATION

Sustainable conservation often involves the effective participation of local communities in the conservation of natural resources and the equitable sharing of benefits (World Bank, 2008). Some of these protected areas such as Mafia Marine Park, engage nearby communities in sustainable resource management through community-based organizations such as the Beach Management Unit; hence promoting practices that benefit both people and the environment (Rocliffe and Udelhoven, 2011; Amin et al., 2018). This system of conservation promotes transparency and accountability in conserving biodiversity and the entire ecosystem.

12.7 CONCLUSION AND RECOMMENDATION

The East African coastal region is home to a diverse small mammal that consists of a highly unique and differentiated group based on habitat heterogeneity. An understanding of the characteristics of small species within a given ecosystem is a key start for conservation and management. Since small mammals are important because of the ecological role they play, conservation and management should be addressed based on the habitat range they occupy. Intensive field surveys and associated morphological and molecular genetics analyses are strongly advised to document the small mammals of East Africa for conservation purposes.

ACKNOWLEDGMENTS

We are grateful to Professor Thammineni Pullaiah of the University of Sri Krishnadevaraya and Dr. Ally J.K Nkwabi of the Tanzania Wildlife Research Institute (TAWIRI) for their valuable comments and edits of this book chapter.

KEYWORDS

- **small mammal**
- **rodents**
- **bats**
- **small primate**
- **coastal forests**

REFERENCES

Amin, R.; Wacher, T.; Bowkett, A. E.; Ogwoka, B.; Morris, M.; Agwanda, B. R. Africa's forgotten forests: The conservation value of Kenya's northern coastal forests for large mammals. *J. East African Nat. Hist.* **2018,** *107*(2), 41–61. https://doi.org/10.2982/028.107.0203

Andanje, S.; Davey, K.; Ogwoka, B.; Risky Agwanda, B.; Ali, A.; Bruce, T.; Wacher, T.; Amin, R. *Mammal Diversity Surveys in the Coastal Forests: Kenya*. The Zoological Society of London, **2010**.

Andrews, G. Mafia Island Marine Park, Tanzania: Implications of applying a Marine Park paradigm in a developing country. *Methods* **1998,** 267–279.

Avenant, N. L.; Cavallini, P. Correlating rodent community structure with ecological integrity, Tussen-die-Riviere nature reserve, free state province, South Africa. *Integr. Zool.* **2007,** *2*(4), 212–219. https://doi.org/10.1111/j.1749-4877.2007.00064.x

Azeria, E. T.; Sanmartín, I.; Ås, S.; Carlson, A.; Burgess, N. Biogeographic patterns of the East African coastal forest vertebrate fauna. *Biodivers Conserv.* **2007,** *16*, 883–912. https://doi.org/10.1007/s10531-006-9022-0.

Baillie, J.; Groombridge, B.; Gärdenfors, U.; Stattersfield, A. *IUCN Red List of Threatened Animals, World Conservation Monitoring Centre., International Union for Conservation of Nature and Natural Resources.* Species Survival Commission., BirdLife International and Conservation International, IUCN: Gland, Switzerland, **1996**.

Barnett, A. *Expedition Field Techniques: Small Mammals (excluding bats)*. Expedition Advisory Centre, Royal Geographical Society: London, **1992**.

Bergmans, W.; Gerlach, J.; Howell, K.; Hutson, A. M.; Mickleburgh, S.; Monadjem, A. *Pteropus seychellensis. The IUCN Red List of Threatened Species*. **2017,** e.T18759A22087745. https://doi.org/10.2305/IUCN.UK.2017-2.RLTS.T18759A22087745.en.

Bloesch, U.; Frank-Klötzli, G. T. Z. *Coastal Forests of the Saadani National Park, Conservation Values and Management Strategies.* Paper No. 37. Wildlife Division in Tanzania Ministry of Natural Resources & Tourism: Dar es Salaam, **2004**.

Burgess, N. D.; Clarke, G. P. *Coastal Forests of Eastern Africa*. IUCN-The World Conservation Union, Publications Services Unit: Gland, Switzerland and Cambridge, UK, **2000,** pp. 69–235.

Burgess, N. D.; Clarke, G. P.; Rodgers, W. A. Coastal forests of eastern Africa: Status, endemism patterns and their potential causes. *Biol. J. Linn. Soc.* 1998, *64*, 337–367.

Burgess, N. D.; Malugu, I.; Sumbi, P.; Kashindye, A.; Kijazi, A.; Tabor, K.; Mbilinyi, B.; Kashaigili, J.; Wright, T. M.; Gereau, R. E.; Coad, L.; Knights, K.; Carr, J.; Ahrends, A.; Newham, R. L. Two decades of change in state, pressure and conservation responses in the coastal forest biodiversity hotspot of Tanzania. *Oryx* **2017,** *51*(1), 77–86. https://doi.org/10.1017/S003060531500099X

Burgess, N. D.; Muir, C. (Eds.). Coastal forests of Eastern Africa: Biodiversity and conservation. In: *Proceedings of a Workshop Held at the University of Dar es Salaam*, Society for Environmental Exploration & Royal Society for the Protection of Birds, UK, August 9–11, **1994**.

Burgess, N. D.; Mwasumbi, L. B.; Hawthorne, W. J.; Dickinson, A.; Doggett, R. A. Preliminary assessment of the distribution, status and biological importance of coastal forests in Tanzania. *Biol. Conserv.* **1992,** *62*(3), 205–218.

Butler, P. M. Fossil macroscelidea. *Mammol. Rev.* **1995,** *25*, 3–14.

Butynski, T. M.; de Jong, Y. A.; Perkin, A. W.; Bearder, S. K.; Honess, P. E. Taxonomy, distribution, and conservation status of three species of dwarf galagos (*Galagoides*) in Eastern Africa. *Primate Conserv.* **2006,** *21*, 63–79.

Bytebier, B. *Taita Hills Biodiversity Project Report*. National Museums of Kenya: Nairobi, **2001,** p. 121.

Cheke, A. S. Was the Flying-Fox *Pteropus comorensis* on Mafia Island (Tanzania) Introduced by Humans. *J. East African Nat. Hist.* **2011,** *100*(1–2), 59–68. https://doi.org/10.2982/028.100.0104

Clarke, G. P. A new regional Centre of Endemism in Africa. In: Huxley, C. R.; Lock, J. M.; Cutler, D. F. (Eds.), *Chrology, Taxonomy and Ecology of the Flora of Africa and Madagascar*, **1998,** pp. 53–65.

Cockle, A.; Kock, D.; Stubblefield, L.; Howell, K. M.; Burgess, N. D. Bat assemblages in Tanzanian coastal forests. *Mammalia*, **1998,** *62*(1), 53–68. https://doi.org/10.1515/mamm.1998.62.1.53

Cooper-Bohannon, R.; Fahr, J.; Jacobs, D. *Kerivoula africana*. The IUCN Red List of Threatened Species **2020,** e.T10966A21975149. https://dx.doi.org/10.2305/IUCN.UK.2020- 2. RLTS.T10966A21975149.en

Denys, C.; Lalis, A.; Lecompte, É.; Cornette, R.; Moulin, S.; Makundi, R. H.; Machang'u, R. S.; Volobouev, V.; Aniskine, V. M. Inventory of fauna in the region of Kingu Pira, southern Tanzania and description of the new karyotypes of small mammals and the new development of rongeur Muridae (Mammalia, Rodentia). *Zoosystema*. **2011,** *33*(1), 5–47. https://doi.org/10.5252/z2011n1a1

Douady, C. J.; Douzery, E. J. P. Hedgehogs, shrews, moles, and solenodons (Eulipotyphla). In: Hedges, S. B.; Kumar, S. (Eds.), *The Timetree of Life*, Oxford University Press, **2009,** pp. 495–498.

Entwistle, A. C.; Juma, J. *Pteropus voeltzkowi*. The IUCN Red List of Threatened Species **2016,** e.T18768A22089205. http://dx.doi.org/10.2305/IUCN.UK.2016- 1. RLTS. T18768A22089205.en

Fitzgibbon, C. D.; Leirs, H.; Verheyen, W. Distribution, population dynamics and habitat use of the lesser pouched rat, *Beamys hindei*. *J. Zool.* **1995,** *236*(3), 499–512.

FitzGibbon, C. D.; Rathbun, G. B. *Rhynchocyon chrysopygus*. The IUCN Red List of Threatened Species **2015,** e.T19705A21287265. http://dx.doi.org/10.2305/IUCN.UK.2015- 2.RLTS.T19705A21287265.en

Francis, J.; Nilsson, A.; Waruinge, D. Marine protected areas in the Eastern African Region: How successful are they? *Ambio*. **2002,** *31*(7–8), 503–511. https://doi.org/10.1579/0044-7447-31.7.503

Gereau, R. E.; Cumberlidge, N.; Hemp, C.; Hochkirch, A.; Jones, T.; Kariuki, M.; Lange, C. N.; Loader, S. P.; Malonza, P. K. Globally threatened biodiversity of the Eastern Arc mountains and coastal forests of Kenya and Tanzania. *J. East African Nat. Hist.* **2016,** *105*(1), 115–201.

Habel, J. C.; Casanova, I. C. C.; Zamora, C.; Teucher, M.; Hornetz, B.; Shauri, H.; Mulwa, R. K.; Lens, L. East African coastal forest under pressure. *Biodiversity and Conservation*. **2017,** *26*(11), 2751–2758. https://doi.org/10.1007/s10531-017-1375-z

Hafner, M. S. Field research in mammalogy: An enterprise in peril. *J. Mammal.* **2007,** *88*(5), 1119–1128, https://doi.org/10.1644/07-MAMM-S-139.1

Happold, D. C. D. *Mammals of Africa Volume III—Rodents, Hares and Rabbits*. Bloomsbury Publishers: London, **2013,** pp. 27–691.

Holly, G.; Rey da Silva, A.; Henderson, J.; Bita, C.; Forsythe, W.; Ombe, Z. A.; Poonian, C.; Roberts, H. Utilizing marine cultural heritage for the preservation of coastal systems in East Africa. *J. Marine Sci. Eng.* **2022,** *10*(5), 693. https://doi.org/10.3390/jmse10050693

Honess, P. E.; Perkin, A. W.; Butynski, T. M. *Galagoides zanzibaricus* Zanzibar dwarf galago. In: Butynski, T. M.; Kingdon, J.; Kalina, J. (Eds.), *Mammals of Africa. Volume II: Primates*, Bloomsbury Publishing: London, UK, **2013,** pp. 447–449.

Kibet, S. Plant communities, species diversity, richness, and regeneration of a traditionally managed coastal forest, Kenya. *For. Ecol. Manag.* **2011,** *261*(6), 949–957. https://doi.org/10.1016/j.foreco.2010.11.027

Kingdon, J. *The Kingdon Field Guide to African Mammals.* A & C Publishers Limited: London, 2nd Edition, **2015,** pp. 204–358.

Kiwia, H. Y. D. Species richness and abundance estimates of small mammals in Zaraninge coastal forest in Tanzania. *Tanzania J. Sci.* **2009,** *32*(2), 51–60.

Koch, A.; Kaplan, J. O. Tropical forest restoration under future climate change. *Nat. Clim. Change* **2022,** *12*(3), 279–283. https://doi.org/10.1038/s41558-022-01289-6

Kock, D.; Csorba, G.; Howell, K. M. *Rhinolophus maendeleo* n. sp. from Tanzania, a horseshoe bat noteworthy for its systematics and biogeography (Mammalia, Chiroptera, Rhinolophidae). *Senckenbergiana Biologica* **2000,** *80*(1/2), 233–240.

Kock, D.; Stanley, W. T. Mammals of mafia Island, Tanzania. *Mammalia*, **2009,** 73(4), 339–352. https://doi.org/10.1515/MAMM.2009.046

Merritt, J. F. *The Biology of Small Mammals*. The Johns Hopkins University Press: Baltimore, Maryland, **2010.**

Milner, J. A.; Gaylard, A. *Dendrohyrax arboreus* Southern Tree Hyrax. In: Kingdon, J. S.; Happold, D. C. D.; Hoffmann, M.; Butynski, T. M.; Happold, M.; Kalina, J. (Eds.), *Mammals of Africa. Vol. I. Introductory Chapters and Afrotheria.* Bloomsbury Publishing: London, UK, **2013,** pp.152–155.

Mulungu, L. S.; Makundi, R. H.; Massawe, A. W.; Machang'u, R. S.; Mbije, N. E. Diversity and distribution of rodent and shrew species associated with variations in altitude on Mount Kilimanjaro, Tanzania. *Mammalia* **2008,** *72*(3), 178–185. https://doi.org/10.1515/MAMM.2008.021

Ndang'ang'a, P. K.; Eshiamwata, G.; Ngari, A.; John, J.; Arinaitwe, J.; Mbwana, S. *Status Report for the Eastern Arc Mountains and Coastal forests of Kenya and Tanzania Region, 2008*. Instituting a standardized sustainable biodiversity monitoring system in the Eastern Arc Mountains and Coastal forests of Kenya and Tanzania region" Critical Ecosystem Partnership Fund, **2008.**

Ngumbau, V. M.; Luke, Q.; Nyange, M.; Wanga, V. O.; Watuma, B. M.; Mbuni, Y. M.; Munyao, J. N.; Oulo, M. A.; Mkala, E. M.; Kipkoech, S.; Itambo, M.; Hu, G. W.; Wang, Q. F. An annotated checklist of the coastal forests of Kenya, East Africa. *PhytoKeys.* **2020,** *147*, 1–191. https://doi.org/10.3897/PHYTOKEYS.147.49602

Nkwabi, A. K.; John, B.; Kija, H.; Otsyina, R. M.; Monjare, J. F.; Kajuni, A. R. Abundance and distribution of small mammals relative to human activities in the wildlife management areas of Ruvuma landscape, southern Tanzania. *Int. J. Fauna Biol. Stud.* **2018,** *5*(2), 156–162.

Ntukey, L. T.; Munishi, L. K.; Treydte, A. C. Land use land/cover change reduces woody plant diversity and carbon stocks in a lowland coastal forest ecosystem, Tanzania. *Sustainability (Switzerland)*. **2022,** *14*(14), 8551. https://doi.org/10.3390/su14148551

Olson, D. M.; Dinerstein, E.; Wikramanayake, E. D.; Burgess, N. D.; Powell, G. V.; Underwood, E. C.; D'amico, J. A.; Itoua, I.; Strand, H. E.; Morrison, J. C.; Loucks, C. J.; Allnutt, T. F.; Ricketts, T. H., Kura, Y.; Lamoreux, J. F.; Wettengel, W. W.; Hedao, P.; Kassem, K. R. Terrestrial ecoregions of the world: A new map of life on earth: A new global map of terrestrial ecoregions provides an innovative tool for conserving biodiversity. *BioScience*, **2001,** *51*(11), 933–938.

Pollard, E. Present and past threats and response on the east coast of Africa: An archaeological perspective. *J. Coastal Conserv.* **2012,** *16*(2), 143–158. https://doi.org/10.1007/s11852-012-0204-5

Rescigno, V.; Maletta, S. Biodiversity hotspots. In: *Biodiversity Hotspots*. Nova Science Publishers, Inc., **2011.** https://doi.org/10.1017/9781009325844.018

Riggio, J.; Jacobson, A. P.; Hijmans, R. J.; Caro, T. How effective are the protected areas of East Africa? *Glob. Ecol. Conserv.* **2019,** *17*, e00573. https://doi.org/10.1016/j.gecco.2019.e00573

Rocliffe, S.; Udelhoven, J. *Protecting East Africa's Marine and Coastal Biodiversity: The Role and feasibility of Marine Conservation Agreements in the Western Indian Ocean*, **2011,** pp. 19–38.

Rodgers, W. A.; Owen, C. F.; Homewood, K. M. Biogeography of East African forest mammals. *J. Biogeogr.* **1982,** *9*(1), 41.

Russell, J. C.; Cole, N. C; Zuël, N.; Rocamora, G. Introduced mammals on Western Indian Ocean islands. *Global Ecol. Conserv.* **2016,** *6*, 132–144. https://doi.org/10.1016/j.gecco.2016.02.005

Sabuni, C. A.; Sluydts, V.; Mulungu, L. S.; Maganga, S. L. S.; Makundi, R. H.; Leirs, H. Distribution and ecology of lesser pouched rat, *Beamys hindei*, in Tanzanian coastal forests. *Integrat. Zool.* **2015,** *10*(6), 531–542. https://doi.org/10.1111/1749-4877.12159

Salehe, J.; Stirton, B. *The Eastern Africa Coastal Forests Ecoregion Strategic Framework for Conservation*, **2006,** pp. 2005–2025.

Shapiro, J. *Mops bakarii*. The IUCN Red List of Threatened Species **2017,** e.T67363010A67363012. http://dx.doi.org/10.2305/IUCN.UK.2017- 2.RLTS.T67363010A67363012.en

Shapiro, J.; Cooper-Bohannon, R.; Jacobs, D.; Cotterill, F. P. D.; Taylor, P. J. *Rhinolophus deckenii*. *The IUCN Red List of Threatened Species* **2020,** e.T19537A21979537. https://dx.doi.org/10.2305/IUCN.UK.2020-2.RLTS.T19537A21979537.en

Stanley, W. T.; Goodman, S. M.; Hutterer, R. Small mammal inventories in the East and West Usambara Mountains, Tanzania, 2. Families soricidae (Shrews) and macroscelididae (Elephant Shrews). *Life Earth Sci.* **2011,** *4*, 18–33.

Tabor, K.; Burgess, N. D.; Mbilinyi, B. P.; Kashaigili, J. J.; Steininger, M. K. Forest and woodland cover and change in coastal Tanzania and Kenya, 1990 to 2000. *J. East African Nat. Hist.* **2010,** *99*(1), 19–45. https://doi.org/10.2982/028.099.0102

Taylor, P. *Myonycteris relicta*. *The IUCN Red List of Threatened Species* **2016,** e.T14098A22046760. http://dx.doi.org/10.2305/IUCN.UK.2016-1.RLTS.T14098A22046760.en

Verheyen, W. N.; Hulselmans, J. L. J.; Dierckx, T.; Mulungu, L. The characterization of the Kilimanjaro *Lophuromys aquilus* TRUE 1892 population and the description of a new *Lophuromys* species (Rodentia, Muridae). *Biologie* **2007,** *77*, 23–75.

Wilson, D. E.; Mittermeier, R. A. *Vespertilionidae, Handbook of the Mammals of the World – Volume 9 Bats*. Lynx Edicions: Barcelona, **2019,** pp. 716–981.

World Bank. *Coastal Profile for Zanzibar 2014 Region Volume II Investment Prioritisation for Resilient Livelihoods and Ecosystems in Coastal Zones of Tanzania*. **2008,** *74*(1), 163–174.

CHAPTER 13

THE COASTAL FORESTS OF EAST AFRICA: INSIGHTS ON THREATS AND CONSERVATION

KELVIN NGONGOLO[1*], JOHN K. BUKOMBE[2], RICHARD D. LYAMUYA[2], NEDUVOTO P. MOLLEL[3], IMMACULATE K. CONSTANTINE[3], PIUS Y. KAVANA[2], SIRI A. ABIHUDI[4], ATUHOMBYE SIGALA[1], ASANTERABI LOWASSA[2], HAMZA K. KIJA[2], BARAKA NAFTAL[2], WILFRED MAREALLE[2], LAZARO MANGEWA[5], SAMUEL MTOKA[2], NAZA MMBAGA[1], FRANCO MBISE[1], MOSES KYANDO[1], FRANCIS CHEBBY[1], ALLY J. K. NKWABI[2], and EMMANUEL F. NZUNDA[6]

[1]*Department of Biology, College of Natural and Mathematical Sciences, The University of Dodoma, Dodoma, Tanzania*

[2]*Tanzania Wildlife Research Institute (TAWIRI), Arusha, Tanzania*

[3]*National Herbarium of Tanzania, Tanzania Plant Health and Pesticides Authority (TPHPA), Arusha, Tanzania*

[4]*Institute of Traditional Medicine, Muhimbili University of Health and Allied Sciences, Dar es Salaam, Tanzania*

[5]*Department of Wildlife Management, College of Forestry, Wildlife and Tourism, Sokoine University of Agriculture (SUA), Morogoro, Tanzania*

[6]*Department of Forest Resources Assessment and Management, College of Forestry, Wildlife and Tourism, Sokoine University of Agriculture (SUA), Morogoro, Tanzania*

*Corresponding author

ABSTRACT

This chapter provides insights on the threats and possible ways to preserve coastal forest, which form one of the most distinctive and threatened ecosystems in East Africa region and the world. Due to the high diversity of rare and endemic flora and fauna species, the East African coastal forests are ranked among the biodiverse regions of the entire globe. Coastal forests save as a rich sources of food, fuel wood, building materials, and medicinal plants. They help to maintain a regular water supply for both humans and livestock. However, this valuable biodiversity is seriously disintergrated by a number of threats, including habitat

Biodiversity Hotspot of the Coastal Forests of Eastern Africa. T. Pullaiah and Ally J. K. Nkwabi (Eds.)

DOI: 10.1201/9781998511099-13

degradation, fragmentation, invasive species, climate change, and deforestation. This chapter examines the myriad threats leading to the destruction of coastal forests, as well as the wide-ranging regional and global implications. By analyzing the dynamics of the interplay between human actions and environmental systems, it elicits critical challenges to sustainable conservation including the unstable legal and institutional frameworks along with socioeconomic problems, hinder these efforts. This chapter provides immediate solutions to reduce the fast deterioration of forest resources while underlining the necessity of research and public awareness campaigns, and enhancing global collaboration. In the end, the chapter highlights the fundamental notion that safeguarding biodiversity in East African coastal forests requires collaborative efforts at local, regional, and global levels.

13.1 INTRODUCTION

The coastal forests of East Africa sustain an astonishing array of species, many of which are utterly unique to our planet. These lush, biologically diverse environments, which may be found along the shores of Somalia, Kenya, Tanzania, and Mozambique, are attributed to the glories of nature. But beneath the colorful canopy is a tale of tremendous beauty and impending danger. Nestled along the Eastern shores of Africa, the coastal forests of East Africa stand as sentinel guardians of an irreplaceable treasure of unparalleled biodiversity richness and significance. For instance, a study in the Namatimbili forest in the southern coastal forests of Tanzania revealed a total of 312 plant species in 62 families dominated by Fabaceae (51 species) of which 26 species are cited under the International Union of Conservation of Nature (IUCN) red list and 36 are endemic species of East African Coastal Forests (Mligo, 2015). These ancient woodlands, stretching from the southern tip of Somalia to the lush hills of northern Mozambique, have long captivated the imaginations of biologists, conservationists, and explorers alike (Tabor et al., 2010). Within their emerald embrace lies a mosaic of lifeforms, many of which are found nowhere else on our planet. Taking another example from a study conducted in the coastal forests of northeastern Tanzania, it was reported that 57 species of birds are found within 27 families whereby 13 species were forest specialists 15 were forest generalists, 14 were forest visitors, and 15 were savannah/woodland species (Modest and Hassan, 2016).

The Eastern African coastal forest serves ecosystem provisioning and cultural services in the form of social and economic benefits to the inhabitants who reside adjacent to these ecological assets. In accordance with a study carried out in Kichi Forest among coastal forests in Tanzania, the forest provided habitat for pollinating for sesame crops planted alongside (Ngongolo et al., 2015a).

13.1.1 HABITAT RESTORATION

Additionally, in Wazo Hill coastal forest in Tanzania, studies demonstrated increased productivity through seed setting for *Leucaena leucocephala* plants, which in turn provided potential homes for butterfly survival and created spaces for butterfly viewing (aesthetic value) (Ngongolo et al., 2014a, 2014b; Ngongolo and Mtoka, 2013). Mangrove forests have been observed to be degrading in coastal forests regardless of their importance, such as, ecosystem service provision, nutrient recycling, flood control, sediment trapping, and erosion barriers that demand restoration (Ngongolo et al., 2015b). In Dar-Es-Salaam, the restoration of mangroves was reported to be taking place with emphasis on women to participate in restoration and conservation efforts (Sabai, 2019). Human activities are the major threat to the Mangrove ecosystem, for instance, in Kunduchi and Mbweni. Major threats, such as, salt works and settlement in Kunduchi and settlement and trampling in Mbweni, were identified and 40% of Kunduchi's mangroves and 31% of Mbweni's were vulnerable to human pressures (Mabula et al., 2017). All these demand human initiative and provide opportunities for restoration programs for the sustainable conservation of mangroves (Ngongolo et al., 2015b). Mangrove restoration gained global attention for its ecological importance. In Kenya, it was observed that there is a significant mangrove loss of 600 ha/year. This harms the local economy, agriculture, and marine resources (Kairo, 2001a; 2001b). Gazi Bay's community-led efforts planted 300,000 trees, but over-exploitation persists due to awareness and economic challenges. Urgent action and attitude shifts are needed (Kairo, 2001a, 2001b). On the other hand, the conservation and restoration of the coastal forest at Gede National Monument in Kenya, which surrounds the historic stone city ruins, has been taking place with the aim of restoring a 5-hectare degraded forest, conducting enrichment plantings, and training local staff in conservation techniques (Robertson et al., 2002). However, it has been reported that Achieving ambitious Coastal forest-like mangrove restoration goals demands a holistic, evidence-driven approach that integrates multiple disciplines and stakeholders to ensure success in preserving these critical coastal ecosystems (Friess et al., 2022).

13.1.2 ECOLOGICAL AND SOCIOECONOMIC BENEFITS OF COASTAL FOREST

The coastal forests of East Africa have important plant species that are used for pharmacological and health benefits. A good example was provided from a study on the medicinal plants used by people from Duruma, Giriama, and Digo tribes of the Midzichenda (also known as Mijikenda) in Kenya's coastal forest (Githitho, 2003). The study found that 56% of all plant species in that particular forest community were utilized for both routine medical health care and cultural rituals (Pakia et al., 2003). Due to the important role of plants in human beings, a number of plant species are overharvested for medicinal uses. Additionally, logging and the exploitation of wood for fuel in the form of fuel wood, charcoal, timber, and construction poles are other dubious benefits (Burgess et al., 2017). Coastal forests provide benefits such as areas for biodiversity hotspots, habitat for endangered species, carbon sequestration, erosion control, and water regulation. On the other hand, they play a key role in socioeconomic benefits like a source of food, medicine, and materials for traditional crafts, employment opportunities in the tourism and forestry sectors, the tourism sector through aesthetic value, traditional knowledge about forest resources, preserving cultural heritage, fisheries support, buffering against the impacts of climate change by reducing storm damage, maintaining stable microclimates, and ensuring a sustainable supply of freshwater. These ecosystems serve as living laboratories for ecological research, conservation efforts, and environmental education, promoting awareness and understanding of biodiversity and natural processes (de la Torre-Castro and Rönnbäck, 2004). Coastal forests offer opportunities for diversifying local economies beyond traditional sectors, promoting sustainable development, and reducing dependence on resource extraction. These forests hold cultural and spiritual importance for many indigenous and local communities, playing a role in traditional ceremonies and belief systems (Pakia et al., 2003). A study to understand the Links between humans and seagrasses from tropical East Africa showed that Fishing grounds, seaweed cultivation areas, and bait collection sites played pivotal roles for fishermen and farmers. Seagrass-supported trap fisheries proved the most lucrative, offering the highest daily income per fisherman while supplying vital animal protein. Seagrasses served multiple purposes, including traditional medicine and fertilizers, and provided aesthetic, instrumental, spiritual, and religious value. Local ecological knowledge among fishermen was extensive, but no specific institution managed seagrasses (de la Torre-Castro and Rönnbäck, 2004). A study on Mangroves that are found in major deltas of Africa and Asia including the Rufiji delta of Tanzania, showed that, between 2000 and 2016, around 1.6% (12,270 hectares) of total mangrove area in river deltas vanished, mainly due to erosion and agriculture conversion (Lagomasino et al., 2019). However, some regions saw rapid mangrove expansion, resulting in taller, carbon-dense forests that offset 44% of the total carbon loss (5,332,843 Mg C) (Lagomasino et al., 2019). New mangrove growth, regardless of net area change, stored 84%–160% of aboveground carbon lost (Lagomasino et al., 2019).

Despite the ecological and socioeconomic benefits of coastal forest biodiversity resources, the environment is nevertheless facing land use and climate change that necessitate conservation intervention efforts from various stakeholders. This chapter sets out on a voyage across this magical world in an effort to sort through the complex changes that pose threats to its biodiversity. We will show some of the impacts of land use and climate change that these forests face and highlight the tireless efforts of communities and conservationists.

13.1.3 SCOPE AND PURPOSE

Within the larger context of ecological and conservation studies, this fulfills a crucial function. Its scope includes a thorough investigation of the particular difficulties that East African coastal forests, which are among the most biodiverse and vulnerable ecosystems in the world, face. An in-depth study of the several threats that these forests face, including habitat loss and climate change, as well as how they affect the region's abundant biodiversity, is covered in this chapter. The chapter delves into an in-depth analysis of the multitude of threats these forests encounter, ranging from habitat loss to climate change, and their cascading impacts on the region's rich biodiversity. Furthermore, the chapter aims to highlight and propose effective conservation strategies and management practices to safeguard these invaluable ecosystems for future generations. Overall, it provides an essential resource for researchers, policymakers, and conservationists working to protect the ecological treasures of East African coastal forests.

13.2 MATERIALS AND METHODS

A detailed literature review on threats to biodiversity and conservation efforts in East African coastal forests was conducted using a database search including PubMed using Boolean operators and Google Scholar. A combination of keywords related to coastal forest biodiversity, risks, and conservation in the East African context were searched (Table 13.1). Only peer-reviewed articles, reports, and reputable publications were taken into consideration, and studies pertinent to the East African coastal forest region were given precedence. The qualitative data obtained was analyzed by using content analysis. The analysis was initially done by reading the information obtained from various searches followed by highlighting the text that appeared to describe key themes (Lund et al., 2011; Stemler, 2015). The keywords were written in the margin of the text (open coding). After coding all the transcripts, data within a particular code were examined, leading to a process of combing some codes, and others were split into subcategories, including habitat loss, climate change, invasive species, conservation strategies, recurring trends, and research gaps. By adhering to these methods, we present a comprehensive and evidence-based analysis of the threats and conservation strategies pertaining to biodiversity in East African coastal forests (Figure 13.1).

TABLE 13.1 Keywords Were Used to Search Information of Possible Threats and Conservation Strategies and Possible Combination for Searching PubMed Using Boolean Operators.

Sr. No.	Keywords Used for Searching	Combining Keywords Using Boolean Operators (AND, OR) for Searching in PubMed Website (https://pubmed.ncbi.nlm.nih.gov/)
1	Coastal forest biodiversity	1. ("Coastal forest biodiversity" OR "East Africa conservation") AND ("Biodiversity threats" OR "Conservation challenges")
2	East Africa Conservation	
3	Biodiversity threats	2. "Coastal forests of East Africa" OR "East African coastal forests" OR "East Africa's coastal biodiversity"
4	Coastal forest ecosystem	3. "Biodiversity conservation" OR "Conservation strategies" OR "Ecosystem protection"
5	East African biodiversity	
6	Conservation challenges	4. "Threats to biodiversity" OR "Ecological challenges" OR "Conservation issues"
7	Biodiversity preservation	
8	Coastal forest conservation	5. ("Coastal forest biodiversity" OR "East Africa conservation") AND ("Biodiversity threats" OR "Conservation challenges")
9	East African ecosystems	
10	Threatened species East Africa	6. ("Coastal forests" OR "coastal woodland") AND ("East Africa" OR "Eastern Africa" OR "East African coast") AND ("Biodiversity conservation" OR "conservation of biodiversity" OR "biodiversity threats").

13.3 THREATS TO BIODIVERSITY ALONG THE COASTAL FORESTS

The coastal forests of East Africa face a multitude of threats to their biodiversity, which collectively jeopardize the delicate balance of these unique ecosystems. Some of the primary threats include the following: increasing infrastructures and other new economic developments mainly attributed by the ever-increasing human population for East African countries and global overseas economic demands, invasive nonnative species, overharvesting or overexploitation of particular species, pollution, climate change, and unsubscribed fire.

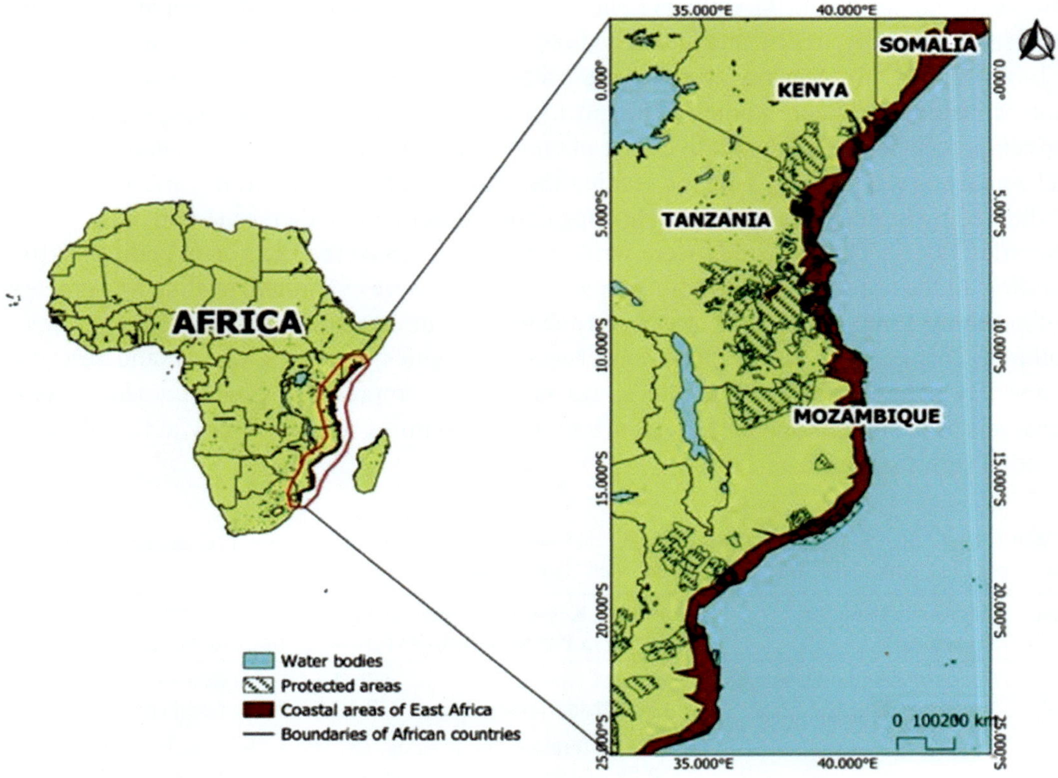

FIGURE 13.1 Map showing waterbodies including Indian Ocean and coastal areas of Eastern Africa, also shows the conservation actions along the coastal forest by showing protected areas along the coastal ecosystem of Eastern Africa.

13.3.1 INFRASTRUCTURES AND NEW ECONOMIC DEVELOPMENTS

Coastal infrastructure and other economic development threaten the biodiversity resources in the East African coastal areas. The rapid growth of coastal cities and tourism infrastructure often leads to land reclamation and alteration of coastal habitats (Abuodha and Kairo, 2001). These changes can result in habitat alteration, pollution, and increased human–wildlife conflicts as the development in the coastal area of East Africa persists (Thoya et al., 2022). For instance, Kenya and Tanzania are doing mega development along the coastal line, one being the construction of ports to support economic growth and expansion of the ocean economy in East Africa. Furthermore, little is known about how coastal human communities are negatively impacted by the increasing coastal and marine development due to habitat change (Thoya et al., 2022). According to a study done in Kenya, the absence of huge corals near new resorts, was related to human activities and proof of coral damage and clearing of associated coral rag forests resulting from constructions of buildings. Pollution and sewage are a few of the detrimental repercussions (Cowburn et al., 2018). Additionally, it was alleged that 100 ha of the Lamu mangrove ecosystem had been lost. Mangrove lands in Ngomeni and Karawa in the coast of Kenya had been bulldozed for salt production (Abuodha and Kairo, 2001). Gazi Bay on the southern coast of Kenya has been cleared for fuelwood, about 100 ha have been repurposed for aquaculture. In Makupa Creek (Kenya) 10 hectares perished as a result of oil pollution (Abuodha and Kairo, 2001). However, the need for development increases as the population increases, and new threats to biodiversity result from human pressure (LeMarie et al., 2006). In accordance with other report, urbanization, agriculture, and aquaculture are causing mangrove stands on coastal line of East Africa to be highly reduced (Martin and Burgess, 2023). Mangrove vegetation is important as it gathers sewage, industrial pollutants, sediment, and pesticide runoff, which causes a change in the seagrass habitat and prevents the silting of the coral reef and coastal erosion. However, their survival is threatened by land use change in the name of economic development and climate change, which is causing the rising of sea levels (Martin and Burgess, 2023).

13.3.2 INVASIVE NONNATIVE SPECIES

The protection of biodiversity in the coastal forests of East Africa is also negatively impacted by invasive species. Flora and fauna that are found thriving in a certain region or nation can be either indigenous to the area or those that are introduced from another area. On one hand, some of these introduced plant species can be beneficial to the human community when they provide food, shelter, medicine, and aesthetic enjoyment. While on the other hand, they can have significant impacts on both human activities and the environment. Native ecosystems undergo disruption when invasive species are introduced into them. Native species can become endangered or go extinct as a result of invasive plants, animals, and viruses that disrupt food networks and outcompete native species. In a report for Anon, the annotated checklist of Kenya's coastal woodland in East Africa, found about 72 alien plant species were listed (Ngumbau et al., 2020). Also, a study in Sacred sites, which are hotspots for biodiversity in coastal Kenya revealed a total of 121 plant species in 43 families and 97 genera, including invasive species like *Azadirachta indica* (Metcalfe et al., 2010). Research from Kenya and Tanzania has identified ten significant invasive plant species that have an impact on the dry lands and coastal ecosystem include: *Arenga pinnata, Rubus pinnatus, R. alba, Maesopsis eminii, Lantana camara, Senegalia polyacantha, Senegalia nigrescens, Cedrela odorata, Datura stramonium,* and *Argemone mexicana* (Lyimo et al., 2009). Their impact on human activities can be associated with livestock production, including interfering with grazing and farming practices, lowering yield and quality of forage, increasing costs of managing and producing livestock, slowing animal weight gain, reducing the quality of meat and milk as well as poisoning livestock, they can kill livestock by poisoning and ruining their foliage. Moreover, invasive alien species can expedite biodiversity loss through the elimination of native plants, and increase disease by providing a breeding ground for mosquitoes and other insects that spread diseases like Trypanosomiasis (Obiri, 2011). The introduction of House Crow (*Corvus splendens*) in Zanzibar provided an example of the impact of an invasive animal species on native biodiversity along the coastal areas of East Africa (Nkwabi et al., 2018; Ndimuligo et al., 2022). There is a notable higher density and distribution of house crows which resulted into suppression of native crows and songbirds along coastal areas of East Africa. House Crow is detrimental to indigenous bird species in which they predate or provide competition and displace other birds by attacking, mobbing and destroying eggs as well as nests (Nkwabi et al., 2018).

13.3.3 HUMAN POPULATION GROWTH

With an average growth rate of 6.7% between 2013 and 2017—double the African average—East Africa is one of the fastest-growing regions in the world, which goes with fast social enhancements (UNECA, 2018). Human population increase rates and the associated expansion of subsistence agriculture, may permanently convert natural habitats to farmland is a major threat to coastal forests biodiversity (Gachanja and Kanyanya, 2004). Population increases are also linked to habitat degradation associated with increased demand for firewood, charcoal, timber, and bushmeat. The above is compounded when rural people out of employment are involved in commercial activities such as industrial fuel wood collection, commercial pole cutting, and charcoal production as is observed in Kwale District along the Coast of Kenya (Gachanja and Kanyanya, 2004). For example, in Kenya, it was found that there was the highest rate of forest loss in coastal Kenya with the potential for species extinctions occurring in Kilifi County (Karyn et al., 2010).

13.3.4 AGRICULTURAL PRACTICES

Much of the Coastal Forests of Eastern Africa, biodiversity Hotspot's original forests have been lost to agricultural conversion and urbanization; only about 10% of the original vegetation remains in pristine condition (Bullock et al., 2021). Most agricultural development involves short-term shifting cultivation concentrating on food crops such as cassava, maize, banana, pawpaw, and coconut. The human population is increasing, and the demand for additional farmland increases every year. Commercial agricultural development, in the form of coconut, sisal, clove, cardamom, and cashew nut plantations, has also led to the loss of lowland coastal forests and other natural habitats. This can be exemplified by agriculture expansion along the coastal forests of Kenya which is still the biggest industry and is the main source of income for the majority of the population.

13.3.5 CHARCOAL BURNING AND FUEL WOOD

Wood fuel is the dominant source of energy for the coastal people in East Africa, and accounts for 90% of rural household energy, and 85% of urban household energy consumption volumes (Gachanja and Kanyanya, 2004). During a study along the coastal areas, Gachanja and Kanyanya (2004) observed that in the urban centers, charcoal is the most used fuel for cooking, while kerosene is used by both urban and rural households for cooking and lighting (Godoy et al., 2012). According to Godoy et al. (2012), within the Dar es Salaam and Coastal regions, forest use and degradation are highly related to distance to Dar es Salaam city. In this regard, burning of woody plants for charcoal production causes major habitat loss near coastal towns and alongside main roads in Tanzania and other parts of the Eastern African coastal forests. In addition, collection of firewood poses a threat in areas away from towns and roads. For example, charcoal burning seriously threatened the Coastal Forests because of the high demands for charcoal in urban centers and for the illegal but lucrative export market (Gachanja and Kanyanya, 2004). According to Gachanja and Kanyanya (2004), in Kwale, it had been estimated from District Forest Officer permit records that 45,000 bags of charcoal were legally transported from Kwale to Mombasa every month in 2001. Therefore, illegal movements are estimated to be more than three times as high. However, forests close to tourist areas, such as Arabuko-Sokoke Forest near Malindi and Watamu in Kenya, suffer from the high demand for wood carving (*Brachylaena huillensis*) and timber for the construction of hotels, private residences, and tourist attractions. Uncontrolled burning to clear farmland, to drive animals for hunting, to collect honey, and to reduce tsetse flies also threatens lowland coastal forests and thicket patches, often replacing rare, endemic coastal forest species with more common wide-ranging, fire-adapted species.

13.3.6 UNPRESCRIBED FIRE

Fire is a threat to all coastal forests and mangroves to some extent especially during the dry seasons (Gachanja and Kanyanya, 2004). The coastal traditional fire-use system in land clearing is destructive to biodiversity and a cause of uncontrolled wild fires that indiscriminately destroys forest and bushland ecosystems. Ecological imbalance arises when a high frequency of uncontrolled wild fires tends to maintain certain types of vegetation, to the detriment of narrow endemic Coastal Forest specialist species (Gachanja and Kanyanya, 2004). Apart from arsonists, forest fires were caused by honey hunters, grazers, smokers, and from adjoining farmers clearing land for crop production (Gachanja and Kanyanya, 2004). A study in Selous-Rufiji coastal landscapes showed that wildfires in the Selous Game Reserve are primarily caused by human activities such as poaching, charcoaling, agriculture, pastoralism, fishing, and residential protection (Ngongolo and Bruno, 2015). Socioeconomic activities such as charcoal production and pastoralism contribute to wildfire occurrence (Ngongolo and Bruno, 2015). Fire may negatively impact diversity, but not the insects species abundance (Ngongolo and Bruno, 2015). The local communities' attitudes vary due to socioeconomic differences.

13.3.7 POACHING AND OVERHARVESTING OF FOREST RESOURCES

Poaching of animals for bush meat and trophies, particularly in protected areas, has resulted into depletion of animal population, and this poses another major threat to biological diversity. African elephants and rhinoceros have, particularly, been very vulnerable to poaching. Consequently, the black rhinoceros have been reduced to quite a few, mainly in Nyerere–Selous protected area of the coast. Coastal region in East Africa are also in danger of losing the elephant as a species of significant landscaping and commercial value. The fauna of coastal forests is seriously threatened by poaching for bushmeat, traditional medicine extraction, and the illegal wildlife trade; iconic species like elephants, primates, and big cats are especially at risk. Human activities are also threatening aquatic resources; dynamite fishing destroys not only coral reefs, which are crucial breeding grounds for many fish and other marine species, but also nontargeted fish species.

The findings in Saadani National Park within the coastal ecosystem of Tanzania have shown that poaching is rampant because communities in that place have a tradition of hunting mainly for bush-meat (Muganda, 2019). Poaching has also been reported in Arabuko Sokoke Forest, Coast-Kenya, as among the illegal activities in Kenya's coastal forest (Kaleb and Humphrey, 2020). In Arabuko Sokoke Forest, Coast-Kenya, poachers avoid major roads to avoid confrontations with rangers and forest guards. Vehicle passage through the forest significantly affects activities, with lower illegal tree-cutting rates near major roads and higher values near paths. The distance inside the forest increases tree cutting, making the distances from the boundary crucial for controlling illegal activities. Different sites have different poaching rates due to availability of raw materials and poachers. Forest guards are few and poorly equipped, making arrests difficult. Design strategies for curbing a combination of these factors are crucial for controlling illegal activities in the forest (Kaleb and Humphrey, 2020). This chapter also proposes strategies for curbing poaching or overharvesting of forest resources along the coastal ecosystems of Eastern Africa.

An increase in illegal anthropogenic activities including cultivation in most of the coastal forest in East Africa has caused an increase in human–wildlife conflict as the area has diverse wildlife species like African Elephant, African buffalo, samango monkeys, vervet monkeys, bushbabies, duikers, and birds such as African fish eagles, palm-nut vultures, brown-headed parrots, and various species of hornbills and turacos. As human populations encroach on wildlife habitats, conflicts between humans and wildlife can escalate, leading to retaliation against wildlife and further threatening certain species. In Saadani National Park along the coastal regions of Tanzania, it was observed that, to resolve the conflict between humans and wildlife, engagement of local communities adjacent to the protected area is vital. Strengthened law enforcement and increased community awareness through conservation education are also important (Muganda, 2019). In Kichi Forest, where illegal activities like poaching, and lumbering are reported, enhancing community participation through exposing the benefits obtained from the forest, such as pollination services were considered to be essential (Ngongolo et al., 2015a). The study conducted in the Shimba Hills ecosystem in Kenya concluded that distances to water, fences, roads, and settlements were related to conflict sites and nonconflict sites between humans and elephants. This called for increased community awareness on early detection of Human Elephant Conflicts and the necessary mitigation measures (Wanyingi et al., 2013).

As human populations' increases, forests are cleared to make way for agriculture, transportation and communication infrastructures, settlements, and industries, resulting in the loss of critical habitat for many species. A study done in Arabuko Sokoke forest revealed that deforestation and biodiversity loss driven by illegal logging, charcoal burning, and poaching of wild animals (Habel et al., 2017). Also, it was noted that agriculture in low-land coastal forests in Tanzania, reduced wood plant diversity and stocking density, which suggests that, the expansion of Agricultural activities in and around the coastal forests of East Africa has detrimental effects on the conservation of biodiversity resources (Ntukey et al., 2022).

13.3.8 ILLEGAL LOGGING

Over-exploitation of plant resources in the coastal forests of Eastern Africa can have significant negative effects on biodiversity. Collection of plant resources like timber, medicinal plants, or nontimber forest products, can lead to habitat alteration and fragmentation. This directly reduces the available living space for many species, which can lead to population declines or even local extinctions. These can lead to the loss of genetic diversity within these species. Reduced genetic diversity can make plants more susceptible to diseases, pests, and environmental changes, making them less resilient. A good example of over-exploitation of plant resources is Illegal logging and deforestation which is another threat in most of the coastal forests of East Africa (Nzunda and Midtgaard, 2019). The illegal logging of valuable timber species remains a pervasive problem, depleting forest resources and disrupting ecosystems. Illegally harvested wood is often exported to international markets, contributing to the black-market timber trade in Tanzania. In addition to agriculture, pole cutting, and logging for timber and fuel were reported to threaten biodiversity in coastal forests (Burgess et al., 1992). A study conducted in Mozambique's coastal forest found that logging was one potential threat to the preservation of biodiversity that impacted the distribution of tree species (Joaquim-Meque et al., 2023). The East Coast Akalat (*Sheppardia gunning*), Reichenow's Batis (*Batis mixta reichenowi*), the Spotted Flat-lizard (*Platysaurus maculatus*), the Woodland Toad (*Mertensophryne micranotis*), and the shrub (*Gardenia transvenulosa*) are just a few of the endemic species that are threatened by forest disturbances in the Coastal Forests of Mtwara, Tanzania (Wegner et al., 2009). The disturbances were associated with clearing the vegetation through logging, lumbering, Agriculture practices, and fire burning (Wegner et al., 2009). In addition, population increase has catalyzed the increase in anthropogenic activities such as logging which threatens more species that are found in the coastal forests in East Africa, including Kenya (Fungomeli et al., 2020). Tropical African hardwood species like *Scorodophloeus fischeri* (Fabaceae, subfamily Caesalpinoideae) and *Manilkara sulcata* (Sapotaceae) are native to East African coastal forests. Human activities like logging and farming in some woods past the point of natural regeneration pose a threat to some plant species (Mligo et al., 2011). It was determined that *Scorodophloeus fischeri* has a high potential for spontaneous regeneration in Zaraninge forest in Saadani National Park with the least amount of human disturbance. The variations in *Scorodophloeus fischeri* and *Manilkara sulcata*'s potential for natural regeneration in these woods were influenced by disturbances caused by human activities (Mligo et al., 2011).

13.3.9 WILDLIFE POISONING

Wildlife poisoning poses a severe threat to the delicate ecosystems of coastal forests in East Africa (Ogada et al., 2022). This clandestine practice involves the use of toxic substances, such as, Carbofuran pesticides and poisoned bait, to eliminate perceived threats like predators or pests. Tragically, it often results in unintended casualties among nontargeted wildlife, including threatened species like lions, leopards, and vultures. In Dar es Salaam, Tanzania, for instance, the House Crow (*Corvus splendens*) has been controlled by poisoning to minimize the threats it poses to humans (Shimba and Jonah, 2017). In Kenya, it was reported that common reasons for using poison among the communities living within the coastal ecosystem area were to control damage-causing animals, harvest fish, harvest traditional medicine, poach for wildlife products, and kill sentinels, which led to recommendations such as banning use of pesticides, awareness campaigns, improving regulations, enhancing enforcement, increasing international support, community engagement to promote coexistence, and developing regional pesticide centers (Ogada, 2014). Coastal forests in East Africa are biodiversity hotspots, harboring unique flora and fauna. Poisoning disrupts the natural balance, leading to imbalances in predator-prey relationships and the decline of vital species as the poison passes through the food chain. A study in Tanzania examined trace metals in 60 sediment samples and 160 giant tiger prawns from the Tanzanian coast, assessing their distribution and potential threats to mangrove fauna. High levels of chromium, cobalt, copper, iron, manganese, nickel, and vanadium were found in the mangroves of the rivers Pangani, Wami, and Rufiji, primarily from weathering and erosion (Rumisha et al., 2016). This not only threatens local wildlife but also disrupts the delicate ecological web, impacting forest health and the livelihoods of communities dependent on these resources.

13.3.10 LAND USE LAND COVER CHANGES

Land use and land cover changes were among the results of agricultural operations that boosted soil erosion in Kenya's Taita Hill; new croplands are projected to be planted mostly in the lowlands, which include regions with lower soil erosion potential (Maeda et al., 2010). Relating finding in Kenya and Tanzania coastal forests have revealed that, these anthropogenic activities have caused loss of habitats to different biodiversity species thus threatening different species in the ecosystem (Gereau et al., 2016). For instance, in the year 2015, the number of threatened class for animals like Amphibia, Aves, Gastropoda, Insecta, Malacostraca, Mammalia, Reptilia, and class for plants such as Bryopsida, Cycadopsida, Jungermanniopsida, Liliopsida, Lycopodiopsida, Magnoliopsida, Pinopsida, Polypodiopsida. Due to deforestation and habitat loss in coastal forests were; 54 in Bagamoyo District, 34 in Dar es Salaam, 34 in Handeni District, 9 in Jozani Forest Reserve of Zanzibar, 33 in Kilifi District in Kenya, and 32 in Kwale District in Kenya (Gereau et al., 2016).

13.3.11 POLLUTION AND HABITAT DEGRADATION

Pollutants from agricultural runoff, industrial activities, and urban areas can contaminate water sources and disrupt ecosystems. This pollution affects both terrestrial and aquatic species in coastal areas. In Tanzania and Mozambique, the study revealed that floating anthropogenic litter, including meso- and micro-litter, is pervasive in the coastal waters of East Africa, particularly in Tanzania and northern Mozambique. The litter load increases with distance from major coastal cities, with 95% of meso- and micro-litter being plastic, and only 6% of microfibres (Weideman et al., 2023). The study by Shilla (2016) investigated the distribution and environmental implications of heavy metals in water, sediment, suspended particulate matter, and biota from two marine coasts in Zanzibar, Tanzania. The results show higher metal concentrations in Malindi compared to Fumba coast, suggesting pollution from anthropogenic activities. The study emphasizes the importance of understanding that metal concentrations in coastal environments depend on industrial and household waste inputs and the geochemical composition of the area (Shilla, 2016). A study in Zanzibar showed that, Coastal pollution in Zanzibar is causing serious environmental and economic damage. Untreated municipal wastewater is affecting mangroves, seagrass beds, and coral reefs, threatening livelihoods and causing waterborne diseases. Sewage pollution also threatens the tourism trade. A holistic approach is suggested, integrating sewage management within integrated coastal zone management initiatives to address the pollution problem (Mohammed, 2002). A study in Tanzania revealed that mangrove was found to be highly polluted with heavy metals, including lead, mercury, and zinc, due to urbanization and industrial activities. The metal pollution index was high in areas like Pangani, Saadani, and Rufiji, indicating human activities (Rumisha et al., 2016).

13.3.12 CLIMATE CHANGE

East African coastal forests are significantly impacted by higher temperatures, changing precipitation patterns, and frequency of extreme weather events. This involves modifications to the timing of flowering and fruiting as well as increased susceptibility to pests and diseases. The vulnerability of coastal resources to climatic change in Mozambique paves a need for integrated coastal zone management as an illustration of this (Chemane et al., 1997). Also, in Mozambique, it has been observed that the expansion of savannas at the expense of forests in the coastal lowlands of Mozambique has been attributed to anthropogenic influence; additionally, climate change, which has caused an increase in drought, has been shown to affect the biodiversity resources in coastal forests in Mozambique; this suggests that the forest fragments currently present on the Mozambique coast are naturally vulnerable to climatic stress and as such are highly sensitive areas to future climate change (Ekblom, 2008). The majority of the Rondo dwarf galago's habitat and distribution range will be exposed to a temperature increase of 1–1.3°C by 2050, which when combined with other threats is likely to further endanger the species survival. The Rondo dwarf galago (*Paragalago rondoensis*) is an endangered primate found in the coastal forests of eastern Tanzania (Haji and Bakuza, 2023). Temperature was found to have an impact on species distribution in an evaluation to predict the spread of the Critically Endangered (CR) *Karomia gigas* tree species under climate change in Tanzania (Mapunda and Andrew, 2023). The strongest predictive contribution and most helpful data to forecast the likely future distributions of *K. gigas* in the forest reserves were found in the annual temperature range. According to the study, between 2050 and 2070, Tanzania's habitat suitability distribution for *K. gigas* will dramatically rise (Mapunda and Andrew, 2023). Addressing these threats requires a multifaceted approach, including habitat protection, sustainable resource management, community involvement, and global cooperation. Conservation initiatives must work to balance the needs of both human communities and the rich biodiversity of the coastal forests of East Africa to ensure their long-term survival.

13.4 IMPACT ANALYSIS OF THE THREATS TO BIODIVERSITY

The delicate flora ecosystems and the fauna that call them home are severely impacted by the challenges to biodiversity along East Africa's coastal forests. These far-reaching repercussions include ecological, economic, social, and cultural effects. Threats to biodiversity have caused species local extinction and loss of genetic diversity in the ecosystem. In these woods, many species have declined and even gone extinct as a result of habitat loss, fragmentation, and poaching. The risk of extinction for many rare and endemic species is rising as their habitats disappear. For example, a rare plant species *Solanum ruvu* is thought to be extinct along the coastal areas of East Africa, and such disappearances were related to prolonged human activities toward areas where it was found (Vorontsova and Mbago, 2010). These species have not been seen again although there is great effort by scientist to search them from different habitats (Vorontsova and Mbago, 2010). A new species, *Solanum ruvu*, was described from only one collection, with unusually long inflorescences and dense prickles. It likely shares affinities with another coastal forest species, *Solanum zanzibarense.* The Ruvu Forest that is the only locality for *Solanum ruvu* is increasingly populated with rice and sesame farming, making recollection efforts unsuccessful (Vorontsova and Mbago, 2010). Another endangered species included the CR *Aloe boscawenii* Christian, missing since 1953, was rediscovered in 2019 in Tanga region (Figure 13.2a). Threatened by salt mines and flooding, conservation efforts are crucial (Ojija, et al., 2022). Other plant species of conservation concern include *Uvaria puguensis* D.M. Johnson, listed as CR and *Uvaria pandensis* Verdc listed as endangered and *Dalbergia melanoxylon* which is Near threatened as per IUCN Red Listed (EAMCFCPAP, 2020). The native range of both *Uvaria puguensis* (Figure 13.2b) and *Uvaria pandensis* (Figure 13.2c) is Tanzania and grow primarily in the wet tropical biome.

Findings on the eastern tree hyrax (*Dendrohyrax validus*) populations in coastal Kenya's woodlands revealed that conservation efforts are essentials to save these species isolated *D. validus* from local extinction and loss of genetic diversity (Rosti et al., 2023).

Coastal forests provide essential ecosystem services, such as carbon sequestration, water purification, and pollination. Their degradation can result in decreased capacity to provide these services, impacting human well-being and livelihoods. For instance, if mangroves are threatened and destroyed, ecosystem services which are likely to be lost includes regulatory services of mangroves such as erosion control and sediment accumulation (Friess, 2016). This suggests that, biodiversity loss may lead to increased Vulnerability to Natural Disasters as Coastal forests provide natural buffers against storm surges and coastal erosion. Their degradation can leave communities more vulnerable to the impacts of natural disasters.

FIGURE 13.2 A= *Aloe boscawenii* classied as Critically Endangered by IUCN red list found along the coast in Boma ward, Tanga region, B= *Uvaria pugunensis* IUCN Red Listed Critical Endangered (CR) from Pande Game reserve (Field Photo by Frank Mbago)**, C=** *Uvaria pandensis* IUCN Red Listed Critical Endangered (CR) from Pande Game reserve, D= *Dalbergia melanoxylon* IUCN Red Listed Near threatened (*Field Photos by Siri A. Abihudi, Frank Mbago, and Bukombe John*).

The loss of biodiversity resources in Wazo Hill coastal forest of Tanzania was observed to cause a loss of habitats for pollinating agents as a result of clearing plants like *Leucaena leucocephala* which support the lives of pollinating insects and birds (Ngongolo et al., 2014b). It is also predicted that the destruction occurring in the Kichi forest reserve, the coastal forest will lead to the loss of pollination services for sesame crop and decline of other services like food security, modification of microclimate, income from tourism sector, employment, and conservation of important wildlife species like elephants (*Loxodonta africana*) (Ngongolo et al., 2015a). This shows that biodiversity loss in coastal forests of East Africa has Economic Impacts as well. The coastal forests support various industries, including tourism and fisheries. Degradation of these ecosystems can result in reduced tourism opportunities, declining fish populations, and economic hardships for local communities. Also, loss of biodiversity of coastal forest of East Africa may lead to climate change amplification as deforestation and habitat loss contribute to climate change by releasing stored carbon dioxide. The loss of these forests exacerbates global warming, with implications for regional weather patterns and sea-level rise. The Rufiji delta in Tanzania has a changing environment due to less rainfall, rising temperatures, coastal flooding, and sea level rise, according to a study on community views of climate change and ecosystem-based resilience. The perceived pressures include weakened coastal protection, harmed fish breeding grounds, and altered climate regulation. Declines in agriculture, fish, and honey output have an impact on community livelihoods, while there are potential solutions (Nyangoko et al., 2022).

Also, the increasing in biodiversity loss in this important ecosystem, resulted into human–wildlife conflict and increased health risks. As habitats shrink and human populations expand, conflicts between humans and wildlife escalate. Crop raiding and livestock predation by wildlife can lead to retaliatory killings and negative perceptions of wildlife. Habitat degradation and encroachment into wildlife habitats increases the risk of zoonotic diseases spilling over from wildlife to humans, as seen with certain viral outbreaks. In Saadani National Park Wildlife–human conflict is a common phenomenon as the interaction between wildlife and people increase to loss of biodiversity (Franco et al., 2021). This has been reported to be handled through benefit-sharing and collaborative management with local communities adjacent the protected areas (Muganda, 2019). The study in Shimbi Hill showed the presence of vector and parasite which can affect both human and animal. For instance, parasitological surveys revealed presence of *Glossina brevipalpis* and *Glossina austeni, Glossina pallidipes* which are vector of *Trypanosoma* species that causes trypanosomiasis in human and animals (Ebhodaghe et al., 2022). In this case, the interaction between the livestock keepers and wildlife in this manner will lead to human-livestock-wildlife conflict while at the same time the individuals in the ecosystem being at risk of getting infection from one host.

In overall, the consequences of the threats to biodiversity in East African coastal forests extend beyond the loss of individual species. They encompass ecological imbalances, economic setbacks, cultural shifts, and increased vulnerabilities to environmental changes. Addressing these consequences requires concerted efforts in biodiversity conservation, sustainable resource management, and community engagement to protect these invaluable ecosystems and the life they support.

13.5 CONSERVATION EFFORTS ALONG THE COASTAL FOREST OF EAST AFRICA

Conservation efforts aimed at preserving the biodiversity of the coastal forests of East Africa have become increasingly critical in the face of mounting threats. Numerous organizations, governments, and communities are actively engaged in a range of initiatives to protect and sustain these invaluable ecosystems. Some of the key conservation efforts include developing robust policies, Program and projects in conservation (Figure 13.3).

Different coastal forests in East Africa have been gazetted as Protected Areas and Reserves. The establishment and management of protected areas and reserves play a fundamental role in safeguarding biodiversity. These areas provide a safe haven for a wide range of species and their habitats. Examples include the Arabuko-Sokoke Forest Reserve and national parks with diverse species which are threatened established or gazette in 1943. The International Union for Conservation of Nature (IUCN) Red List categorizes threats into CR, endangered (EN), near threatened (NT), vulnerable (VU), and least concern (LC). Along the coastal forest of Eastern Africa some of the threatened species that have been identified includes:

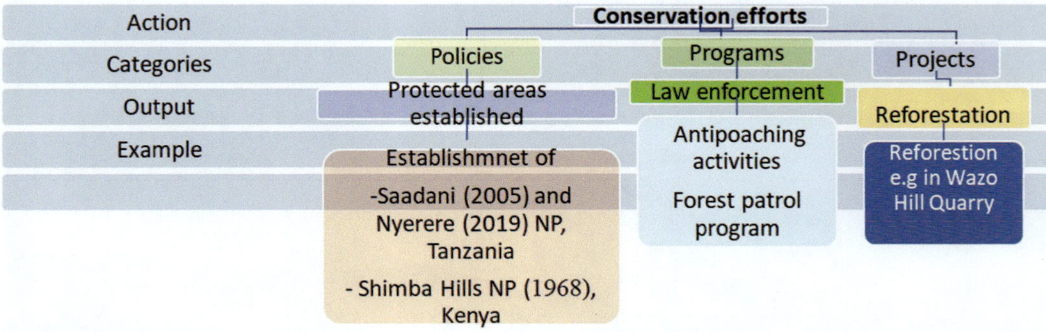

FIGURE 13.3 The conservation efforts for coastal forests of Eastern Africa.

(a) Fauna

Anthreptes pallidigaster (EN), *Anthus sokokensis* (EN), *Cephalophus adersi* (EN), *Loxodonta africana* (EN), *Otus ireneae* (EN), *Paraxerus palliatus* (VU), *Ploceus golandi* (EN), *Rhynchocyon chrysopygus* (EN), *Sheppardia gunningi* (VU), *Zoothera guttata* (VU), *Trigonoceps occipitalis* (CR), *Gyps africanus* (CR), and *Lycaon pictus* (EN) (Figure 13.4) (Kaleb and Humphrey, 2020).

(b) Flora

Aristogeitonia monophylla (VU), *Buxus obtusifolia* (VU), *Canthium kilifiense* (VU), *Canthium pseudoverticillatum* (VU), *Cynometra webberi* (VU), *Gardenia transvenulosa* (VU), *Newtonia paucijuga* (VU), and *Rothmannia macrosiphon* (VU) (Kaleb and Humphrey, 2020).

With the help of the Forest Conservation Programme, the Kenya Wildlife Services, National Museums of Kenya, Kenya Forest Research Institute, IUCN, and BirdLife International are continuing conservation efforts to safeguard the diverse forest biodiversity (Matiku, 2005). In Tanzania, a good example is Saadani National Park established in 2005, by combining the Saadani Game Reserve, Zaraninge Forest Reserve, and Mkwaja ranch to form a National Park, which is overseen by Tanzania National Park. Baixo Pinda Forest Reserve is among the coastal forests in Mozambique, established in 1957, a 19,600-ha area was set aside primarily for the conservation and restoration of *Androstachys johnsonii* a plant species in the family Picrodendraceae. The designation of Baixo Pinda Forest Reserve as means of conservation and recovering of *Androstachys johnsonii* shows the important of restoration of habitat as means of conservation efforts that are taking place in the coastal forests of Eastern Africa.

FIGURE 13.4 Some of the threatened fauna species can be observed in Selous Game Reserve within coastal forests ecosystem: A = *Loxodonta africana,* B = *Gyps africanus,* C = *Lycaon pictus,* and D = *Zoothera guttata* (*Field photo by Samuel Mtoka and Ally J. K. Nkwabi*).

Habitat restoration is the intentional and planned process of restoring a degraded natural ecosystem, aiming to restore its original ecological functions, biodiversity, and sustainability. This involves removing invasive species, replanting native vegetation, improving water quality, and addressing other factors contributing to habitat degradation. Habitat restoration is among the conservation efforts taking place in the coastal forests of East Africa so as to enhance the health and functioning of biodiversity. In Wazo Hill in Dar es Salaam, where quarrying is taking place for cement production, restoration by planting native trees has been observed. A number of studies showed that the restored area harbored more diversity of butterflies and other wildlife species with higher pollination services compared to unrestored areas (Ngongolo et al., 2014a, 2014b; Ngongolo and Mtoka, 2013). Similar efforts have been observed in Shimba Hills Nature Reserve in Kenya particullary in the areas where Bamburi Cement Ltd is working, there has been restoration program taking place through WWF initiatives (TFCG, 2006). These efforts are made successful when conservationists and local communities collaborate to restore degraded habitats within coastal forests. Other efforts may include afforestation and the removal of invasive species to enhance the resilience of ecosystems. Engaging local communities in conservation efforts is crucial. Community-based conservation programs involve communities in decision-making processes, provide alternative livelihoods programs, and raise awareness about the importance of biodiversity conservation (Franco et al., 2021; Moshi, 2016; Muganda, 2019). These has been observed in different coastal forests of Eastern Africa to have fruitful in the conservation efforts and enable the local communities to support conservation of biodiversity resources in the area willingly (Franco et al., 2021; Moshi, 2016; Muganda, 2019). In Saadani National Park, it was observed that if the local community is benefiting from the conserved biodiversity resources, there is a likelihood of them supporting positively the conservation efforts taking place (Franco et al., 2021). Another study on mangrove of Eastern Africa in three communities namely Kisakasaka village in Zanzibar, Mida Creek in Kenya, and Inhaca Island in Mozambique found that the mangroves in Inhaca are exploited for household needs, while in Kisakasaka and Mida Creek, the demand for mangrove products leads to an ecologically unsustainable rate of exploitation (Zorini et al., 2004). However, the study indicated that a participatory approach among stakeholders for such local communities, provision of alternative livelihoods program, are acceptable methods for biodiversity conservation (Zorini et al., 2004). Conservation organizations work to raise awareness about the significance of coastal forests and the threats they face. Educational programs target local communities, schools, and the broader public to foster a culture of conservation.

Antipoaching measures while ensuring better legislation and policy also helps to address the threats on biodiversity in the coastal forests. Initiatives to combat poaching and illegal wildlife trade include the deployment of antipoaching patrols, community-based wildlife management and monitoring, transboundary campaign and the use of technology such as camera traps, to deter illegal activities and protect endangered species (Lotter and Clark, 2014). Advocacy efforts focus on strengthening legislation and policies related to biodiversity conservation. These include enforcing anti-poaching laws, regulating land use, and promoting sustainable resource management. In the ecosystem between Tanzania and Mozambique which connect Nyerere-Selous protected areas in southern coast of Tanzania and Niassa National Reserve in northern coast of Mozambique, it was observed that, the population of elephants are enhanced when the antipoaching practices are implemented in collaboration with local communities adjacent the areas. The activities involved in the antipoaching efforts, includes among others joint patrols and operations; and intelligence-led operations within and outside the protected areas (Lotter and Clark, 2014). Also, the efforts involved different institution like international organization like PAMS Foundation, Thin Green Line Foundation and member of IUCN and Natural resources, and the World Commission on Protected Areas (Lotter and Clark, 2014). This indicates that international organizations and donors contribute to biodiversity conservation efforts through funding, technical assistance, and capacity building while respective countries can prove conducive polices and Legislation that support sustainable biodiversity conservation in coastal forests of East Africa (TFCG, 2006; Thoya et al., 2022).

Efforts to mitigate climate change, such as Reducing emissions from deforestation and forest degradation in developing countries (REDD+), help protect coastal forests by preserving their carbon-storing capacity (Beymer-Farris and Bassett, 2012). The project of assessing environmental knowledge in Tanzania's Rufiji Delta, focusing on the "REDD-readiness" argued that proponents of REDD+ underestimate the agency of forest-reliant communities, who have played a significant role in shaping the delta landscape. The shift from community-based resource management to fortress conservation is seen as an adaptive strategy to counter global climate change effects (Beymer-Farris and Bassett, 2012). To compare the REDD+ payments needed to compensate for opportunity costs (OC) of converting montane forest and miombo woodlands into cropland in Morogoro Region, Tanzania. Data from 250 households was used for OC estimation. The median compensation ranged from USD 1 $tCO2e^{-1}$ for montane forest to USD 39 $tCO2e^{-1}$ for degraded miombo woodlands. Incentives in the form of sustainable harvests could offset up to 55% of the total median OC. Implementing the policy in villages around the coastal miombo area would require high compensation levels (Beymer-Farris and Bassett, 2012).

All these efforts need to be done in institutionalized planning while the countries maintain sovereignty, enhance stakeholder awareness, make gender-appropriate considerations, and do more research for successful conservation efforts. For instance, a study by Nzunda (2022) revealed that most plans are not institutionalized in gender considerations, lack recent national inventories, and have low knowledge and skills. These conservation efforts represent a multifaceted approach to address the threats facing the biodiversity of coastal forests in East Africa. While challenges persist, these initiatives exemplify the dedication and commitment of stakeholders to protect these vital ecosystems and the remarkable species they harbor. It should be clear that there are strengths, weaknesses, opportunities, and challenges for the conservation efforts being undertaken in the coastal forests of East Africa (Tables 13.2–13.4).

13.6 WAY FORWARD AND CONCLUSION

The biodiversity along East African coastal forests is a national, regional, and global treasure, containing unique species and ecosystems worth conserving for sustainability. These forests contribute to global climate regulations, ecological balance, and human and wildlife well-being. The forests continue facing various threats including threats like deforestation, habitat loss, climate change, poaching, and illegal wildlife trade, which lead to deterioration of the conditions within. Although, ongoing conservation efforts, political supports, and community engagement initiatives provide hope in addressing the threats, the future of the biodiversity in these areas depends on collective efforts and well focused strategies, including empowering local communities, as well as enhancing scientific research, legal protections, and international cooperation. Protecting the biodiversity within need addressing issues of poaching, and environmental pressures. Promotion of public awareness and integration with technology may make a strong opportunity toward human behavioural change which may enhance climate change adaptation.

The SWOC of conservation efforts in East Africa coastal forests reveals strengths, weaknesses, opportunities, and challenges. Protected areas, such as, National parks, game reserves, can effectively preserve coastal forests. However, enforcement challenges, human activities, climate change, and land rights can hinder conservation efforts. Programs like law enforcement, antipoaching activities, and forest patrol programs can deter illegal activities but face resource constraints, corruption, and political interference. Antipoaching activities can protect endangered species but require significant resources and ongoing adaptation.

TABLE 13.2 Strengths, Weaknesses, Opportunities, and Challenges (SWOC) on Policies of the Conservation Efforts

S/n	Category	Strengths	Weaknesses	Opportunities	Challenges
2.1	*Protected Areas*	Protected areas can effectively preserve the unique biodiversity of coastal forests in East Africa.	Monitoring and enforcing protection can be difficult due to limited resources and manpower.	Involving local communities in protected area management can improve conservation efforts.	Balancing conservation with the rights and needs of local communities can be challenging.
		Provides a legal basis for the protection of these vital ecosystems.	Encroachment, illegal logging, and poaching can still occur within protected areas.	Seek funding from international organizations and carbon offset programs to support maintenance.	Coastal forests are vulnerable to climate change, requiring adaptive management strategies.
		Can generate revenue through ecotourism, supporting local economies and conservation efforts.	Protected areas may not be sufficient to address the impacts of climate change on coastal forests.	Enhance scientific research to adapt strategies to changing conditions.	Ensuring consistent support from governments and stakeholders over the long term can be a challenge

TABLE 13.3 Law Enforcement Archieved in Conservation of Biodiversity Along the Coastal Areas of East Africa

S/n	Category	Strengths	Weaknesses	Opportunities	Challenges
3.1.	*Law Enforcement*	Effective law enforcement can deter illegal activities like logging and poaching.	Limited funding and personnel can hinder enforcement efforts.	Utilize modern technology like drones and GIS for monitoring and patrolling.	Some illegal activities may involve well-organized criminal networks.
		Collaborative efforts with other agencies and countries can strengthen enforcement.	Corruption within law enforcement agencies can undermine effectiveness.	Training and equipping enforcement personnel to be more effective.	Political pressure can sometimes compromise enforcement actions.
		Supported by existing legal structures and regulations.	Coastal forests cover large areas, making comprehensive enforcement challenging.	Promote awareness of the importance of forest protection to gain public support.	Ensuring that local communities are not adversely affected by enforcement efforts.
3.1.	*Antipoaching Activities*	Targeted efforts to prevent poaching help preserve endangered species.	Antipoaching efforts can require significant resources.	Enhance antipoaching teams with modern training and technology.	Poachers may adapt and develop new strategies, requiring ongoing adaptation.
		Can work in tandem with law enforcement programs.	Poachers may pose physical risks to antipoaching teams.	Coordinate efforts with law enforcement and conservation organizations.	Legal processes for prosecuting poachers can be complex and lengthy.
		Antipoaching activities often enjoy public sympathy and support	Antipoaching alone may not address broader conservation challenges.	Engage local communities in antipoaching efforts.	The global wildlife trafficking network poses a significant challenge.
3.3.	*Forest Patrol Program*	Patrols can provide consistent oversight of forest areas.	Limited funding and manpower can restrict patrol coverage.	Use GPS, camera traps, and other tech for more efficient patrols.	Corruption within patrol teams can undermine efforts.
		Quick reaction to illegal activities can prevent extensive damage.	Rugged terrain and remote locations can impede patrols.	Involve local communities as forest stewards and information sources.	Patrollers may face threats or violence from illegal actors.
		Patrols can gather valuable data on forest health and threats	Ensuring effective coordination among patrol teams can be challenging.	Analyze patrol data to identify trends and inform conservation strategies.	Addressing natural threats like wildfires alongside illegal activities is complex

TABLE 13.4 Efforts Archived by Projects in Restoring Habitats in Coastal Forest of East Africa

S/n	Category	Strengths	Weaknesses	Opportunities	Challenges
4.1.	*Forest Landscape Restoration Projects*	Can restore degraded coastal forest ecosystems. Contributes to climate change mitigation through carbon capture. Can provide jobs and resources to local communities.	Restoration projects can take many years to show significant results. Reliance on grants and funding sources can be unstable. Restoring a balanced ecosystem can be challenging.	Involve governments, NGOs, and local communities. Develop comprehensive, long-term restoration plans. Incorporate sustainable forest management into restoration efforts.	Land tenure issues can hinder restoration efforts. Invasive plants and pests can disrupt restoration progress. Changing climate conditions may affect restoration success.
4.2.	*Reforestation Projects*	Reforestation can yield visible results in a relatively short time. Helps mitigate climate change by absorbing CO_2. Provides habitats for wildlife and preserves biodiversity.	Restoration projects can take many years to show significant results. Reliance on grants and funding sources can be unstable. Restoring a balanced ecosystem can be challenging.	Raise public awareness and engage volunteers in reforestation. Integrate reforestation with sustainable agricultural practices. Collaborate with businesses for funding and land use.	Ensuring high tree survival rates can be difficult, especially in harsh conditions. Securing consistent funding for reforestation initiatives can be a hurdle. Unpredictable weather patterns can affect reforestation outcomes

Training and technology can enhance antipoaching teams, while legal processes can be complex. Forest patrol programs provide consistent oversight and data collection, but face resource constraints, coordination, and environmental pressures. Forest landscape restoration projects can restore ecosystems and contribute to climate change mitigation through carbon capture. However, they are time-intensive, rely on funding dependence, and have ecological complexity. Reforestation projects can yield visible results in a short time but require long-term planning, sustainable practices, land ownership, invasive species, and climate change.

For the resilience and long-term health of the East African coastal forests and biodiversity there in, integrated approaches are essential in addressing various threats and drivers. Extending protected areas, building wildlife corridors, enhancing community involvement, putting climate resilience strategies into practice, encouraging transboundary collaboration, promoting sustainable resource management, exploring novel conservation finance mechanisms, investing in capacity building, raising public awareness, promoting stricter environmental policies, bolstering international collaboration, and placing education at the top of the agenda are just a few of the important directions for the future. These tactics will aid in preserving these ecosystems' distinctive biodiversity and long-term health and resilience. A concerted effort to address existing dangers and prepare for upcoming difficulties is necessary to ensure the biodiversity along East Africa's coastal forests is preserved.

KEYWORDS

- **biodiversity hotspot**
- **potential threats**
- **coastal forest conservation**
- **management strategies**
- **challenges**

REFERENCES

Abuodha, P. A. W.; Kairo, J. G. Human-induced stresses on mangrove swamps along the Kenyan coast. *Hydrobiologia* **2001,** *458*(1/3), 255–265. https://doi.org/10.1023/A:1013130916811

Beymer-Farris, B. A.; Bassett, T. J. The REDD menace: Resurgent protectionism in Tanzania's mangrove forests. *Glob. Environ. Change* **2012,** *22*(2), 332–341. https://doi.org/10.1016/j.gloenvcha.2011.11.006

Bullock, E. L.; Healey, S. P.; Yang, Z.; Oduor, P.; Gorelick, N.; Omondi, S.; Ouko, E.; Cohen, W. B. Three decades of land cover change in East Africa. *Landb* **2021,** *10*(2), 150.

Burgess, N. D.; Malugu, I.; Sumbi, P.; Kashindye, A.; Kijazi, A.; Tabor, K.; Mbilinyi, B.; Kashaigili, J.; Wright, T. M.; Gereau, R. E.; Coad, L., Knights, K.; Carr, J.; Ahrends, A.; Newham, R. L. Two decades of change in state, pressure and conservation responses in the coastal forest biodiversity hotspot of Tanzania. *Oryx* **2017,** *51*(1), 77–86. https://doi.org/10.1017/S003060531500099X.

Burgess, N. D.; Mwasumbi, L. B.; Hawthorne, W. J.; Dickinson, A.; Doggett, R. A. Preliminary assessment of the distribution, status and biological importance of coastal forests in Tanzania. *Biol. Conserv.* **1992,** *62*(3), 205–218. https://doi.org/10.1016/0006-3207(92)91048-W.

Chemane, D.; Motta, H.; Achimo, M. Vulnerability of coastal resources to climate changes in Mozambique: A call for integrated coastal zone management. *Ocean Coast. Manag.* **1997,** *37*(1), 63–83. https://doi.org/10.1016/S0964-5691(97)00073-2

Cowburn, B.; Moritz, C.; Birrell, C.; Grimsditch, G.; Abdulla, A. Can luxury and environmental sustainability co-exist? Assessing the environmental impact of resort tourism on coral reefs in the Maldives. *Ocean Coast. Manag.* **2018,** *158*, 120–127. https://doi.org/10.1016/j.ocecoaman.2018.03.025.

de la Torre-Castro, M.; Rönnbäck, P. Links between humans and seagrasses—an example from tropical East Africa. *Ocean Coast. Manag.* **2004,** *47*(7–8), 361–387.

Eastern Arc Mountains & Coastal Forests CEPF Plant Assessment Project (EAMCFCPAP) 2020. *Uvaria pandensis* (amended version of 2009 assessment). The IUCN Red List of Threatened Species **2020,** https://dx.doi.org/10.2305/IUCN.UK.2020-. Accessed 05 October 2023.

Ebhodaghe, F. I.; Bastos, A. D. S.; Okal, M. N.; Masiga, D. K. Entomological assessment of tsetse-borne trypanosome risk in the Shimba Hills human-wildlife-livestock interface, Kenya. *Front. Vet. Sci.* **2022,** *9*, 931078. https://doi.org/10.3389/fvets.2022.931078.

Ekblom, A. Forest–savanna dynamics in the coastal lowland of southern Mozambique since c. AD 1400. *Holocene* **2008,** *18*(8), 1247–1257. https://doi.org/10.1177/0959683608096599

Franco, P. M.; Beatrice, M.; Eivin, R. Impact of protected areas on the livelihood of locals: A case study in Saadani National Park, Tanzania. *Int. J. Biodivers. Conserv.* **2021,** *13*(3), 98–108. https://doi.org/10.5897/IJBC2021.1474.

Friess, D. A.; Gatt, Y. M.; Ahmad, R.; Brown, B. M.; Sidik, F.; Wodehouse, D. Achieving ambitious mangrove restoration targets will need a transdisciplinary and evidence-informed approach. *One Earth* **2022,** *5*(5), 456–460.

Friess, D. Ecosystem services and disservices of mangrove forests: Insights from historical colonial observations. *Forests* **2016,** *7*(12), 183. https://doi.org/10.3390/f7090183.

Fungomeli, M.; Githitho, A.; Frascaroli, F.; Chidzinga, S.; Cianciaruso, M.; Chiarucci, A. A new vegetation-plot database for the coastal forests of Kenya. *Veg. Classif. Surv.* **2020,** *1*, 103–109. https://doi.org/10.3897/VCS/2020/47180.

Gachanja, M.; Kanyanya, E. *Conservation of Kenya's Coastal Forests: Social Economic Resources,* WWF East African Coastal Forest Ecoregion Programme, Nairobi, Kenya. **2004**.

Gereau, E. R.; Cumberlidge, N.; Hemp, C.; Hochkirch, A.; Jones, T.; Kariuki, M.; Lange, N. C.; Loader, P. S.; Malonza, K. P.; Menegon, M.; Ndang'ang'a, K. P.; Rovero, F.; Shirk, P. Globally threatened biodiversity of the Eastern Arc Mountains and Coastal Forests of Kenya and Tanzania. *J. East Afr. Nat, Hist.* **2016,** *105*(1), 115–201.

Githitho, A. N. The sacred Mijikenda Kaya forests of coastal Kenya and biodiversity conservation. In: *The Importance of Sacred Natural Sites for Biodiversity Conservation* **2003,** pp. 27–35.

Godoy, F. L.; Tabor, K.; Burgess, N. D.; Mbilinyi, B. P.; Kashaigili, J. J.; Steininger, M. K. Deforestation and CO_2 emissions in coastal Tanzania from 1990 to 2007. *Environ. Conserv.* **2012,** *39*(1), 62–71. https://doi.org/10.1017/S037689291100035X

Habel, J. C.; Casanova, I. C. C.; Zamora, C.; Teucher, M.; Hornetz, B.; Shauri, H.; Mulwa, R. K.; Lens, L. East African coastal forest under pressure. *Biodivers. Conserv.* **2017,** *26*(11), 2751–2758. https://doi.org/10.1007/s10531-017-1375-z.

Haji, M.; Bakuza, J. S. Climate change vulnerability assessment for the rondo dwarf galago in coastal forests, Tanzania. *Environ. Manag.* **2023,** *71*(1), 145–158. https://doi.org/10.1007/s00267-022-01605-y.

Joaquim-Meque, E.; Lousada, J.; Liberato, M. L. R.; Fonseca, T. F. Forest in mozambique: Actual distribution of tree species and potential threats. *Land* **2023,** *12*(8), 1519. https://doi.org/10.3390/land12081519.

Kairo, J. G. Ecology and restoration of mangrove systems in Kenya PhD dissertation, Vrije Universiteit Brussel: Brussels, Belgium, **2001a**.

Kairo, J. G. Mangrove restoration in Kenya: Constraints and prospects. *ACP–EU Fish. Res. Rep. Number* **2001b**, *10*, 135.

Kaleb, A. M.; Humphrey, A. Influence of forest roads, distance from the boundary and location of sites on illegal activities in Arabuko Sokoke Forest, Coast-Kenya. *Amer. J. Agric. Forestry* **2020,** *8*(6), 235. https://doi.org/10.11648/j.ajaf.20200806.11.

Karyn, T.; Neil, D. B.; Boniface, P. M.; Japhet, J. K.; Marc, K. S. Forest and woodland cover and change in coastal Tanzania and Kenya, 1990 to 2000. *J. East African Nat. Hist.* **2010,** *99*(1), 19–45. https://doi.org/10.2982/028.099.0102.

Lagomasino, D.; Fatoyinbo, T.; Lee, S.; Feliciano, E.; Trettin, C.; Shapiro, A.; Mangora, M. M. Measuring mangrove carbon loss and gain in deltas. *Environ. Res. Lett.* **2019,** *14*(2), 025002.

LeMarie, M.; van der Zaag, P.; Menting, G.; Baquete, E.; Schotanus, D. The use of remote sensing for monitoring environmental indicators: The case of the Incomati estuary, Mozambique. *Phys. Chem. Earth, Parts A/B/C* **2006,** *31*(15–16), 857–863. https://doi.org/10.1016/j.pce.2006.08.023

Lotter, W.; Clark, K. Community involvement and joint operations aid effective anti-poaching in Tanzania. *Parks* **2014,** *20*(1), 19–28.

Lund, J. F.; Shackleton, S.; Luckert, M. Getting quality data. Measuring livelihoods environmental dependence: Methods for research fieldwork **2011,** 175–89.

Lyimo, J. G.; Kangalawe, R. Y.; Liwenga, E. T. Status, impact and management of invasive alien species in Tanzania. *Tanzania J. For. Nat. Conserv.* **2009,** *79*(2), 28–42.

Mabula, M. J.; Mangora, M. M.; Muhando, C. A. Peri-urban mangroves of Dar es Salaam-Tanzania are highly vulnerable to anthropogenic pressures. *Adv. Ecol. Environ. Res.* **2017,** 141–172.

Maeda, E. E.; Pellikka, P. K. E.; Siljander, M.; Clark, B. J. F. Potential impacts of agricultural expansion and climate change on soil erosion in the Eastern Arc Mountains of Kenya. *Geomorphology* **2010,** *123*(3–4), 279–289. https://doi.org/10.1016/j.geomorph.2010.07.019

Mapunda, K. K.; Andrew, S. M. Predicting the distribution of critically endangered tree species Karomia gigas under climate change in Tanzania. *Ecol. Eng.* **2023,** *195*, 107065. https://doi.org/10.1016/j.ecoleng.2023.107065

Martin, E.; Burgess, N. East African Mangroves. *One Earth,* **2023,** https://www.Oneearth.Org/Ecoregions/East-African-Mangroves/. Accessed 11 September 2023.

Matiku, P. The Coastal Forests of Kenya: Forests data, threats, socio-economic issues, values, stakeholders, challenges, strategies, investment and enabling environment--A national synthesis report for the development of the WWF-EARPO Eastern Africa Coastal Forests Ecoregion Programme, Technical Reports, Nature Kenya **2005**. https://aquadocs.org/handle/1834/8857. Accessed 27 September 2023.

Metcalfe, K.; Ffrench-Constant, R.; Gordon, I. Sacred sites as hotspots for biodiversity: The three sisters cave complex in coastal Kenya. *Oryx* **2010,** *44*(01), 118. https://doi.org/10.1017/S0030605309990731

Mligo, C. Conservation of plant biodiversity of Namatimbili forest in the southern coastal forests of Tanzania. *Int. J. Biodivers. Conserv.* **2015,** *7*(3), 148–172. https://doi.org/10.5897/IJBC2014.0771

Mligo, C.; Lyaruu, H.;d Ndangalasi, H. The effect of anthropogenic disturbances on population structure and regeneration of *Scorodophloeus fischeri* and *Manilkara sulcata* in coastal forests of Tanzania. *Southern For.: J. For. Sci.* **2011,** *73*(1), 33–40. https://doi.org/10.2989/20702620.2011.574830.

Modest, R. B.; Hassan, S. N. Species composition of tropical understory birds in threatened East African coastal forests based on capture data. *Int. J. Zool.* **2016,** *2016*(4), 1–9. https://doi.org/10.1155/2016/1390364.

Mohammed, S. M. Pollution management in Zanzibar: The need for a new approach. *Ocean Coast. Manag.* **2002,** *45*(4–5), 301–311. https://doi.org/10.1016/S0964-5691(02)00060-1.

Moshi, B. S. Impacts of protected areas on local livelihood: A case study of Saadani National Park. *Master's Thesis in Natural Resources Management (Biology).* Norwegian University of Science and Technology, Faculty of Natural Science and Technology **2016,** vol. 1, no. 1, pp. 1–35.

Muganda, M. Protected area-people relationships in Tanzania: A case study of Saadani National Park and its neighboring communities. *J. Soc. Polit. Sci.* **2019,** *2*(3), 21. https://doi.org/10.31014/aior.1991.02.03.114.

Ndimuligo, S. A.; Mbwambo, B. N.; Kavana, P. Y.; and Nkwabi, A. K. Predicting the impacts of climate change on the potential suitable habitat distribution of house crows (*Corvus splendens*) in Tanzania. *Open Access Library J.* **2022,** *9*, e9014. https://doi.org/10.4236/oalib.1109014 not quoted in text.

Ngongolo, K.; Mtoka, S. Using butterflies to measure biodiversity health in Wazo hill restored quarry. *J. Entomol. Zool. Stud.* **2013,** *1*(4), 81–86.

Ngongolo, K.; Bruno, N. A. Impacts of wildfire on insect diversity in the selous game reserve, Tanzania. In: *TAWIRI Biennual Conference Proceedings.* Arusha, **2015,** pp. 58–69.

Ngongolo, K.; Mtoka, S.; Mahulu, A. Wet season diversity of butterflies in restored mine of Wazo Hill Tanzania. *Int. J. Fauna Biol. Stud.* **2014a**, *1*(3), 1–3.

Ngongolo, K.; Mtoka, S.; Mahulu, A.; Sigala, A. The *Leucaena leucocephala* floral visitors, pollinators and their predators in the restored Wazo Hill Quarry, Tanzania. *Entomol. Appl. Sci. Lett.* **2014b**, *1*(3), 36–42.

Ngongolo, K.; Mtoka, S.; Rubanza, C. D. Floral visitors and pollinators of sesame (*Sesamum indicum* L) from Kichi forest to the adjacent local communities' farms. *Entomol. Appl. Sci. Lett.* **2015a** *2*(2), 32–39.

Ngongolo, K.; Mtoka, S.;d Mahulu, A. Challenges and opportunities for restoring the threatened mangroves. *J. Sci. Res. Rep.* **2015b**, *5*(5), 352–360.

Ngumbau, V. M.; Luke, Q.; Nyange, M.; Wanga, V. O.; Watuma, B. M.; Mbuni, Y. M.; Munyao, J. N.; Oulo, M. A.; Mkala, E. M.; Kipkoech, S.; Itambo, M.; Hu, G.-W.; Wang, Q.-F. An annotated checklist of the coastal forests of Kenya, East Africa. *PhytoKeys* **2020**, *147*, 1–191. https://doi.org/10.3897/phytokeys.147.49602

Nkwabi, A. K.; Bukombe, J.; Maliti, H.; Liseki, S.; Lesio, N.; Kija, H. An overview of biodiversity of Tanzania and conservation efforts. In: Pullaiah, T. (Ed.), *Global Biodiversity Volume 3, Selected Countries in Africa.* Apple Academic Press: Waretown, USA, **2018**, pp. 297–342.

Ntukey, L. T.; Munishi, L. K.; Treydte, A. C. Land use land/cover change reduces woody plant diversity and carbon stocks in a lowland coastal forest ecosystem, Tanzania. *Sustainability* **2022,** *14*(14), 8551. https://doi.org/10.3390/su14148551.

Nyangoko, B. P.; Berg, H.; Mangora, M. M.; Shalli, M. S.; Gullström, M. Community perceptions of climate change and ecosystem-based adaptation in the mangrove ecosystem of the Rufiji Delta, Tanzania. *Clim. Dev.* **2022,** *14*(10), 896–908. https://doi.org/10.1080/17565529.2021.2022449.

Nzunda, E. F.; Midtgaard, F. Deforestation and loss of bushland and grassland primarily due to expansion of cultivation in mainland Tanzania (1995–2010). *J. Sustain. For.* **2019,** *38*(6), 509–525.

Nzunda, E. *Assessment of Forest Planning and Management in Different Forest Types in Selected African Countries*. AFF Working Paper. African Forest **2022.** https://afforum.org/wp-content/uploads/2022/09/Vol-5-Issue-5_ENG-1.pdf. Accessed 27 September 2023.

Obiri, F. J. Invasive plant species and their disaster-effects in dry tropical forests and rangelands of Kenya and Tanzania. *J. Disaster Risk Stud.* **2011,** *3*(2).

Ogada, D. L. The power of poison: Pesticide poisoning of Africa's wildlife. *Ann. New York Acad. Sci.* **2014,** 1322(1), 1–20. https://doi.org/10.1111/nyas.12405

Ogada, D.; Virani, M. Z.; Thiollay, J. M.; Kendall, C. J.; Thomsett, S.; Odino, M.; Kapila, S.; Patel, T.; Wairasho, P.; Dunn, L.; Shaw, P. Evidence of widespread declines in Kenya's raptor populations over a 40-year period. *Biol. Conserv.* **2022,** *266*, 109361.

Ojija, F., Manyanza, N. M., Mataba, G. R. (2022). Distribution, habitat and conservation status of critically endangered aloes in Tanzania. *S. Afr. J. Bot. 147*, 1051–1057.

Pakia, M.; Cooke, J. A.; van Staden, J. The ethnobotany of the Midzichenda tribes of the coastal forest areas in Kenya: 2. Medicinal plant uses. *South African J. Bot.* **2003,** *69*(3), 382–395. https://doi.org/10.1016/S0254-6299 (15)30321-5

Robertson, A.; Hankamer, C.; Ngonyo, M. Restoration of a small tropical coastal forest in Kenya: Gede National Monument Forest Restoration Project. In: *Plant Conservation in the Tropics: Perspectives and Practice* **2002,** pp. 241–268.

Rosti, H.; Pihlström, H.; Rottcher, N.; Bearder, S.; Mwangala, L.; Maghenda, M.; Rikkinen, J. Species identity and behavior of cave-dwelling tree hyraxes of the Kenyan coast. *Ecol. Evol.* **2023,** *13*(1), e9693. https://doi.org/10.1002/ece3.9693

Rumisha, C.; Mdegela, R. H.; Kochzius, M.; Leermakers, M.; Elskens, M. Trace metals in the giant tiger prawn *Penaeus monodon* and mangrove sediments of the Tanzania coast: Is there a risk to marine fauna and public health? *Ecotoxicol. Environ. Saf.* **2016,** *132*, 77–86. https://doi.org/10.1016/j.ecoenv.2016.05.028

Sabai, D. The role of women in re-orienting mangrove-based practices into practice in the eastern coast of Tanzania. *Tanzania J. Popul. Stud. Dev.* **2019,** *26*(1), 49–67.

Shilla, D. A. Distribution of Pb, Cr, Cu and Zn in the marine-coastal region of Zanzibar (Tanzanian archipelago, East Africa). *Chem. Ecol.* **2016,** *32*(8), 774–785. https://doi.org/10.1080/02757540.2016.1178727

Shimba, M. J.;Jonah, F. E. Nest success of the Indian House Crow *Corvus splendens* : An urban invasive bird species in Dar es Salaam, Tanzania. *Ostrich* **2017,** *88*(1), 27–31. https://doi.org/10.2989/00306525.2016.1223766

Stemler, S. E. Content analysis. In: Scott, R. and Kosslyn, S. (Eds.), *Emerging Trends in the Social and Behavioral Sciences: An Interdisciplinary, Searchable, and Linkable Resource,* John Wiley & Sons, Inc., **2015,** 1–14. https://www.researchgate.net/profile/Steven-Stemler/publication/279917349_Emerging_Trends_in_Content_Analysis/links/5b38b9a64585150d23ea2d4f/Emerging-Trends-in-Content-Analysis.pdf. Accessed 27 Semptember 2023.

Tabor, K.; Burgess, N. D.; Mbilinyi, B. P.; Kashaigili, J. J.; Steininger, M. K. Forest and woodland cover and change in coastal Tanzania and Kenya, 1990 to 2000. *J. East Afr. Nat. Hist.* **2010,** *99*(1), 19–45. https://doi.org/10.2982/028.099.0102

TFCG, T. F. C. G *Coastal forest of Kenya and Tanzania* **2006**. http://Coastalforests.Tfcg.Org/Tz-Coast.Html.

Thoya, P.; Horigue, V.; Möllmann, C.; Maina, J.; Schiele, K. S. Policy gaps in the East African blue economy: Perspectives of small-scale fishers on port development in Kenya and Tanzania. *Front. Mar. Sci.* **2022,** *9*. https://doi.org/10.3389/fmars.2022.933111

UNECA [United Nations Economic Commission for East Africa] Media Briefs | United Nations Economic Commission for Africa **2018,** https://archive.uneca.org/ea-ice22/pages/media-briefs#:~:text=With%20an%20average%20growth%20rate,accompanied%20by%20rapid%20social%20improvements. Accessed 27 September 2023.

Vorontsova, M. S.; Mbago, F. M. New *Solanum* species from Tanzanian coastal forests may already be extinct. *J. East African Nat. Hist.* **2010,** *99*(2), 227–234. https://doi.org/10.2982/028.099.0202

Wanyingi, N. J.; Mwasi, B.; Kairu, K. J.; Kimanzi, K. J. Predictive modelling of human - elephant conflicts in Shimba hills ecosystem. *Afr. J. Educ. Sci. Technol.* **2013,** *1*(2), 163–172.

Wegner, G.; Howell, K. M.; Davenport, T. R. B.;Burgess, N. D. The Forgotten 'Coastal forests' of Mtwara, Tanzania: A biologically impoverished and yet important ecosystem. *J. East African Nat. Hist.* **2009,** *98*(2), 167–209. https://doi.org/10.2982/028.098.0202

Weideman, E. A.; Perold, V.; Donnarumma, V.; Suaria, G.;d Ryan, P. G. Proximity to coast and major rivers influence the density of floating microplastics and other litter in east African coastal waters. *Mar. Pollut. Bull.* **2023,** *188*, 114644. https://doi.org/10.1016/j.marpolbul.2023.114644

Zorini, L. O.; Contini, C.; Jiddawi, N.; Ochiewo, J.; Shunula, J.; Cannicci, S. Participatory appraisal for potential community-based mangrove management in East Africa. *Wetl. Ecol. Manag.* **2004,** *12*(2), 87–102. https://doi.org/10.1023/B:WETL.0000021672.15252.54

CHAPTER 14

REMOTE SENSING APPLICATIONS IN MIOMBO AND THE COASTAL MOSAIC FORESTS OF TANZANIA: ECOLOGICAL CONDITIONS AND ANTHROPOGENIC DISTURBANCES IN THE SELOUS-NIASSA WILDLIFE CORRIDOR, TANZANIA

HAMZA KIJA[1*], WILFRED MAREALLE[1], BUKOMBE JOHN[1], ASANTERABI LOWASSA[1], BARAKA NAFTALI[1], MACHOKE MWITA[1], LAZARO MANGEWA[2], RICHARD KATONDO[3], ALMAS KASHINDYE[3], DENIS IKANDA[3], ROB HARRIS[3], and EMMANUEL F. NZUNDA[4]

[1]*Tanzania Wildlife Research Institute (TAWIRI), Arusha, Tanzania*

[2]*Department of Wildlife Management, College of Forestry, Wildlife and Tourism, Sokoine University of Agriculture (SUA), Morogoro, Tanzania*

[3]*World Wildlife Fund for Nature (WWF), Tanzania Country Office, Dar es Salaam, Tanzania*

[4]*Department of Forest Resources Assessment and Management, College of Forestry, Wildlife and Tourism, Sokoine University of Agriculture (SUA), Morogoro, Tanzania*

*Corresponding author

ABSTRACT

This chapter explores the application of remote sensing technology to assess ecological conditions and anthropogenic disturbances in the Miombo and Coastal Mosaic Forest of Tanzania, with a specific focus on the Selous-Niassa Wildlife Corridor. This critical corridor facilitates wildlife movement between the Selous Game Reserve, Nyerere National Park (Tanzania), and the Niassa Special Reserve (Mozambique). Remote sensing proves indispensable for monitoring and comprehensively evaluating the coastal forest's health and changes, providing nonintrusive and accurate data collection over vast and inaccessible areas. The findings highlight significant land use changes, stable vegetation conditions from 2001 to 2012, and subsequent improvements from 2013 to 2020, while

Biodiversity Hotspot of the Coastal Forests of Eastern Africa. T. Pullaiah and Ally J. K. Nkwabi (Eds.)

DOI: 10.1201/9781998511099-14

rainfall patterns and fire incidences fluctuate. To safeguard the corridor's ecological integrity, conservation and management strategies that prioritize maintaining connectivity and mitigating human–wildlife conflicts and regulating land use practices are essential. This chapter offers a comprehensive analysis of the corridor's ecological conditions, emphasizing the significance of well-informed conservation and management strategies that balance conservation goals with the well-being of the communities within the Ruvuma landscape.

14.1 INTRODUCTION

14.1.1 BACKGROUND

Miombo and Coastal mosaic forests in Tanzania are ecologically significant regions that hold biodiversity and play a crucial role in supporting various ecosystems. These forests, situated along the eastern shores of Tanzania, encompass a diverse range of vegetation types, including mangroves, lowland rainforests, and coastal woodlands (Schabel, 2006, Tabor et al., 2010). Their unique geographical location between land and sea makes them valuable habitats for a wide array of plant and animal species, contributing to the overall ecological balance and providing essential ecosystem services such as carbon sequestration, coastal protection, and support for local livelihoods (Kalaba, 2014; Lupala et al., 2014; Filho et al., 2021). However, the increasing human pressures on these woodlands and coastal mosaic forests, including deforestation, land-use changes, and wildlife conflicts, have led to significant challenges in maintaining their ecological integrity.

One of the vital ecological links in the region is the Selous-Niassa Wildlife Corridor (SNWC), a critical ecological link within the Ruvuma landscape of Tanzania. Spanning vast areas between the Selous Game Reserve, Nyerere National Park (NP), and the Niassa Special Reserve (NSR) in Mozambique, this corridor serves as a crucial pathway for wildlife movement spanning vast areas of the Ruvuma landscape. This corridor serves as a critical migration pathway for diverse wildlife species, facilitating gene flow, and maintaining the genetic diversity essential for the long-term population viability of various wildlife populations (Baldus et al., 2003; Bluwstein and Lund, 2018). This genetic connectivity is vital for maintaining the long-term viability and genetic diversity of various species, enhancing their resilience to environmental changes, and ensuring healthy populations (Mpanduji, 2004; Picard, 2010). Moreover, the corridor supports seasonal migrations, enabling wildlife to access key resources such as water, food, and breeding grounds, promoting the overall health and stability of the entire ecosystem. Preserving the integrity of the SNWC is of utmost importance to safeguard the region's biodiversity and support sustainable wildlife conservation (Kija et al., 2021).

Despite its ecological significance, the miombo and coastal mosaic forests of Tanzania, including those in the SNWC, face numerous challenges resulting from anthropogenic activities (Zella, 2016; Bluwstein and Lund, 2018; Botey et al., 2022). The expansion of human settlements, unsustainable land use practices, illegal logging, and poaching are among the primary threats that contribute to deforestation and habitat degradation within the region. These disturbances can disrupt wildlife movement patterns, fragment habitats, and exacerbate human–wildlife conflicts, impacting both ecological integrity and the well-being of local communities that rely on the forest's resources.

In recent years, remote sensing technology has emerged as a powerful tool in ecological studies and conservation efforts worldwide. Remote sensing involves the acquisition of information about the Earth's surface from a distance, utilizing various sensing platforms such as satellites and drones. This technology allows researchers and conservationists to assess large geographical areas with reduced fieldwork requirements, making it ideal for monitoring vast and remote coastal forest regions like the SNWC. By collecting multispectral and hyperspectral data, remote sensing enables the analysis of land use and land cover changes, evaluation of vegetation health, identification of anthropogenic disturbances, and mapping of fire incidences, among other ecological parameters.

Given its nonintrusive and comprehensive data collection capabilities, remote sensing has revolutionized studies on coastal forests and their ecological dynamics (Phinn et al., 2006; Jackson et al., 2020). By integrating remote sensing techniques into their studies, scientists can gain valuable insights on ecological conditions and changes over time of the coastal forests (including within the SNWC) and identify human-induced disturbances that may impact the corridor's integrity. Remote sensing techniques offer a wide range of applications in ecological studies, including land use and land cover mapping, vegetation health assessment, wildfire detection, and identification of human-induced disturbances (Svoray et al., 2013; Willis, 2015; Gao et al., 2020). This evidence-based approach to conservation allows for the development of informed and effective strategies to protect and manage the coastal mosaic forest ecosystems, safeguarding their ecological balance for the benefit of wildlife, local communities, and the sustainable future of the Ruvuma landscape.

This chapter aims to explore the applications of remote sensing technology in assessing the ecological conditions and anthropogenic disturbances in the Coastal Forest of Tanzania, with a specific focus on the SNWC. By integrating remote sensing techniques into ecological studies, the chapter seeks to provide a comprehensive understanding of the current state of the coastal forests and the corridor. Additionally, it aims to emphasize the significance of well-informed conservation and management strategies to safeguard the integrity of the SNWC and preserve the ecological balance for the benefit of wildlife and local communities within the Ruvuma landscape. The integration of remote sensing technology with ecological studies enhances our ability to address the complex challenges facing coastal forests and wildlife corridors, contributing to the sustainable conservation of these critical ecosystems in Tanzania.

14.1.2 REMOTE SENSING STUDIES IN THE COASTAL EASTERN AFRICA ECOREGION

Remote sensing studies along the Coastal Eastern Africa Ecoregion have been at the forefront of environmental research and conservation initiatives. This ecologically diverse region, spanning across countries like Tanzania, Kenya, Mozambique, and Somalia, is characterized by a unique blend of terrestrial and marine habitats, including mangroves, coastal mosaic forests, savannas, and marine ecosystems. Remote sensing techniques have played an important role in mapping and monitoring land cover and land use patterns, allowing to assess changes in natural and human-induced landscapes (Chughtai et al., 2021; Roy et al., 2022). By employing satellite imagery and advanced sensors, scientists have gained valuable insights into the spatial distribution of different land cover types, enabling the identification of areas under threat and facilitating evidence-based conservation planning.

Among the prominent applications of remote sensing in the Coastal Eastern Africa Ecoregion is the monitoring of mangrove ecosystems. Mangroves are important in coastal protection, carbon sequestration, and supporting marine biodiversity (Bouillon et al., 2009; Howard et al., 2017; Sievers et al., 2023). Remote sensing studies have allowed for the reliable mapping and monitoring of mangrove forests, providing critical data on their health and extent. By detecting changes in mangrove cover over time, can assist in identifying areas at risk of degradation or deforestation, supporting conservation efforts to safeguard these invaluable coastal habitats and their ecosystem services (Estoque et al., 2018; Bryan-Brown et al., 2020; Nyangoko et al., 2022).

Coastal erosion and sea level rise are significant environmental challenges faced by the Coastal Eastern Africa Ecoregion. Remote sensing technologies, including LiDAR and satellite imagery, have enabled to monitor shoreline changes and assess the impact of rising sea levels (Gray et al., 2018; Adebisi et al., 2021). By analyzing the rate of coastal erosion and identifying vulnerable areas, remote sensing studies contribute to the formulation of adaptation strategies to mitigate the potential impacts on coastal communities, infrastructure, and ecosystems. This information is vital for sustainable coastal zone management and climate change resilience planning.

Biodiversity conservation is a key focus of remote sensing studies in the Coastal Eastern Africa Ecoregion. Satellite-based technologies have been used to identify biodiversity hotspots, critical habitats for endangered species, and wildlife corridors for movement between protected areas (Hansen and DeFries, 2007; Marvin et al., 2016; Mathur et al., 2017). Remote sensing data aids in the mapping of important conservation areas, assisting in the design and implementation of effective management plans to preserve the region's rich biodiversity. Moreover, remote sensing tools have also been instrumental in monitoring changes in wildlife populations and their habitats, supporting efforts to protect and sustainably manage this diverse and delicate ecosystem (Kaplan et al., 2019; Gabriele et al., 2022).

Remote sensing studies also extend to monitoring marine and coastal environments within the Coastal Eastern Africa Ecoregion. Ocean color sensors on satellites enable the observation of marine ecosystem health, including coral reefs and seagrass beds (Mumby et al., 2004; Eakin et al., 2010). These data can provide valuable insights into the status and dynamics of these marine habitats, assisting in marine conservation and fisheries management efforts. By understanding the health of marine ecosystems, scientists and policymakers can implement measures to protect these critical habitats and ensure the sustainability of marine resources for the benefit of both biodiversity and coastal communities.

Studies in remote sensing in the SNWC have emerged as vital tools for comprehending the ecological conditions and anthropogenic disturbances. Remote sensing techniques have been employed to assess the corridor's land use and land cover dynamics, fire incidences, vegetation health, and spatiotemporal changes over the years (Zella, 2016; Riggio and Caro, 2017; Zella et al., 2017; Mamboleo, 2018; Mwakoba, 2021; Mango et al., 2022). Analyzing remote sensing data can provide insights into the corridor's habitat suitability for various wildlife species, and the potential threats to biodiversity and ecological integrity. These studies contribute in informing effective conservation and management strategies, prioritizing connectivity maintenance, mitigating human–wildlife conflicts, and safeguarding the long-term viability of the corridor.

14.1.3 SELOUS-NIASSA WILDLIFE CORRIDOR

The SNWC is situated within the Ruvuma landscape, spanning from 9° 52′52″ S to 11° 46′ 30″ S latitude and 35° 47′ 34″ E to 37° 18′ 20″ E longitude in the Ruvuma Region of Tanzania. It shares its borders with Nyerere NP in the north and the NSR in Mozambique in the south (Figure 14.1. A = SNWC in relation to B = Tanzania, C = Villages in the Selous-Niassa, D = Wildlife Management Areas (WMAs) in the SNWC). The SNWC serves as a vital link connecting Nyerere NP/SGR in Tanzania and the NSR in Mozambique, with the Ruvuma River dissecting the corridor (Figure 14.2).

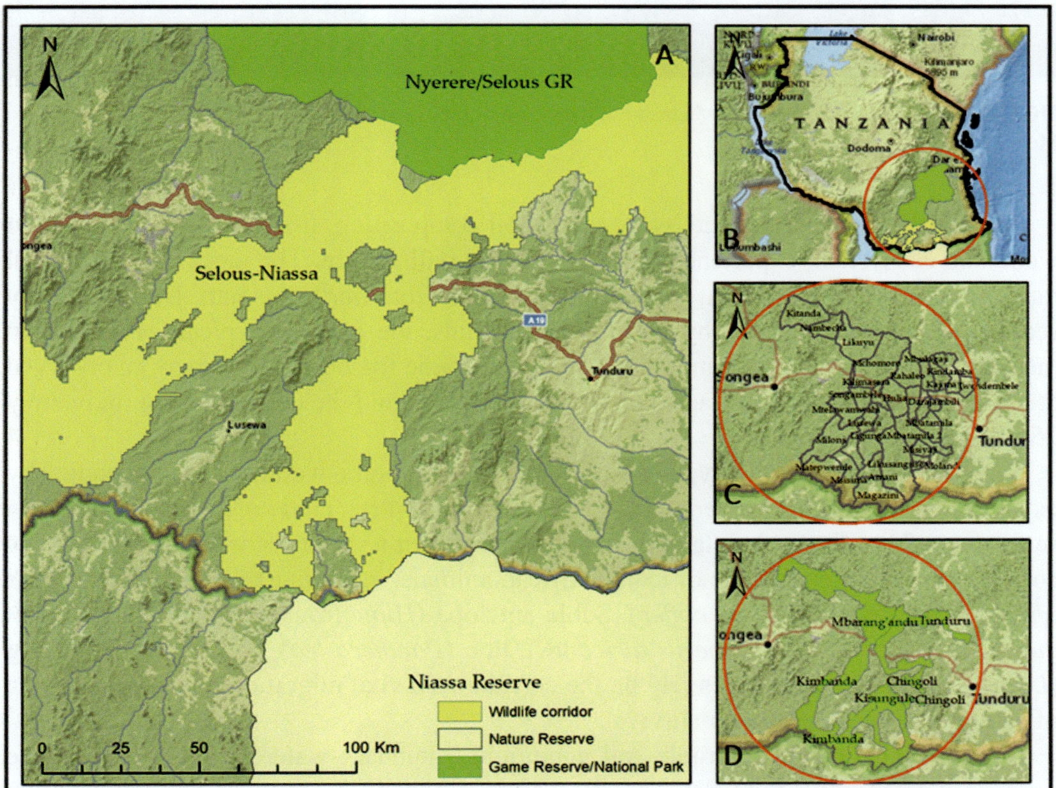

FIGURE 14.1 Map of the Selous-Niassa Wildlife Corridor, Tanzania.

FIGURE 14.2 Ruvuma river (at Magazini Village), an attractive natural feature separating Tanzania and Mozambique.

The SNWC is characterized by its sandy soil composition. It falls within the Miombo woodland ecosystem, experiencing a unimodal rainfall pattern. The mean annual rainfall ranges from 800 to 1200 mm, gradually decreasing from north to south. During the dry season, which is influenced by the northeast winds, the corridor typically receives no measurable rain for at least 5 months, although sporadic fog occurrences may be observed at higher elevations. The average annual temperature within the corridor is approximately 25°C.

The dominant vegetation type in the SNWC is the miombo woodland, accompanied by extensive grasslands, lush riverine forests, meandering rivers, cultivated lands, and human settlements. This diverse habitat provides a sanctuary for a variety of wildlife species. The corridor supports populations of iconic African animals, including Elephants (*Loxodonta africana*), Buffalo (*Syncerus caffer*), Sable antelope (*Hippotragus niger*), Zebra (*Equus quagga*), Warthog (*Phacochoerus africanus*), lion (*Panthera leo*), and African wild dog (*Lycaon pictus*). These species rely on the corridor as a vital migration route and a source of essential resources for their survival.

The combination of favorable soil conditions, abundant water sources, and diverse vegetation types makes the SNWC a critical ecological zone that sustains a rich and vibrant biodiversity. The presence of iconic and keystone wildlife species underscores the significance of this corridor for maintaining gene flow, biodiversity, and the ecological balance within the larger Ruvuma landscape. Preserving the integrity of the SNWC is essential for ensuring the continued survival of these species and promoting sustainable ecological coexistence between wildlife and human communities in the region.

14.2 ECOLOGICAL CONDITIONS AND ANTHROPOGENIC DISTURBANCES

14.2.1 ECOLOGICAL CONDITIONS

14.2.1.1 LAND USE AND LAND COVER MAPPING

The thematic classifications conducted for the years 2001 and 2020 demonstrated high accuracy, with all per-class accuracies exceeding 70% and an overall accuracy of ≥85%. These results indicate that the classification was ideal for use, as supported by existing literature (Thomlinson et al., 1999; Foody, 2002; Shalaby and Tateishi, 2007). Additionally, the KIA (Kappa Index of Agreement) values obtained were acceptable, further confirming the reliability of the classification. KIA values greater than 0.8 indicate a perfect classification, while values between 0.6 and 0.8 indicate a good classification (Shalaby and Tateishi, 2007).

The Landsat thematic analysis conducted for the years 2001 and 2020 in the SNWC revealed changes in land cover types. Mosaic and woodland cover types dominated the corridor in 2001, occupying 56.5% and 26.3% of the area, respectively, while cultivation covered 3.3%. However, by 2020, mosaic and woodland cover types constituted 63% of the area, while cultivation expanded significantly to 13.4%. The proportion of grassland increased from 10.5% in 2001 to 13.4% in 2020, potentially due to land conversion for agriculture or deforestation. Shrubland cover also increased from 2% in 2001 to 9.5% in 2020, possibly due to natural vegetation succession after disturbances. The change detection matrix showed shifts between different land cover types, with some areas converted to cultivated land or shrubland. Burnt areas declined over time, potentially due to changes in land use practices. Understanding these land cover changes is important in guiding sustainable land management practices in the corridor. The confusion matrix, as shown in the appendices, assessed the performance of the classification algorithm used in the analysis (Figures 14.3–14.6).

Over the past two decades, there has been a slight increase in the proportion of land classified as bare land within the corridor. However, this increase remains relatively small compared to the overall land cover. The rise in bare land may be attributed to an increase in the human population, including some illegal settlement activities occurring within the WMAs, which often requires the clearing of land for cultivation and settlement purposes (Figure 14.7). The increase in bare land indicates reduced vegetation or habitat availability for wildlife species. This decline in vegetation can lead to a loss of biodiversity and have negative impacts on the populations of both flora and fauna in the coastal forests that depend on vegetated habitats. Moreover, the increase in bare land also exposes the soil to erosion and surface run-off, further exacerbating environmental challenges in the corridor.

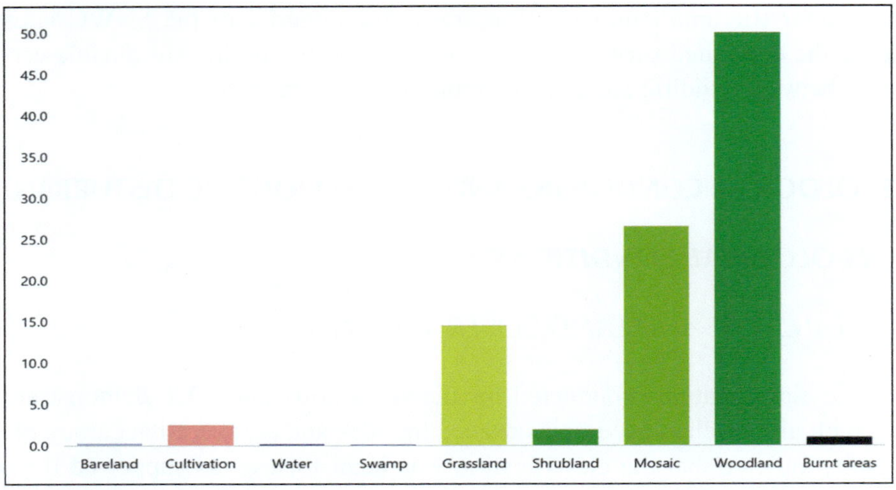

FIGURE 14.3 Areal coverage of land use and land cover types in the Selous-Niassa Wildlife Corridor for the year 2001.

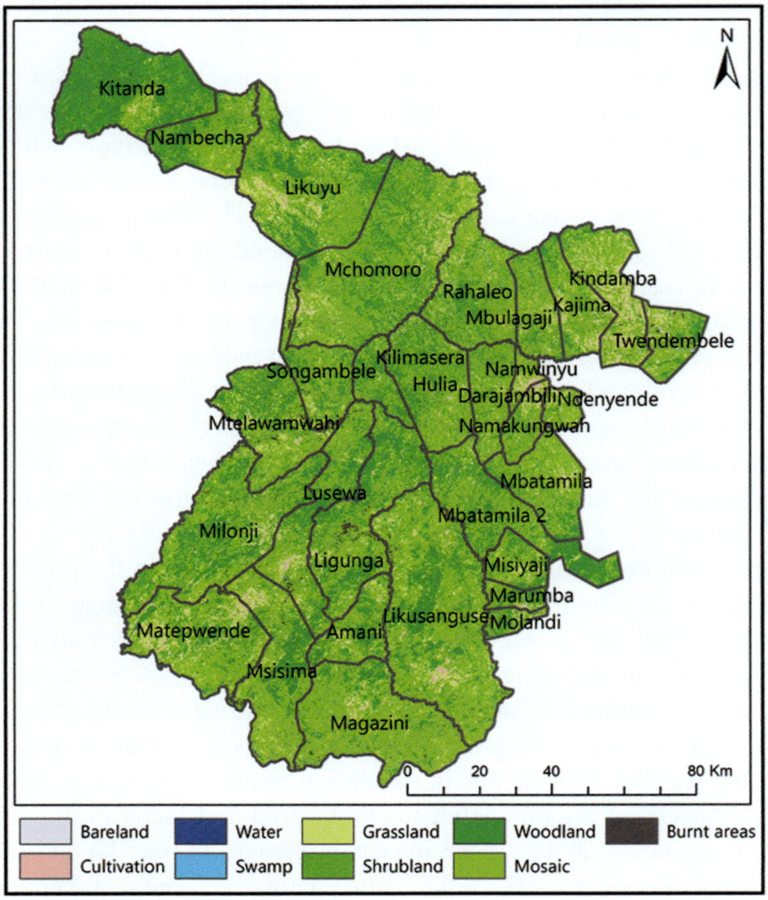

FIGURE 14.4 Land use and land cover map in the Selous-Niassa Wildlife Corridor in 2001.

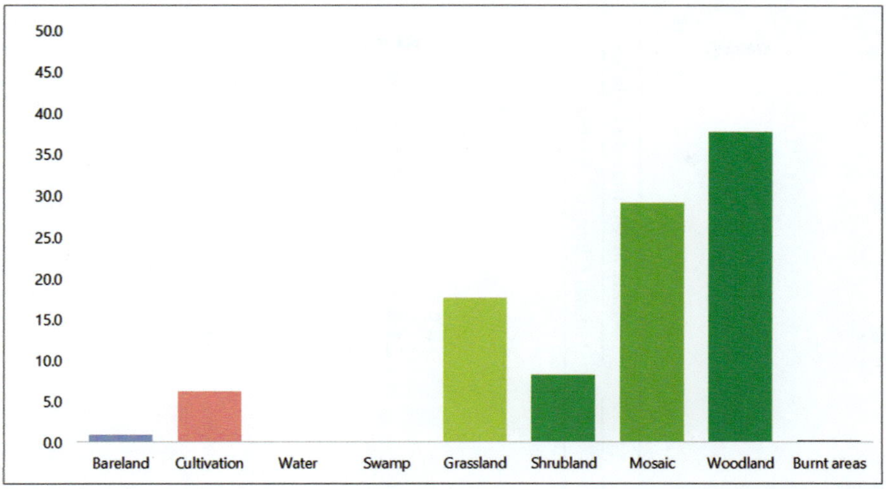

FIGURE 14.5 Areal coverage of land use and land cover types in the Selous-Niassa Wildlife Corridor in 2001.

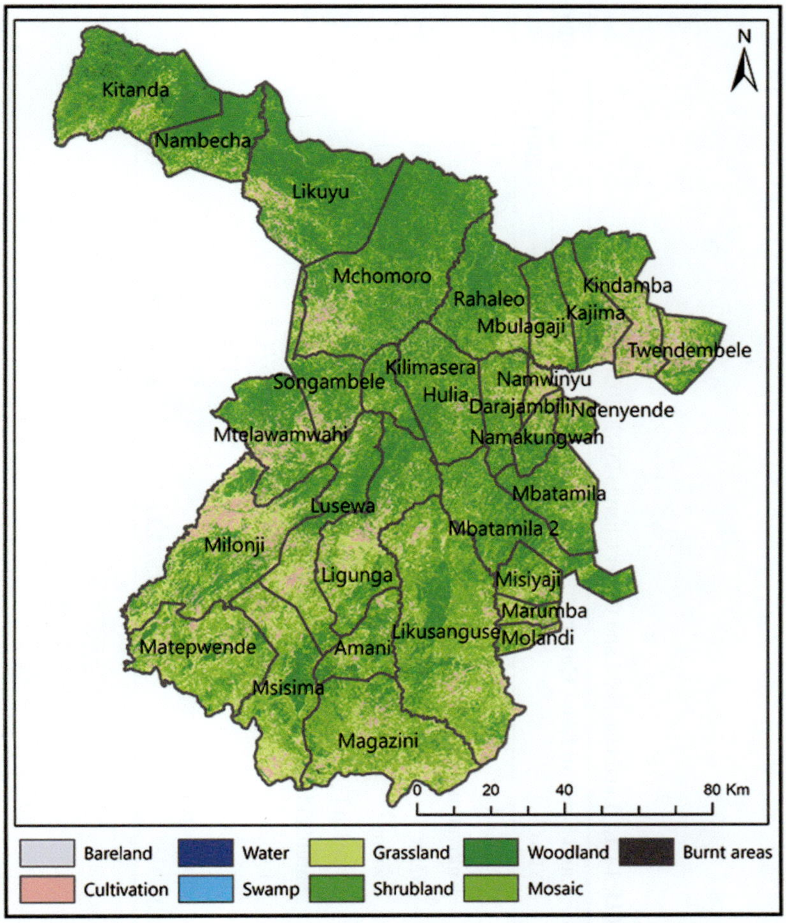

FIGURE 14.6 Land use and land cover map in the Selous-Niassa Wildlife Corridor in 2020.

TABLE 14.1 Land Use and Land Cover Change 2001–2020.

SN	Land Use and Land Cover Type	Areal Coverage (Ha)		Proportion (%)		Rate of Change and Status	
		2001	2020	2001	2020	Annual Change	Status
1	Bareland	19.6	60.0	0.1	0.4	2.0	Increasing
2	Cultivation	559.0	2269.0	3.3	13.4	85.5	Increasing
3	Water	16.0	13.5	0.1	0.1	−0.1	Decreasing
4	Swamp	21.5	4.1	0.1	0.0	−0.9	Decreasing
5	Grassland	1790.2	2279.8	10.5	13.4	24.5	Increasing
6	Shrubland	344.2	1605.4	2.0	9.5	63.1	Increasing
7	Mosaic	4463.3	5574.8	26.3	32.8	55.6	Increasing
8	Woodland	9589.7	5126.4	56.5	30.2	−223.2	Decreasing
9	Burnt areas	179.6	50.1	1.1	0.3	−6.5	Decreasing
Total		**16,983.21**	**16,983.21**	**100.0**	**100.0**	**0.0**	

FIGURE 14.7 Informal settlement in Mbaranga'andu WMA.

In the last two decades, there has been a notable increase in cultivated land within the corridor. Cultivation expansion is a widespread occurrence in human-dominated landscapes and has significant implications for natural resources (Gebrehiwot et al., 2021; Yaynemsa, 2023). The rise in cultivated area may be attributed to growing anthropogenic demands for food, fuelwood, building materials, and other livelihood necessities, leading to encroachment on natural areas and habitat degradation through deforestation (Figure 14.8). Similar trends have been reported in other studies, such as in the Kitendeni wildlife corridor between Amboseli NP and Mount Kilimanjaro, where areas were used for grazing and wildlife (Noe, 2003), as well as in the Kitendeni wildlife corridor (Kikoti et al., 2010) and in the Selous-Niassa corridor (Ntongani et al., 2010).

Land use and land cover changes profoundly influence wildlife habitats and connectivity. When natural habitats are fragmented or lost can lead to a decline in biodiversity and ecosystem functions. On the other hand, proper land-use planning can help maintain ecological corridors, enabling wildlife to access vital migration routes and dispersal opportunities. For instance, studies in the USA have revealed that converting grassland to cropland and urban areas negatively impacts habitat connectivity and biodiversity (Saunders et al., 1991; Haddad et al., 2015). Similarly, in Europe, land-use fragmentation poses a significant threat to wildlife habitats, especially for large carnivores (Kaim et al., 2019). Degraded cover can result in reduced food and shelter availability for wildlife, intensify competition, and increase vulnerability to predators and diseases. Disrupted habitat connectivity also disrupts wildlife migration and dispersal patterns, leading to genetic isolation and reduced population viability (Haddad et al., 2015).

FIGURE 14.8 Cultivation in the Chingoli WMA in the Selous-Niassa Corridor.

In the SNWC, a comparison of land use and land cover types between 2001 and 2020 shows significant changes. Cultivation has increased, while woodland, shrubland, and mosaic areas have decreased, indicating a considerable conversion of natural habitats into agricultural land (Figure 14.9). This transformation could impact the availability of food and shelter for wildlife, alter competition dynamics, and expose them to greater risks from predators and diseases. Moreover, the loss of natural habitats can disrupt wildlife migration and dispersal patterns, leading to genetic isolation and potential threats to population viability (Haddad et al., 2015).

14.2.2 RAINFALL PATTERN

Over the past two decades, the corridor experienced rainfall ranging from 853 to 1294 mm/year, with some years having rainfall below 1000 mm/year while others had higher rainfall. Spatially, the southern half of the corridor received more rainfall compared to the northern half, although some years showed even distribution patterns. The rainfall pattern exhibited variability, indicating periods of both higher and lower rainfall over the 20-year timeframe (Figures 14.10 and 14.11). The corridor's woodlands are transitioning between wet and dry categories, with rainfall between 700 and 1000 mm/year suggesting dry miombo and rainfall beyond 1000 mm/year indicating wet miombo (Frost, 1996). Higher rainfall was typically observed on the eastern side of the corridor, aligns with common rainfall patterns in miombo woodlands.

FIGURE 14.9 Conversation of natural vegetation to cultivation/agriculture in Likuyu village.

The corridor's location in the tropics makes it susceptible to seasonal rainfall influenced by the Inter-Tropical Convergence Zone. The variability in miombo woodland rainfall has significant implications for the ecosystem, affecting vegetation, wildlife, and overall functioning. Factors like climate cycles, topography, and human activities contribute to rainfall variability, including fluctuations caused by sea surface temperatures in the Indian Ocean Dipole. Changes in land use, such as deforestation and urbanization, can also impact rainfall patterns. Studies have shown that deforestation in the Kilimanjaro region of Tanzania decreased rainfall in certain areas (Fairman Jr et al., 2011). Additionally, climate change is causing shifts in temperature and rainfall patterns, leading to more frequent and intense droughts and floods in some regions of Tanzania (Mati et al., 2008; Guzha et al., 2018). Rainfall is important in influencing wildlife distribution, abundance, and movement within ecosystems. It affects the availability of food, water, and shelter for wildlife, determining their survival and reproductive success. Rainfall variability is a significant driver of habitat suitability and connectivity for wildlife corridors in both savannah and miombo landscapes. For example, in Kenya's Maasai Mara National Reserve, rainfall variability affects the availability of food for herbivores, influencing the distribution and movements of predators such as lions and hyenas (Ogutu et al., 2016). Similarly, in Namibia's Zambezi region, rainfall variability impacts the availability of water resources, influencing the spatiotemporal distribution of large herbivores like elephants (Barnes et al., 2018).

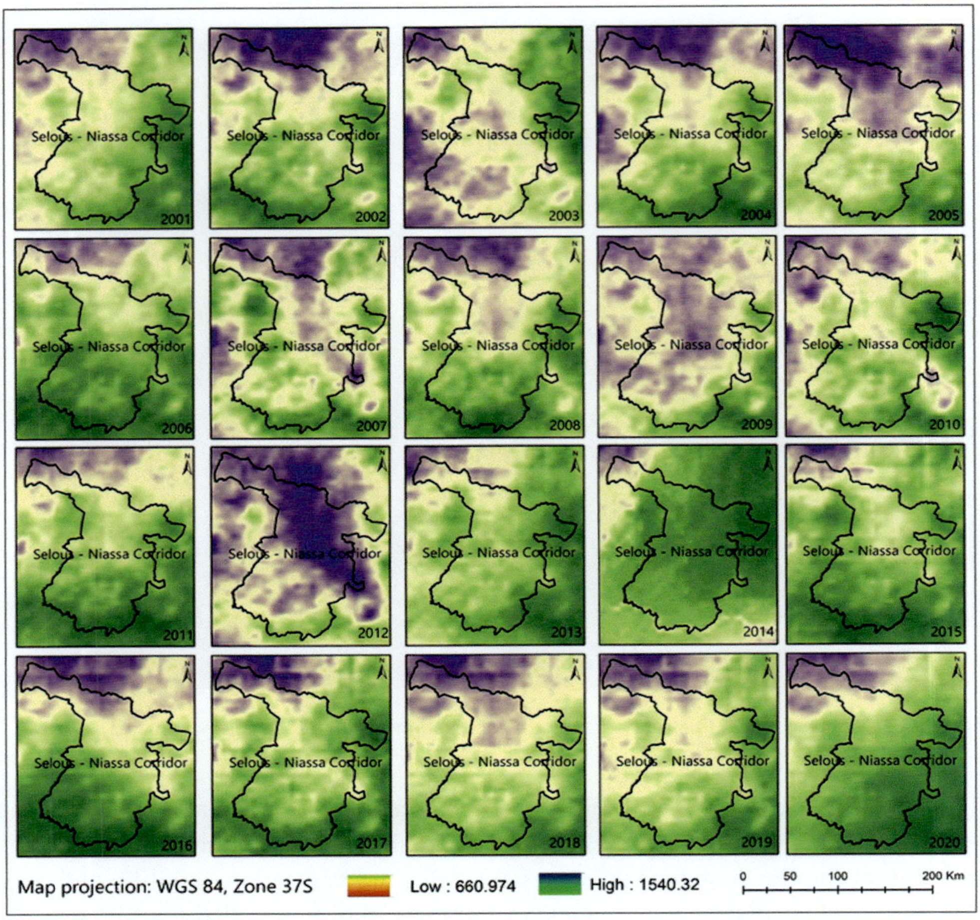

FIGURE 14.10 Rainfall patterns in the Selous-Niassa Wildlife Corridor for the years 2001–2020.

14.3 ENHANCED VEGETATION INDEX (EVI)

Figures 14.12 and 14.13 illustrate the spatial distribution of the 20-year mean EVI from 2001 to 2020. The EVI in the study area remained relatively constant with slight yearly fluctuations for about 10 years (2001–2012), followed by a substantial increase in 2013 and 2014, before stabilizing around 0.4 from 2015 to 2020. The corridor's EVI ranged from a minimum of 0.25 in 2001 to a maximum of 0.43 in 2014 and 2019. However, the average EVI over the two decades was lower than the highest decadal EVI. When comparing the EVI between the two time periods (2001–2010 and 2011–2020), the first slice had an EVI of 0.26, which was 0.12 lower than the second slice.

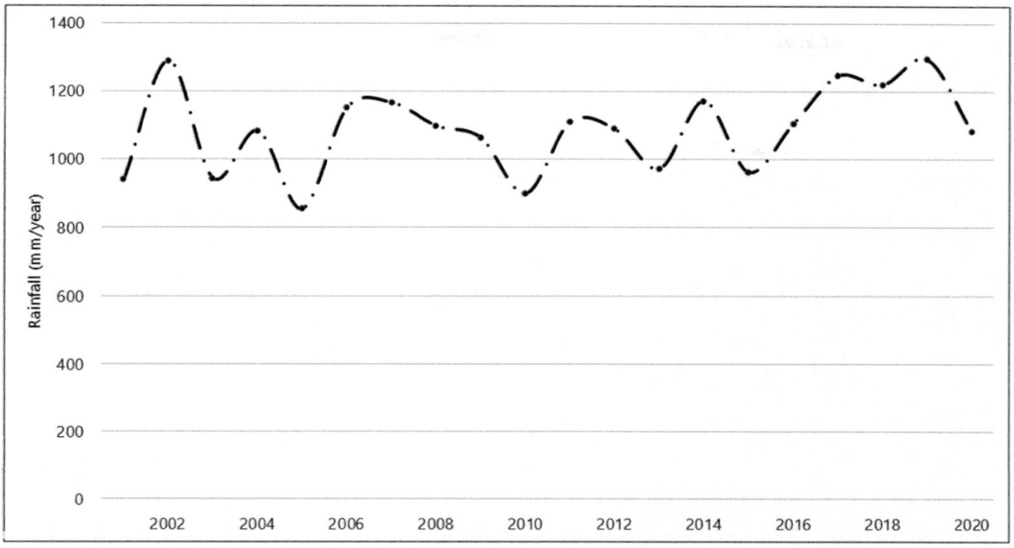

FIGURE 14.11 Rainfall trend in the Selous-Niassa Wildlife Corridor for the years 2001–2020.

The EVI values serve as indicators of habitat conditions. According to Mangewa et al. (2022), EVI ranges can be interpreted as follows: An EVI between 0.6 and 1 suggests very good habitat conditions, indicating a high density of healthy vegetation and favorable environmental conditions for plant growth. An EVI between 0.4 and 0.61 signifies good habitat conditions, suggesting a moderate density of vegetation and relatively favorable environmental conditions for plant growth. An EVI between 0.2 and 0.4 indicates poor habitat conditions with a low density of vegetation and less favorable environmental conditions for plant growth. An EVI less than 0.2 implies very poor habitat conditions, suggesting a significant lack of vegetation and unfavorable environmental conditions for plant growth. Analyzing the EVI values helps assess the overall health and density of vegetation, providing insights into the habitat conditions of an area.

The observed changes in EVI between 2001 and 2020 can be attributed to fluctuations in rainfall and disturbances in the corridor. During the first slice (2001–2010), the EVI remained low due to factors like fluctuating rainfall and disturbances such as fires. In the second slice (2011–2020), there was an increase in EVI, which corresponds to higher rainfall and possibly controlled or reduced fire incidents. These changes in EVI align with the fluctuations in rainfall in the corridor. Increased rainfall has been linked to higher vegetation productivity (Xiao et al., 2015; Zeng et al., 2022).

Low EVI scores indicate lower vegetation vigor or density, which can result from factors like drought, deforestation, habitat degradation, or land-use changes. On the other hand, high EVI scores suggest healthier and denser vegetation, indicating good habitat quality and positive ecological functioning. Identifying areas with high EVI scores can help prioritize conservation efforts, focusing on preserving these ecologically significant areas and implementing measures to prevent habitat fragmentation, encroachment, and disturbances.

In miombo woodlands, low EVI values can be attributed to various human activities such as logging, charcoal production, and farming, which lead to land degradation and deforestation, ultimately reducing vegetation cover (Alemu, 2015).

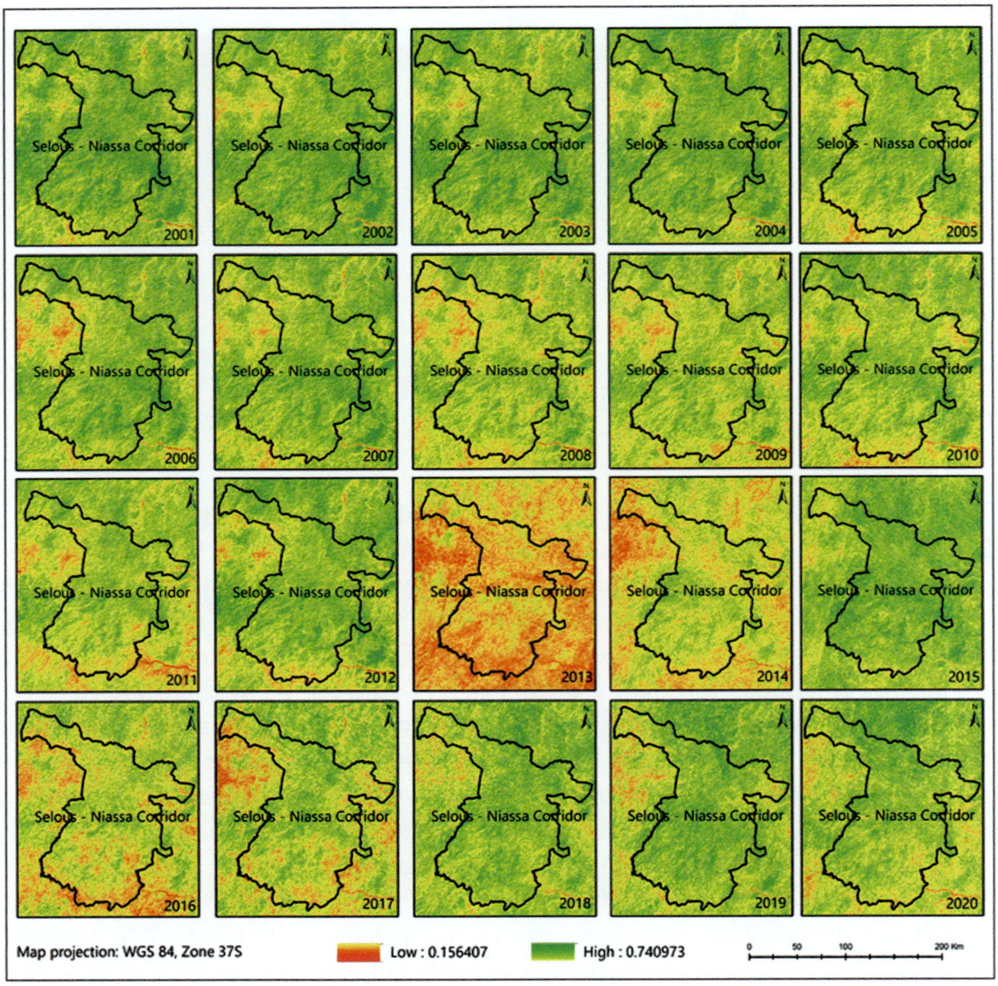

FIGURE 14.12 Enhanced Vegetation Index (EVI) for the years 2001–2020.

Transforming miombo woodlands into croplands or other land uses can lead to a decline in vegetation cover and EVI. A study conducted in Tanzania by Schneibel et al. (2016) demonstrated that land use changes in miombo woodlands caused a substantial reduction in vegetation cover, resulting in lower EVI values. Similarly, in Mozambique's miombo woodlands, areas experiencing high deforestation rates exhibited lower EVI values (Schneibel et al., 2016). These findings indicate that such land use changes can negatively impact the health and density of vegetation, highlighting the importance of sustainable land management and conservation measures to preserve miombo woodlands' ecological integrity.

The observed EVI pattern between 2001 and 2012 may be attributed to favorable local climate conditions with stable rainfall, providing a reliable source of moisture for plant growth. Additionally, increased conservation knowledge from various stakeholders may have contributed to improved EVI.

The visual estimate of EVI shows that areas outside the corridor (western, southern, and eastern) have poorer EVI values compared to inside the corridor. This indicates that nonprotected lands outside the traditional corridor boundaries, excluding the Niassa Reserve, experience lower EVI, likely due to disturbances from fire, human activities, and relatively poor climatic conditions (Alvarado et al., 2017).

High EVI indicates increased biomass production and available biomass for wildlife species. The obtained EVI range values align with other studied areas (Pandey et al., 2019). Several studies have shown the significance of EVI or NDVI as indicators of vegetation dynamics and their implications for wildlife habitats in coastal forests. They emphasize the need for effective land use management practices to maintain or enhance forage resources for wildlife in miombo-dominated landscapes like the Ruvuma landscape.

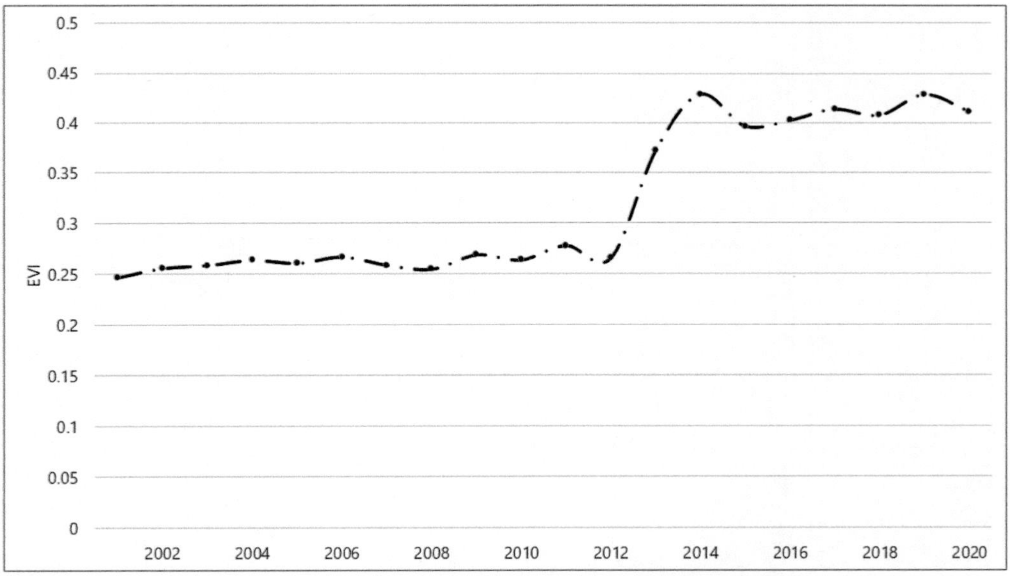

FIGURE 14.13 EVI trend in the Selous-Niassa Wildlife Corridor for the years 2001–2020.

14.4 SOIL MOISTURE

Soil moisture is an important indicator for drought conditions. In addition to soil parameters and management, it depends on both precipitation and evapotranspiration as well as land surface temperature (LST). The soil moisture is expressed as a percentage representing the volumetric water content in the soil. The range of soil moisture values typically falls between 0% (completely dry) and 100% (fully saturated).

Soil moisture is an indicator for drought conditions and depends on precipitation, evapotranspiration, and LST. It is expressed as a percentage representing water content in the soil, with values ranging between 0% (completely dry) and 100% (fully saturated). Over two decades, the corridor's average soil moisture stood at 26.10%, with the minimum recorded at 6.10% in 2007 and the maximum at 34.40% in 2001 and 2002. Soil moisture displayed an up-and-down trend between 2001 and 2012, but from 2013 to 2017, it increased, suggesting more rainfall in the corridor (Figures 14.14 and 14.15). However, reduced moisture in 2018 could be linked to increased surface temperature. Human activities, such as agriculture, can affect water resources, thus impacting soil moisture. Fluctuations in soil moisture along the corridor were observed with changes in the eastern and western halves.

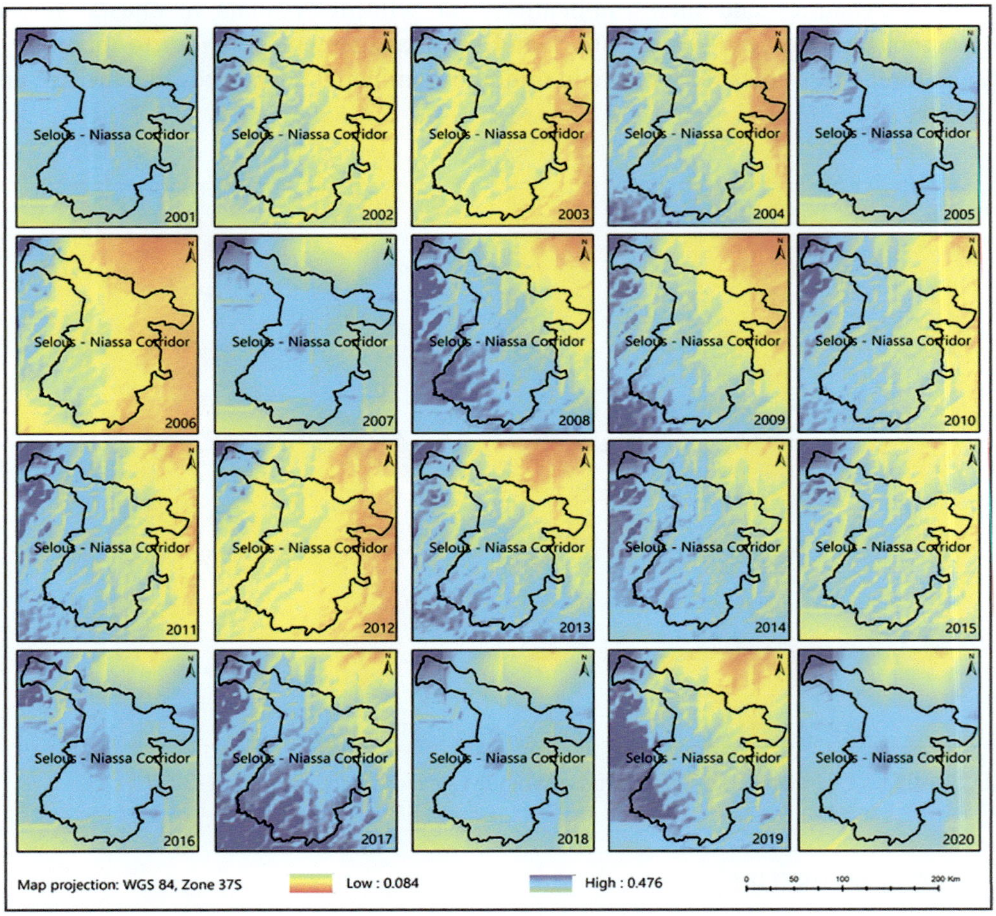

FIGURE 14.14 Soil moisture pattern in the Selous-Niassa Wildlife Corridor for the years 2001–2020.

Changes in precipitation patterns significantly drive soil moisture fluctuations. Heavy rainfall events lead to increased soil moisture, while dry spells result in drying out. Human activities like agriculture and water extraction can also alter the water balance and impact soil moisture (Zhang et al., 2012). Fluctuations in soil moisture influence vegetation growth and wildlife habitat, supporting different vegetation communities (Kidane et al., 2012). Soil moisture also affects land use and habitat types and influences plant richness in grassland, shrubland, and woodland cover types (Moeslund et al., 2013; Kija et al., 2020). Vegetation growth depends primarily on temperature and moisture.

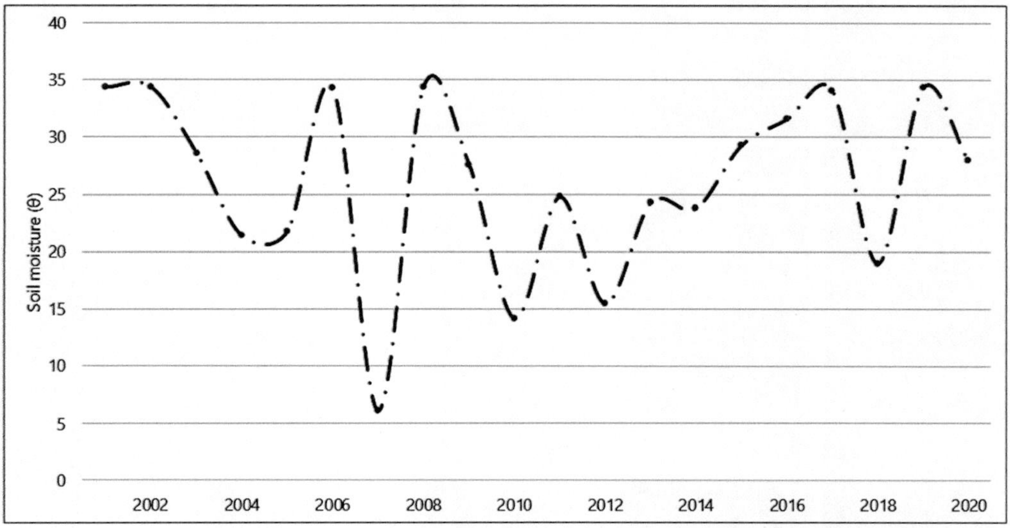

FIGURE 14.15 Pattern of soil moisture in the Selous-Niassa Wildlife Corridor for the period 2001–2020.

14.5 TEMPERATURE PATTERN

LST typically range from the minimum temperature recorded during night time to the maximum temperature observed during daytime. The analysis of LST (2001–2020) depicts an average of 30.19°C temperature, while the minimum and maximum temperature varied between 28.24°C and 30.19°C, respectively (Figures 14.16 and 14.17).

The lowest temperature was recorded in 2004, and the highest in 2010. Visual comparisons of LST patterns show that the northern side of the corridor has lower temperatures compared to the south (Figure 14.16). Conversely, the far southern area outside the corridor exhibits relatively higher temperature patterns in all study years (Figure 14.17).

Temperature fluctuations over years are closely linked to variations in rainfall in the area. Increased temperature aligns with the national temperature records over the past 30 years (1901–2020). Temperature variations can also be influenced by factors like deforestation, desertification, and urbanization, which can alter surface characteristics and heat storage, thus temperature changes.

Climate change, driven by greenhouse gas accumulation, is causing global temperature increases and fluctuations. These variations can impact wildlife distribution, particularly species seeking suitable habitats and resources. As temperatures warm, some species may shift their ranges to higher elevations or within their preferred temperature ranges.

Temperature also affects wildlife physiology and behavior, including reproduction, migration, and hibernation timing, as well as plant species' distribution and abundance.

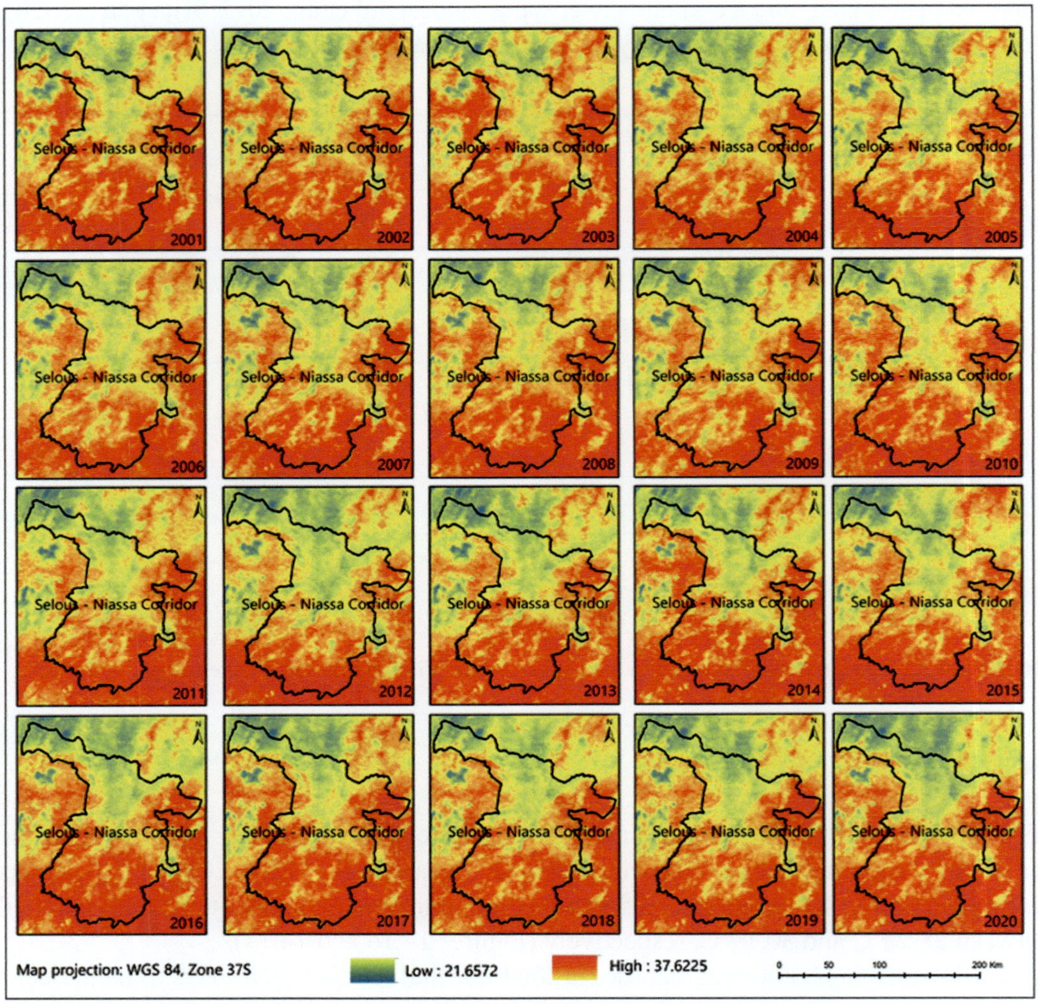

FIGURE 14.16 LST pattern in the Selous-Niassa Wildlife Corridor for the years 2001–2020.

14.6 ANTHROPOGENIC DISTURBANCES

14.6.1 WILDFIRE (FIRE FREQUENCY AND BURNT AREA)

Two decadal analysis (2001–2020) are shown in Figure 14.18. Analysis indicates a total of 201,011 fire incidences in the corridor. Incidences ranged from 3046 (in 2001) to 16,081 (in 2020) and a mean of 10,051 incidents. Fire incidences relate to burnt areas, analysis from burnt areas indicate an average of 273,406 ha (16.11%) are burned per year. The fire frequency and proportions of burnt (in brackets) were 96,998 (5.71%) ha and 398,460 ha (23.48%), respectively (Figures 14.19 and 14.20). Years 2008, 2010, and 2015 had the most extensive annual area burned, whereas, 2001, 2002, and 2004 were the years with least annual area burned. Trend analysis shows fluctuation of fire occurrences and frequency of burn (Figures 14.18 and 14.19), this could be related to increased human activities such as land clearing, agricultural practices, and logging.

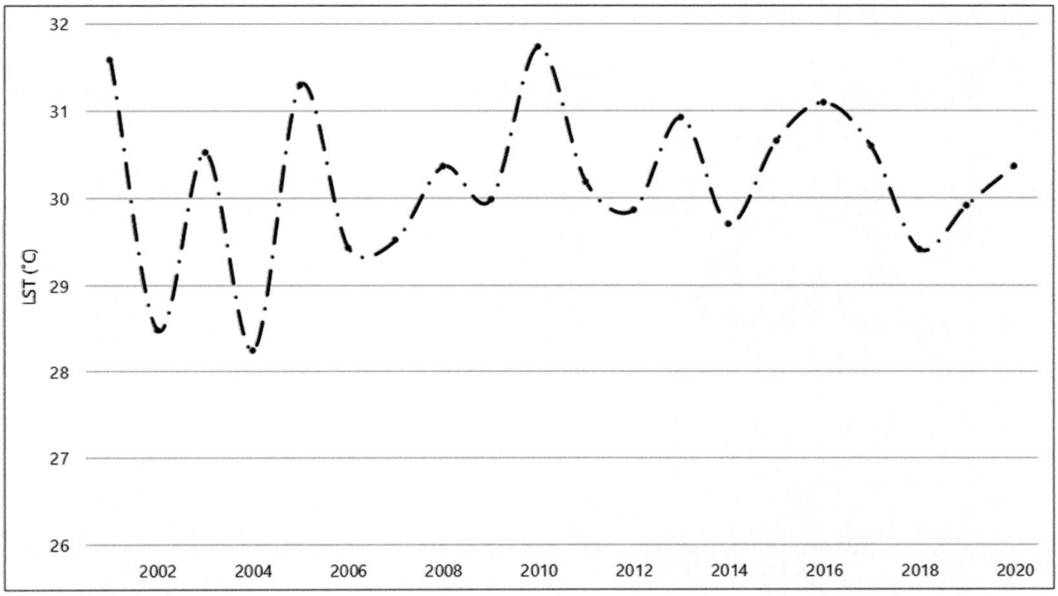

FIGURE 14.17 Land surface temperature (LST) trend in the Selous-Niassa Wildlife Corridor for the period 2001–2020.

Fire is an important parameter in ecology and affects ecological processes in terrestrial ecosystems. Fire incidence is a common phenomenon in farmed-human dominated landscapes, especially in savannah and woodland ecosystems (Frost, 1996). The observed one to two-years fire return area has also been reported in previous studies (Frost, 1996; Tarimo et al., 2015). The observed high amount of burnt area in 2008 could be linked to high amounts of available flammable materials (dry herbaceous layer and dry litter and top soil humus) resulting from El-Nino Oscillations (rainfall) and land use and land cover changes. The observed reduced extent of burnt area has also been reported in other miombo woodlands (Tarimo et al., 2015).

General trend visually indicates that the northern part of the corridor is not much burned compared to the middle and southern parts. High fire incidences in the southern part of the corridor (northern part of the NSR in Mozambique) has also been reported in previous study (Ribeiro et al., 2020). However, for the years 2003, 2008, 2009, 2010, and 2011–2015 burned patches spread evenly across the entire corridor. The quantified burnt areas provide an indication of the distribution of fire incidences in the study area, and can be considered as a wildfire metric for trend analysis, as it quantifies the portion of an area that has been burnt (Figure 14.20).

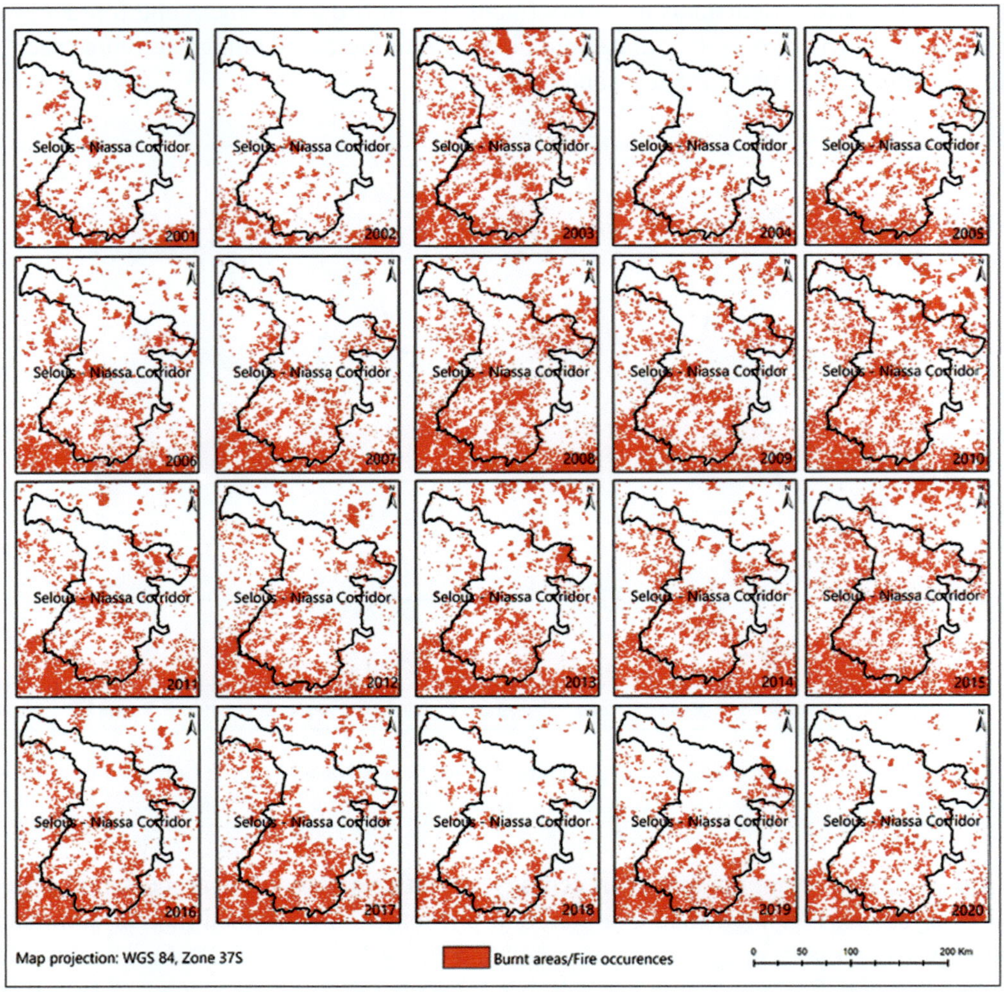

FIGURE 14.18 Pattern of burnt areas in the Selous-Niassa Wildlife Corridor for the years 2001–2020.

The fluctuating fire frequencies and burnt area observed in the corridor can have various causes and ecological implications. For example, they can be attributed to fuel accumulation and changes in vegetation composition. An increase in fire-prone vegetation types or the presence of highly flammable species can lead to more frequent fires. Conversely, shifts toward less flammable vegetation communities may result in reduced fire frequencies. Fires play a role in maintaining ecosystem health, promoting nutrient cycling, and shaping vegetation communities through the removal of accumulated biomass. Anthropogenic activities, including land use practices such as agricultural burning, slash-and-burn farming, or deliberate fire setting, can significantly influence fire frequencies as well, thus more frequent fires, impacting both natural ecosystems and wildlife habitats.

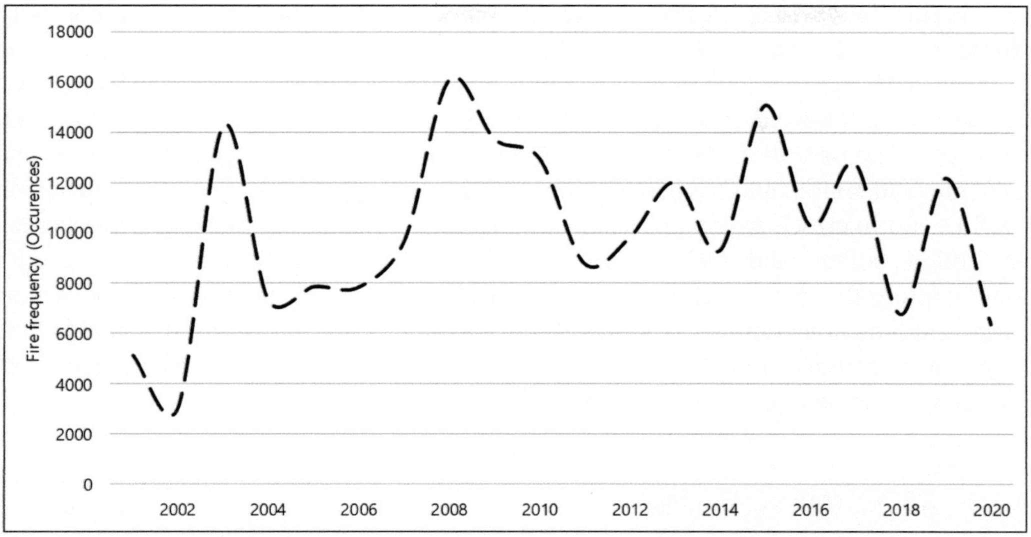

FIGURE 14.19 Fire frequencies in the Selous-Niassa Wildlife Corridor for the years 2001–2020.

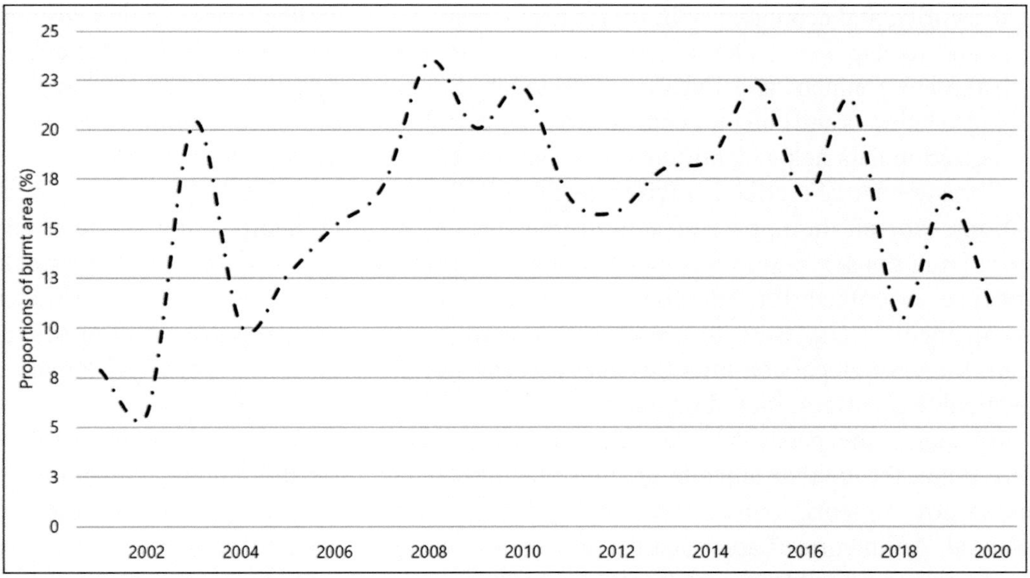

FIGURE 14.20 Burnt area in the Selous-Niassa Wildlife Corridor for the years 2001–2020.

14.7 CONCLUSIONS AND RECOMMENDATIONS

14.7.1 CONCLUSIONS

The study observed significant land use changes in the corridor, leading to the conversion of woodland to grassland and shrubland due to settlement and agricultural intensification. These changes have adverse effects on wildlife habitats and connectivity, causing habitat loss, fragmentation, and decline in wildlife populations. Human–wildlife conflicts, reduced genetic diversity, and ecological resilience are also consequences of these land use changes. Despite variable rainfall patterns, the corridor exhibited a stable EVI between 2001 and 2012, followed by an increase from 2013 to 2020. Factors like temperature, soil moisture, and conservation efforts likely contributed to the EVI stability and improvement. Fluctuations in rainfall and temperature impact vegetation dynamics and wildlife behavior. Wildfires and livestock grazing play crucial roles in ecosystem functioning, while human interventions affect land surface phenology and ecological processes. Future studies should explore the ecological consequences of phenological changes in light of climate change and human activities. To address these challenges, implementing sustainable land use practices, conservation efforts, and involving local communities are recommended to preserve wildlife habitats and connectivity in the corridor.

14.7.2 RECOMMENDATIONS

Implementing Land Use Planning (LUP) can improve human well-being by efficiently using land resources for sustainable agriculture and livestock practices. This boosts productivity, food security, and economic stability for local communities. LUP also encourages sustainable land management, like soil conservation and water resource protection, benefiting both the environment and livelihoods. Habitat restoration programs are vital to restore degraded habitats and enhance connectivity between fragmented areas. Regular monitoring of vegetation data helps identify degraded habitat and track wildlife distribution.

Climate change's effects, seen through rainfall, temperature, and soil moisture fluctuations, worsen the impact of land use changes on wildlife habitats and connectivity. Prioritizing conservation efforts that protect wildlife corridors and promote sustainable land use is crucial for effective conservation and management strategies. Wildfire management is necessary to preserve connectivity and open areas for wildlife movement. Engaging local communities in conservation efforts raises awareness and supports sustainable practices, including eco-tourism and wildlife protection.

Balancing the needs of livestock and wildlife is essential for maintaining healthy ecosystems through sustainable grazing practices. To achieve these goals, collaboration among governments, conservation organizations, communities, and the private sector is essential. An integrated approach that integrates conservation and development activities should reduce habitat fragmentation and enhance connectivity. By prioritizing both wildlife and human needs, we can ensure a sustainable future for our planet.

ACKNOWLEDGMENTS

We express our deep gratitude to the World Wildlife Fund for Nature and the Selous Ecosystem Conservation and Development (SECAD) for funding this work (Project name/number: CMS-SSFA-40000012-402892 & SECAD AWARD 401509). Our sincere thanks go to the Ministry of Natural Resources and Tourism (MNRT), specifically the Director of

Wildlife for their invaluable support throughout this work. We appreciate the unwavering support of the Ruvuma's Regional Office, Namtumbo District Council, Tunduru District Council, Tanzania Forest Services Agency, Tanzania Wildlife Management Authority, and Tanzania National Parks. The technical support from the Director General of the Tanzania Wildlife Research Institute is highly acknowledged. We extend our thanks to the WMAs within the corridor for their cooperation. Lastly, we are thankful to all individuals who contributed to the successful execution of this assignment.

KEYWORDS

- **remote sensing**
- **ecological conditions**
- **anthropogenic disturbances**
- **Coastal Mosaic Forest of Tanzania**
- **Selous-Niassa Wildlife Corridor**
- **Ruvuma landscape**

REFERENCES

Adebisi, N.; Balogun, A.-L.; Mahdianpari, M.; Min, T. H. Assessing the impacts of rising sea level on coastal morpho-dynamics with automated high-frequency shoreline mapping using multi-sensor optical satellites. *Remote Sens.* **2021,** *13,* 3587.

Alemu, B. The effect of land use land cover change on land degradation in the highlands of Ethiopia. *J. Environ. Earth Sci.* **2015,** *5,* 1–13.

Alvarado, S. T.; Fornazari, T.; Cóstola, A.; Morellato, L. P. C.; Silva, T. S. F. Drivers of fire occurrence in a mountainous Brazilian cerrado savanna: Tracking long-term fire regimes using remote sensing. *Ecol. Indic.* **2017,** *78,* 270–281.

Baldus, R.; Hahn, R.; Mpanduji, D.; Siege, L. *The Selous-Niassa Wildlife Corridor. Tanzania Wildlife Discussion Paper No. 34. Rolf D.* Baldus and Ludwig Siege (Eds.). **2003.**

Barnes, R.; Craigie, I. D.; Dudley, N. Understanding the influence of rainfall on elephant population dynamics in the Zambezi region of Namibia. *Peer J.* **2018,** *6,* e5258.

Bluwstein, J.; Lund, J. F. Territoriality by conservation in the Selous–Niassa Corridor in Tanzania. *World Develop.* **2018,** *101,* 453–465.

Botey, M.M.; Soliño, M.; Perea, R.; Martínez-Jauregui, M. Let us give voice to local farmers: preferences for farm-based strategies to enhance human–elephant coexistence in Africa. *Animals* **2022,** *12,* 1867.

Bouillon, S.; Rivera-Monroy, V.; Twilley, R.; Kairo, J. Mangroves. In: Laffoley, D.; Grimsditch, G.D. (Eds.), *The Management of Natural Carbon Sinks in Coastal Ecosystems*. IUCN: Gland, Switzerland, 2009.

Bryan-Brown, D. N.; Connolly, R. M.; Richards, D. R.; Adame, F.; Friess, D. A.; Brown, C. J. Global trends in mangrove forest fragmentation. *Sci. Rep.* **2020,** *10,* 7117.

Chughtai, A. H.; Abbasi, H.; Karas, I. R. A review on change detection method and accuracy assessment for land use land cover. *Remote Sens. Appl.: Soc. Environ.* **2021,** *22,* 100482.

Eakin, C. M.; Nim, C. J.; Brainard, R. E.; Aubrecht, C.; Elvidge, C.; Gledhill, D. K.; Muller-Karger, F.; Mumby, P. J.; Skirving, W. J.; Strong, A. E. Monitoring coral reefs from space. *Oceanography* **2010,** *23,* 118–133.

Estoque, R. C.; Myint, S. W.; Wang, C.; Ishtiaque, A.; Aung, T. T.; Emerton, L.; Ooba, M.; Hijioka, Y.; Mon, M. S.; Wang, Z. Assessing environmental impacts and change in Myanmar's mangrove ecosystem service value due to deforestation (2000–2014). *Glob. Change Biol.* **2018,** *24*, 5391–5410.

Fairman Jr, J. G.; Nair, U. S.; Christopher, S. A.; Mölg, T. Land use change impacts on regional climate over Kilimanjaro. *J. Geophys. Res.*: *Atmospheres* **2011,** *116*.

Filho, W. L.; Azeiteiro, U. M.; Balogun, A.-L.; Setti, A. F. F.; Mucova, S. A.; Ayal, D.; Totin, E.; Lydia, A. M.; Kalaba, F. K.; Oguge, N. O. The influence of ecosystems services depletion to climate change adaptation efforts in Africa. *Sci. Total Environ.* **2021,** *779*, 146414.

Foody, G. M. Status of land cover classification accuracy assessment. *Remote Sens. Environ.* **2002,** *80,* 185–201.

Frost, P. The ecology of miombo woodlands. In: Campbell, B. (Ed.), *The Miombo in Transition: Woodlands and Welfare in Africa*. Centre for International Forestry Research: Bogor, Indonesia. **1996,** pp. 11–57.

Gabriele, M.; Brumana, R.; Previtali, M.; Cazzani, A. A combined GIS and remote sensing approach for monitoring climate change-related land degradation to support landscape preservation and planning tools: The Basilicata case study. *Appl. Geom.* **2022,** *15*, 497–532.

Gao, Y.; Skutsch, M.; Paneque-Gálvez, J.; Ghilardi, A. Remote sensing of forest degradation: A review. *Environ. Res. Lett.* **2020,** *15*, 103001.

Gebrehiwot, K.; Teferi, E.; Woldu, Z.; Fekadu, M.; Desalegn, T.; Demissew, S. Dynamics and drivers of land cover change in the Afroalpine vegetation belt: Abune Yosef mountain range, Northern Ethiopia. *Environ. Dev. Sustain.* **2021,** *23*, 10679–10701.

Gray, P. C.; Ridge, J. T.; Poulin, S. K.; Seymour, A. C.; Schwantes, A. M.; Swenson, J. J.; Johnston, D. W. Integrating drone imagery into high resolution satellite remote sensing assessments of estuarine environments. *Remote Sens.* **2018,** *10,* 1257.

Guzha, A.; Rufino, M. C.; Okoth, S.; Jacobs, S.; Nóbrega, R. Impacts of land use and land cover change on surface runoff, discharge and low flows: Evidence from East Africa. *J. Hydrol. Reg. Stud.* **2018,** *15*, 49–67.

Haddad, N. M.; Brudvig, L. A.; Clobert, J.; Davies, K. F.; Gonzalez, A.; Holt, R. D.; Lovejoy, T. E.; Sexton, J. O.; Austin, M. P.; Collins, C. D. Habitat fragmentation and its lasting impact on Earth's ecosystems. *Sci. Adv.* **2015,** *1*, e1500052.

Hansen, A. J.; Defries, R. Ecological mechanisms linking protected areas to surrounding lands. *Ecol. Appl.* **2007,** *17*, 974–988.

Howard, J.; Sutton-Grier, A.; Herr, D.; Kleypas, J.; Landis, E.; Mcleod, E.; Pidgeon, E.; Simpson, S. Clarifying the role of coastal and marine systems in climate mitigation. *Front. Ecol. Environ.* **2017,** *15*, 42–50.

Jackson, T. D.; Williams, G. J.; Walker-Springett, G.; Davies, A. J. Three-dimensional digital mapping of ecosystems: A new era in spatial ecology. *Proc. Roy. Soc. B* **2020,** *287*, 20192383.

Kaim, D.; Ziółkowska, E.; Szwagrzyk, M.; Price, B.; Kozak, J. Impact of future land use change on large carnivores connectivity in the Polish Carpathians. *Land Degrad. Dev.* **2019,** *8*, 8.

Kalaba, F. K. A conceptual framework for understanding forest socio-ecological systems. *Biodivers. Conserv.* **2014,** *23,* 3391–3403.

Kaplan, G.; Avdan, Z. Y.; Avdan, U. Mapping and monitoring wetland dynamics using thermal, optical, and SAR remote sensing data. In: *Wetlands Management: Assessing Risk and Sustainable Solutions.* InTech Open, **2019**, p. 87.

Kidane, Y.; Stahlmann, R.; Beierkuhnlein, C. Vegetation dynamics, and land use and land cover change in the Bale Mountains, Ethiopia. *Environ. Monit. Assess.* **2012,** *184*, 7473–7489.

Kija, H. K.; Mokheha, S.; Kamuna, M.; Kivuyo, E.; Katondo, R.; Easton, J.; Botey, M. M.; Msoffe, F. Conservation planning and viability in the Selous-Niassa trans frontier conservation area, Southern Tanzania: A safe corridor for people and Wildlife. In: *A Paper Presented at 13th TAWIRI Scientific Conference*, December **2021**.

Kija, H. K.; Ogutu, J. O.; Mangewa, L. J.; Bukombe, J.; Verones, F.; Graae, B. J.; Kideghesho, J. R.; Said, M. Y.; Nzunda, E. F. Land use and land cover change within and around the greater Serengeti ecosystem, Tanzania. *Am. J. Remote Sens.* **2020,** *8*(1), 1–19.

Kikoti, A.; Griffin, C.; Pamphil, L. Elephant use and conflict leads to Tanzania's first wildlife conservation corridor. *Pachyderm* **2010,** *48,* 57–66.

Lupala, Z., Lusambo, L.; Ngaga, Y. Management, growth, and carbon storage in miombo woodlands of Tanzania. *Intern. J. Forestry Res.* **2014,** *2014*, 1–11.

Mamboleo, A. A. *Integration of agent-based and GIS-based modelling for geosimulation of human-elephant interactions in the Bunda district, Tanzania.* A thesis submitted in partial fulfilment of the requirements for the Degree of Doctor of Philosophy at Lincoln University. **2018**.

Mangewa, L. J.; Ndakidemi, P. A.; Alward, R. D.; Kija, H. K.; Bukombe, J. K.; Nasolwa, E. R.; Munishi, L. K. Comparative assessment of UAV and Sentinel-2 NDVI and GNDVI for preliminary diagnosis of habitat conditions in burunge wildlife management area, Tanzania. *Earth* **2022,** *3,* 769–787.

Mango, J.; Valerian-Peter, R.; Kanja, K.; Ngondo, J.; Maobe, A.; Lubida, A.; Ijumulana, J.; Li, X. Evaluating the land cover dynamics in the protected areas using GIS and remote sensing techniques: The case of Nyerere National Park, Tanzania. *Geocarto International* **2022,** *37,* 17361–17381.

Marvin, D. C.; Koh, L. P.; Lynam, A. J.; Wich, S.; Davies, A. B.; Krishnamurthy, R.; Stokes, E.; Starkey, R.; Asner, G. P. Integrating technologies for scalable ecology and conservation. *Glob. Ecol. Conserv.* **2016,** *7,* 262–275.

Mathur, P. K.; Habib, B.; Mathur, P. Technology advancement and integration in the context of wildlife conservation. In: *Human Bond Communication: The Holy Grail of Holistic Communication and Immersive Experience*. Wiley Telecom, **2017,** pp. 115–130.

Mati, B. M.; Mutie, S.; Gadain, H.; Home, P.; Mtalo, F. Impacts of land-use/cover changes on the hydrology of the transboundary Mara River, Kenya/Tanzania. *Lakes Reserv.: Res. Manag.* **2008,** *13,* 169–177.

Moeslund, J. E.; Arge, L.; Bøcher, P. K.; Dalgaard, T.; Ejrnæs, R.; Odgaard, M. V.; Svenning, J.-C. Topographically controlled soil moisture drives plant diversity patterns within grasslands and influences plant richness, especially grassland. *Biodivers. Conserv.* **2013,** *22,* 2151–2166.

Mpanduji, D. G. Population structure, movement and health status of elephants and other wildlife in the Selous-Niassa Wildlife Corridor, southern Tanzania. PhD Thesis. Free University of Berlin. **2004**.

Mumby, P. J.; Skirving, W.; Strong, A. E.; Hardy, J. T.; Ledrew, E. F.; Hochberg, E. J.; Stumpf, R. P.; David, L. T. Remote sensing of coral reefs and their physical environment. *Mar. Pollut. Bull.* **2004.** *48,* 219–228.

Mwakoba, K. *Remote sensing analysis of land cover/use conditions of community-based wildlife conservation areas in Tanzania.* Master Thesis in Geographical Information Science. **2021**.

Ntongani, W. A.; Munishi, P. K.; Mbilinyi, B. P. Land use changes and conservation threats in the eastern Selous–Niassa wildlife corridor, Nachingwea, Tanzania. *African J. Ecol.* **2010,** *48,* 880–887.

Nyangoko, B. P.; Berg, H.; Mangora, M. M.; Shalli, M. S.; Gullström, M. Local perceptions of changes in mangrove ecosystem services and their implications for livelihoods and management in the Rufiji Delta, Tanzania. *Ocean Coast. Manag.* **2022,** *219,* 106065.

Ogutu, J. O.; Piepho, H.-P.; Said, M. Y.; Ojwang, G. O.; Njino, L. W.; Kifugo, S. C.; Wargute, P. W. Extreme wildlife declines and concurrent increase in livestock numbers in Kenya: What are the causes? *PLoS One* **2016,** *11,* e0163249.

Pandey, P. C.; Anand, A.; Srivastava, P. K. Spatial distribution of mangrove forest species and biomass assessment using field inventory and earth observation hyperspectral data. *Biodivers. Conserv.* **2019,** *28,* 2143–2162.

Phinn, S.; Joyce, K.; Scarth, P.; Roelfsema, C. The role of integrated information acquisition and management in the analysis of coastal ecosystem C. In: *Remote Sensing of Aquatic Coastal Ecosystem Processes*. **2006,** pp. 217–249.

Picard, C. H. *The Promise and Peril of Large-Scale Conservation: An Appraisal of the Selous Niassa Wildlife Corridor.* Yale University: Tanzania. **2010**.

Ribeiro, N. S.; Silva De Miranda, P. L.; Timberlake, J. Biogeography and ecology of miombo woodlands. In: *Miombo Woodlands in a Changing Environment: Securing the Resilience and Sustainability of People and Woodlands*. **2020,** pp. 9–53.

Riggio, J.; Caro, T. Structural connectivity at a national scale: Wildlife corridors in Tanzania. *PLoS One.* **2017,** *12,* e0187407.

Roy, P. S.; Ramachandran, R. M.; Paul, O.; Thakur, P. K.; Ravan, S.; Behera, M. D.; Sarangi, C.; Kanawade, V. P. Anthropogenic land use and land cover changes—A review on its environmental consequences and climate change. *J. Indian Soc. Remote Sens.* **2022,** *50,* 1615–1640.

Saunders, D. A.; Hobbs, R. J.; Margules, C. R. Biological consequences of ecosystem fragmentation: A review. *Conserv. Biol.* **1991,** *5,* 18–32.

Schabel, H. G. *Introduction to Tanzania. Forest Entomology in East Africa: Forest Insects of Tanzania.* **2006,** pp. 1–22.

Schneibel, A.; Stellmes, M.; Röder, A.; Finckh, M.; Revermann, R.; Frantz, D.; Hill, J. Evaluating the trade-off between food and timber resulting from the conversion of Miombo forests to agricultural land in Angola using multi-temporal Landsat data. *Sci. Total Environ.* **2016,** *548,* 390–401.

Shalaby, A.; Tateishi, R. Remote sensing and GIS for mapping and monitoring land cover and land-use changes in the Northwestern coastal zone of Egypt. *Appl. Geogr.* **2007,** *27,* 28–41.

Sievers, M.; Brown, C. J.; Mcgowan, J.; Turschwell, M. P.; Buelow, C. A.; Holgate, B.; Pearson, R. M.; Adame, M. F.; Andradi-Brown, D. A.; Arnell, A. Co-occurrence of biodiversity, carbon storage, coastal protection, and fish and invertebrate production to inform global mangrove conservation planning. *Sci. Total Environ.* **2023,** *904,* 166357.

Svoray, T.; Perevolotsky, A.; Atkinson, P. M. Ecological sustainability in rangelands: The contribution of remote sensing. *Intern. J. Remote Sens.* **2013,** *34,* 6216–6242.

Tabor, K.; Burgess, N. D.; Mbilinyi, B. P.; Kashaigili, J. J.; Steininger, M. K. Forest and woodland cover and change in coastal Tanzania and Kenya, 1990 to 2000. *J. East African Nat. Hist.* **2010,** *99,* 19–45.

Tarimo, B.; Dick, Ø. B.; Gobakken, T.;Totland, Ø. Spatial distribution of temporal dynamics in anthropogenic fires in miombo savanna woodlands of Tanzania. *Carbon Balance Manag.* **2015,** *10,* 1–15.

Thomlinson, J. R.; Bolstad, P. V.; Cohen, W. B. Coordinating methodologies for scaling landcover classifications from site-specific to global: Steps toward validating global map products. *Remote Sens. Environ.* **1999,** *70,* 16–28.

Willis, K. S. Remote sensing change detection for ecological monitoring in United States protected areas. *Biol. Conserv.* **2015,** *182,* 233–242.

Xiao, J.; Zhou, Y.; Zhang, L. Contributions of natural and human factors to increases in vegetation productivity in China. *Ecosphere* **2015,** *6,* 1–20.

Yaynemsa, K. G. Ecological connectivity. In: *Plant Biodiversity Conservation in Ethiopia: A Shift to Small Conservation Reserves*. Springer. **2023,** pp. 127–139.

Zella, A. Y. *Impacts of Socio-Economic Activities in Managing Selous-Niassa Wildlife Corridor.* Sokoine University of Agriculture. **2016.**

Zella, A. Y.; Lawi, Y.; Saria, J. Forecasting habitat conversion of eastern corridor of Selous-Niassa Transfrontier Conservation Areas (TFCA). *Int. J. Environ. Sci. Nat. Resour.* **2017,** *8,* 01–06.

Zeng, X.; Hu, Z.; Chen, A.; Yuan, W.; Hou, G.; Han, D.; Liang, M.; Di, K.; Cao, R.; Luo, D. The global decline in the sensitivity of vegetation productivity to precipitation from 2001 to 2018. *Glob. Change Biol.* **2022,** *28,* 6823–6833.

Zhang, A.; Zhang, C.; Fu, G.; Wang, B.; Bao, Z.; Zheng, H. Assessments of impacts of climate change and human activities on runoff with SWAT for the Huifa River Basin, Northeast China. *Water Resour. Manag.* **2012,** *26,* 2199–2217.

CHAPTER 15

THE IMPACT OF CLIMATE CHANGE ALONG THE EASTERN AFRICA COASTAL FORESTS

CECILIA LEWERI*, DAMIAN NGUMA, and RICHARD D. LYAMUYA

Tanzania Wildlife Research Institute (TAWIRI), Arusha, Tanzania

*Corresponding author

ABSTRACT

This chapter explores how climate change has impacted the Eastern African Coastal Forests region. Focusing on the scientific underpinnings of climate change dynamics, and the specific impacts on the Eastern African Coastal Forests much emphasis were put to examine the social-economic and demographic trends and see how local communities are benefiting from its conservation. Emphasizing more on the vulnerability of endemic species and biodiversity hotspots, the chapter discusses the societal and economic ramifications of these environmental changes on local communities. In detailing the direct consequences of rising temperatures, shifting rainfall patterns, and extreme weather events, the chapter examines the challenges faced by coastal populations, particularly those reliant on fisheries and agriculture. It also delves into the potential loss of iconic species and habitats, illustrating the interconnectedness of ecological and human well-being in the region. Additionally, the chapter explores conservation and adaptation strategies, emphasizing the importance of marine protected areas, community-driven initiatives, and sustainable resource management. It calls for a comprehensive and collaborative approach, considering the transboundary nature of many coastal ecosystems and the need for regional cooperation. The chapter concludes with policy recommendations aimed at mitigating climate change impacts and preserving biodiversity.

15.1 INTRODUCTION

15.1.1 SOCIAL-ECONOMIC AND DEMOGRAPHIC TRENDS ALONG THE COAST OF EASTERN AFRICA

The community along the coastal area of Eastern Africa heavily depends on marine and coastal ecosystems for provisioning services including commercial, artisanal, timber/

Biodiversity Hotspot of the Coastal Forests of Eastern Africa. T. Pullaiah and Ally J. K. Nkwabi (Eds.)

DOI: 10.1201/9781998511099-15

wood, charcoal, food, and regulatory services (Ferreira et al., 2012; Herr et al., 2019). They rely on activities like small-scale fishing and charcoal production, but there is a growing trend toward larger scale commercial ventures such as fisheries, oil and gas extraction, mariculture, and urban development, indicating a shift toward more extensive use of the surrounding forest resources which in turn might influence biodiversity conservation and management (Midrar, 2020). The population along Eastern Africa's coast has been rising mainly due to people's movement in search of better economic opportunities (Barbier, 2015). The percentage of people living within the 200 km coastal area varies greatly: it is around 33% in Mozambique, 13.6% in Tanzania, and just 6% in Kenya. Even though Mozambique has the highest coastal population percentage, its population density is the lowest at 113 per km^2, while Kenya and Tanzania have higher densities at 171 and 154 per km^2, respectively (Samoilys et al., 2015). This population growth has led to increased pressure on coastal ecosystems of East Africa, due to a rise in demand for settlement areas, more land for agricultural activities and hence depletion of forest resources as well as different plant and animal species (Diop and Scheren, 2016).

15.1.2 CLIMATE CHANGE (CC) IN EASTERN AFRICA COASTAL FORESTS

CC is a major global threat that has already had an observed impact on biodiversity and natural ecosystems (Sintayehu, 2018). It has been due to the increased human activities since the 1800s, particularly the emission of greenhouse gases by burning fossil fuels like coal, oil, and gas which have been estimated to have caused a significant portion of global warming (Bessou et al., 2009). Moreover, clearing land and cutting down forests can also release carbon dioxide (Wanyama et al., 2019). Additionally, energy, industry, transport, buildings, and agriculture among others are also the main sectors causing greenhouse gases (Bessou et al., 2009). Therefore, CC is an intergovernmental complex challenge globally with its influence over various components of the ecological, environmental, socio-political, and socio-economic disciplines (Adger et al., 2005). It encompasses not only global warming but also long-term shifts in global weather patterns and temperature, melting glaciers, rising sea levels, droughts, and impacts on food production and infectious diseases (Balamurugan et al., 2014; Figure 15.1). For example, according to Sintayehu (2018) it was predicted that CC could result in the extinction of more than a million terrestrial species in the next 50 years. During that period rare species, fragmented ecosystems, and areas already under pressure from pollution and deforestation are the most vulnerable (Sintayehu, 2018). It had been found that fire is a major cause of biodiversity loss in Africa and as global warming increases these fires are likely to get more intense and extensive and may result in significant ecosystem changes that would affect biodiversity through species loss or changes in species composition (Sintayehu, 2018).

CC in coastal forest of Eastern Africa is driven by a combination of natural and anthropogenic factors. One of the primary drivers is the increase in greenhouse gas emissions from human activities, such as burning fossil fuels and deforestation (Payet and Obura, 2004). Eastern Africa coastal forest has experienced a rise in average temperatures, with significant implications for the region's climate (Christy et al., 2009). The warming trend has led to more frequent and severe droughts, as well as changes in rainfall patterns, impacting agricultural productivity and food security among local communities (Choi and Eltahir, 2023; Figure 15.1). CC has impacted wildlife ecosystems and human habitats, it has aggravated poor crop production from prolonged drought (Wassmann et al., 2009; Zhao et al., 2005) and crop damage from wildlife in many rural areas including coastal areas resulting into decreased crop production (Nassary et al., 2022) (Figure 15.1).

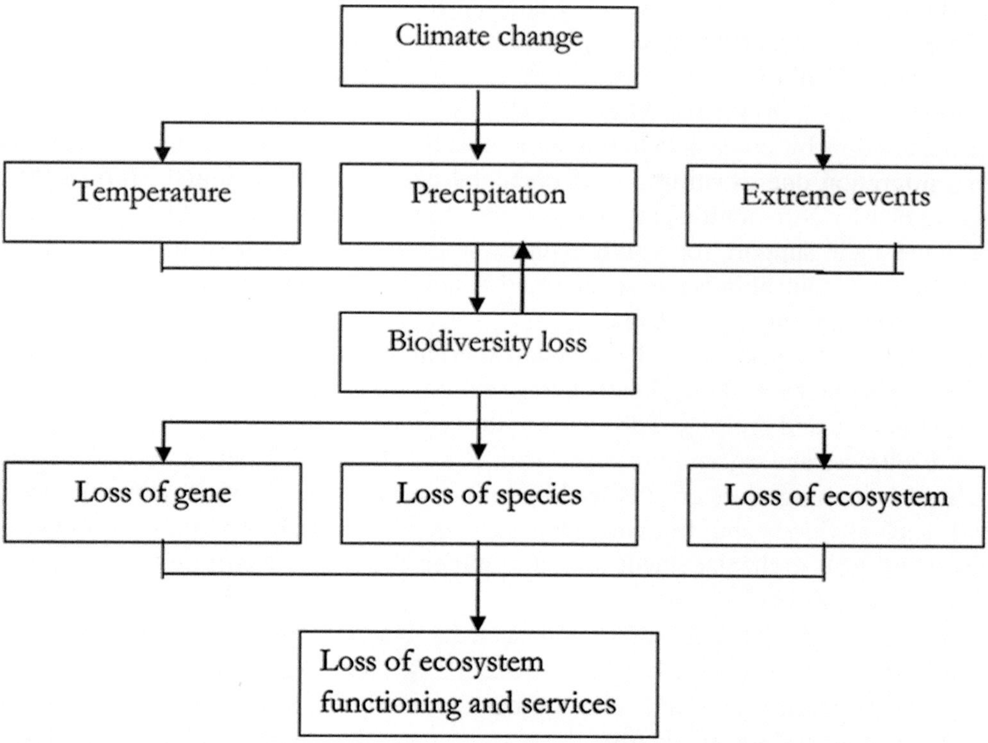

FIGURE 15.1 Link between climate change and its impacts on biodiversity and ecosystem services, and the impact of biodiversity loss on climate change

Source: Reprinted from Sintayehu, 2018. Copyright © 2018 The Author(s). Published by Taylor & Francis Group and Science Press on behalf of the Ecological Society of China. http://creativecommons.org/licenses/by/4.0/

Additionally, the Eastern African region is vulnerable to extreme weather events, like cyclones and floods, which have become more intense due to CC (Palmer et al., 2023). These climatic shifts are affecting not only the environment but also the livelihoods and well-being of the people in coastal forest of Eastern Africa (Choi and Eltahir, 2023). Research suggests that warming sea surface temperatures, especially in the southwest Indian Ocean, in addition to inter-annual climate variability (i.e., El Nin˜o-Southern Oscillation), may play a key role in Eastern African rainfall patterns and may be linked to the change in rainfall across some parts of equatorial and subtropical Eastern Africa (Leal Filho et al., 2021) (Figure 15.1).

15.2 IMPACTS OF CC ALONG THE EASTERN AFRICA COASTAL FORESTS

Eastern Africa has been experiencing various CC trends that have significant implications for the region, increasing temperatures and changing precipitation patterns are some of the key trends affecting Eastern Africa biodiversity (Gbegbelegbe et al., 2018). Also, increasing temperatures and decreased rainfall are anticipated to heighten the loss of biodiversity (Shemsanga et al., 2010), impacting not only the tourism industry but also various other sectors. To mitigate these consequences, it is crucial to establish effective development and management strategies for conservation areas. For example, preventative measures should focus on managing forest fires, particularly in regions expected to become more arid (Orindi and Murray, 2005). Again, preserving animal migration paths from obstruction by other developments is essential, given the likelihood of increased animal migration due to shifting environmental conditions (Orindi and Murray, 2005). Granting nearby communities regulated access to resources, like seed collection, could incentivize their support for wildlife on their lands. Collaborative efforts with local communities to control invasive species and rehabilitate degraded areas also offer viable solutions (Kideghesho et al., 2013).

Eastern Africa coastal areas have been experiencing a warming trend, with temperature increases that exceed the global average (Shongwe et al., 2011). This has contributed to more frequent and prolonged droughts in the region, negatively impacting flora, fauna, water resources, and agriculture (Adhikari et al., 2015). Rainfall patterns in Eastern Africa are becoming more erratic, leading to increased vulnerability to extreme weather events, such as floods and droughts, which can disrupt availability of biodiversity, food security and lead to displacement of communities along the coast forest of East Africa (Muthoni et al., 2019).

Climatic change in the coastal region caused a reduction of yields of food crops with negative effects on food security (Habel et al., 2017). Thus, people are forced to substitute their higher dependency on financial sources frequently derived from illegal harvesting and selling of natural resources from the costal forest, like clearing of forest area for crop cultivation, logging hardwood for poles, timber, carving and charcoal burning (Habel et al., 2017). However, these plant species that are being harvested along the East African-Costal Forest provide key-habitat requirements for many endemic and range-restricted species in the coastal region that they form one of the major centers of endemism in Africa (Azeria et al., 2007).

On biological diversity, CC has the potential impact in reducing species that are unable to track the climate to which they are currently adapted that can be resulted in extinction risk (Samoilys et al., 2015); however, the realized impact can differ greatly between different species. Similarly, CC affects species, and ecosystem composition and function both direct and indirect (Muluneh, 2021). Both species in terrestrial and marine ecosystems are vulnerable to CCs, and die out in their present areas and colonize new sites (Hansen et al., 2001). It predicted that CC in the future will result in the depilation of natural communities and the extinction of species (Condamine et al., 2013). The IPCC (2013) report, projections indicate a rise in global temperatures and precipitation, which will likely become significant contributors to the ongoing loss of biodiversity across various levels in the next

50–100 years. This shifting climate is anticipated to exacerbate the impacts of existing threats, amplifying the already substantial effects on biodiversity decline (Sintayehu, 2018; Figure 15.1).

The shifting climate in costal ecosystem in Eastern African is leading to significant changes in biodiversity (Orindi and Murray, 2005), impacting the way ecosystem's function and the services they provide to human populations. However, there is still much we do not fully understand about how CC affects different aspects of eastern African costal forest biodiversity and how these elements interact within ecosystems (Sintayehu, 2018). This lack of understanding makes it challenging to predict how biodiversity loss, influenced by CC, will affect the stability and overall functioning of ecosystems, subsequently impacting the benefits these ecosystems provide to society (Azeria et al., 2007).

To mitigate these impacts, East African countries are working toward strategies that include sustainable land management, afforestation, and renewable energy adoption (Nyasimi et al., 2013). Implementing these measures can help mitigate the adverse effects of CC and contribute to building resilience in the region.

15.3 ADAPTATION MEASURES TOWARD CC IN EASTERN AFRICA, COASTAL FORESTS

The need for CC adaptation strategies in Africa, particularly in East Africa, is becoming critical not only due to the vulnerability of these regions to CC impacts but also due to crucial role played by the coastal forests in people's lives (Pramova et al., 2012). Decision-oriented metrics have been developed to address key sectors such as agriculture, water supply, fisheries, flood management, urban infrastructure, and urban health in East Africa aiming to achieve adaptation outcomes such as reducing the loss of assets of coastal communities due to extreme events (Bornemann and Weiland, 2019). Some strategies adopted include restoring degraded lands through planting resilient tree species, modifying how forests are used to minimize tree cutting, preserving seed banks, preventing habitat fragmentation, and safeguarding migration routes and buffer areas. Engaging local communities in forest management is essential to sustainably access resources and enhance livelihood opportunities for the region (Orindi and Murray, 2005).

15.4 CONCLUSION, POLICY RECOMMENDATIONS, AND FUTURE PROSPECTS

In conclusion, this chapter underscores the critical interplay between CC and biodiversity in the coastal areas of Eastern Africa. It calls for a holistic approach that combines scientific understanding, community involvement, and policy action to ensure the long-term sustainability of these ecologically significant regions. As Eastern Africa's coastal ecosystems face unprecedented challenges, a concerted effort is required to safeguard both the natural heritage and the livelihoods of the communities depending on these invaluable resources.

We recommend for an integrated research approach that combines ecological studies, climate modeling and social science as well as an establishment of a robust monitoring system that will enable researchers to identify trends, assess the effectiveness of conservation measures, and inform adaptive strategies.

This chapter holds the potential to be a dynamic and enduring resource, influencing a spectrum of stakeholders' researchers and policymakers as well as communities and educators as they strive to understand the complexities of CC and biodiversity conservation along the Eastern African coast. It further seeks to inspire and guide future research agendas by identifying knowledge gaps for further investigations, advance our understanding of the relationship between CC and biodiversity in coastal ecosystems as well as continue to contribute to the ongoing dialogues on CC along the Eastern Africa's coastal areas and hence contributing to the resilience and sustainability of these critical ecosystems.

ACKNOWLEDGMENT

A heartfelt gratitude to the editors, Dr. Ally J. K. Nkwabi and Prof. Pullaiah Thammineni, whose dedication and expertise played a pivotal role in shaping and refining this book chapter. Their insightful guidance, thorough attention to detail, contributions, and collaborative spirit throughout the editorial process and unwavering commitment to excellence have significantly contributed to the overall quality of this publication.

KEYWORDS

- **temperatures**
- **rainfall**
- **biodiversity**
- **coastal areas**
- **habitat loss**
- **socioeconomic impacts**
- **adaptation strategies**

REFERENCES

Adger, W. N.; Arnell, N. W.; Tompkins, E. L. Successful adaptation to climate change across scales. *Glob. Environ. Change*. **2005**, *15*(2), 77–86. https://doi.org/10.1016/j.gloenvcha.2004.12.005.

Adhikari, U.; Nejadhashemi, A. P.; Woznicki, S. A. Climate change and eastern Africa: A review of impact on major crops. *Food Energy Secur*. **2015,** *4*(2), 110–132.

Azeria, E. T.; Sanmartín, I.; Ås, S.; Carlson, A.; Burgess, N. Biogeographic patterns of the East African coastal forest vertebrate fauna. *Biodivers. Conserv.* **2007,** *16*(4), 883–912. https://doi.org/10.1007/s10531-006-9022-0.

Balamurugan, V.; Hemadri, D.; Gajendragad, M. R.; Singh, R. K.; Rahman, H. Diagnosis and control of peste des petits ruminants: a comprehensive review. *VirusDisease*. **2014,** *25*(1), 39–56. https://doi.org/10.1007/S13337-013-0188-2/FIGURES/2.

Barbier, E. B. Climate change impacts on rural poverty in low-elevation coastal zones. *Estuar. Coast. Shelf Sci.* **2015,** *165*, A1–A13.

Bessou, C.; Ferchaud, F.; Gabrielle, B.; Mary, B. Biofuels, greenhouse gases and climate change. *Sustain. Agric.* **2009,** *2*, 365–468. https://doi.org/10.1007/978-94-007-0394-0_20/COVER.

Bornemann, B.; Weiland, S. Empowering people—democratising the food system? Exploring the democratic potential of food-related empowerment forms. *Politics Gov.* **2019,** *7*(4), 105–118. https://doi.org/10.17645/PAG.V7I4.2190.

Choi, Y. W.; Eltahir, E. A. B. Near-term climate change impacts on food crops productivity in East Africa. *Theor. Appl. Climatol.* **2023,** *152*(1–2), 843–860. https://doi.org/10.1007/S00704-023-04408-1.

Christy, J. R.; Norris, W. B.; McNider, R. T. Surface temperature variations in east Africa and possible causes. *J. Climate*. **2009,** *22*(12), 3342–3356. https://doi.org/10.1175/2008JCLI2726.1.

Condamine, F. L.; Rolland, J.; Morlon, H. Macroevolutionary perspectives to environmental change. *Ecol. Lett.* **2013,** *16*(Suppl. 1), 72–85. https://doi.org/10.1111/ele.12062.

Diop, S.; Scheren, P. A. Sustainable oceans and coasts: Lessons learnt from Eastern and Western Africa. *Estuar. Coast. Shelf Sci.* **2016,** *183*, 327–339. https://doi.org/10.1016/J.ECSS.2016.03.032.

Ferreira, M. A.; Andrade, F.; Mendes, R. N.; Paula, J. Use of satellite remote sensing for coastal conservation in the Eastern African Coast: Advantages and shortcomings. *Eur. J. Remote Sens.* **2012,** *45*(1), 293–304. https://doi.org/10.5721/EUJRS20124526.

Gbegbelegbe, S.; Serem, J.; Stirling, C.; Kyazze, F.; Radeny, M.; Misiko, M.; Tongruksawattana, S.; Nafula, L.; Gakii, M.; Sonder, K. Smallholder farmers in eastern Africa and climate change: A review of risks and adaptation options with implications for future adaptation programmes. *Clim. Dev.* **2018,** *10*(4), 289–306). https://doi.org/10.1080/17565529.2017.1374236.

Habel, J. C.; Casanova, I. C. C.; Zamora, C.; Teucher, M.; Hornetz, B.; Shauri, H.; Mulwa, R. K.; Lens, L. East African coastal forest under pressure. *Biodiv. Conserv.* **2017,** *26*(11), 2751–2758. https://doi.org/10.1007/s10531-017-1375-z.

Hansen, A. J.; Neilson, R. P.; Dale, V. H.; Flather, C. H.; Iverson, L. R.; Currie, D. J.; Shafer, S.; Cook, R.; Bartlein, P. J. Global change in forests: Responses of species, communities, and biomes. *BioScience.* **2001,** *51*(9), 765–779. https://doi.org/10.1641/0006-3568(2001)051[0765:GCIFRO]2.0.CO;2.

Herr, D.; Blum, J.; Himes-Cornell, A.; Sutton-Grier, A. An analysis of the potential positive and negative livelihood impacts of coastal carbon offset projects. *J. Environ. Manag.* **2019,** *235*, 463–479. https://doi.org/10.1016/J.JENVMAN.2019.01.067.

IPCC. Climate change 2013: The physical science basis. Contribution of working group I to the fifth assessment report of the intergovernmental panel on climate change. **2013**.

Kideghesho, J. R.; Rija, A. A.; Mwamende, K. A.; Selemani, I. S. Emerging issues and challenges in conservation of biodiversity in the rangelands of Tanzania. *Nat. Conserv.* **2013,** *6*, 1–29. https://doi.org/10.3897/natureconservation.6.5407.

Leal Filho, W.; Matandirotya, N. R.; Lütz, J. M.; Alemu, E. A.; Brearley, F. Q.; Baidoo, A. A.; Kateka, A.; Ogendi, G. M.; Adane, G. B.; Emiru, N.; Mbih, R. A. Impacts of climate change to African indigenous communities and examples of adaptation responses. *Nat. Commun.* **2021,** *12*(1):6224. https://doi.org/10.1038/S41467-021-26540-0.

Midrar, A. Optimizing the utilization of potential fisheries resources in coastal communities. *J. Life. Sci*. **2020,** *1*(1), 24–27. https://doi.org/10.37899/JOURNALLALIFESCI.V1I1.44.

Muluneh, M. G. Impact of climate change on biodiversity and food security: A global perspective—a review article. *Agric. Food Secur.* **2021,** *10*(1), 36. https://doi.org/10.1186/s40066-021-00318-5.

Muthoni, F. K.; Odongo, V. O.; Ochieng, J.; Mugalavai, E. M.; Mourice, S. K.; Hoesche-Zeledon, I.; Mwila, M.; Bekunda, M. Long-term spatial-temporal trends and variability of rainfall over Eastern and Southern Africa. *Theoret. Appl. Climatol.* **2019,** *137*(3–4), 1869–1882. https://doi.org/10.1007/s00704-018-2712-1.

Nassary, E. K.; Msomba, B. H.; Masele, W. E.; Ndaki, P. M.; Kahangwa, C. A. Exploring urban green packages as part of Nature-based solutions for climate change adaptation measures in rapidly growing cities of the Global South. *J. Environ. Manag.* **2022,** *310,* 114786. https://doi.org/10.1016/j.jenvman.2022.114786.

Nyasimi, M.; Radeny, M.; Kinyangi, J. *Climate Change Adaptation and Mitigation Initiatives for Agriculture in East Africa*. CCAFS Working Paper No. 60. **2013.** www.ccafs.cgiar.org

Orindi, V. A.; Murray, L. A. *Adapting to Climate Change in East Africa: A Strategic Approach Series Gatekeeper. Gatekeeper Series 117*. **2005.** https://www.iied.org/9544iied.

Palmer, P. I.; Wainwright, C. M.; Dong, B.; Maidment, R. I.; Wheeler, K. G.; Gedney, N.; Hickman, J. E.; Madani, N.; Folwell, S. S.; Abdo, G.; Allan, R. P.; Black, E. C. L.; Feng, L.; Gudoshava, M.; Haines, K.; Huntingford, C.; Kilavi, M.; Lunt, M. F.; Shaaban, A.; Turner, A. G. Drivers and impacts of Eastern African rainfall variability. *Nat. Rev. Earth Environ.* **2023,** *4*(4), 254–270. https://doi.org/10.1038/s43017-023-00397-x.

Payet, R.; Obura, D. The negative impacts of human activities in the eastern African region: An international waters perspective. *Ambio.* **2004,** 33(1–2), 24–33. https://doi.org/10.1579/0044-7447-33.1.24.

Pramova, E.; Locatelli, B.; Djoudi, H.; Somorin, O. A. Forests and trees for social adaptation to climate variability and change. *Wiley Interdiscip. Rev. Clim. Change*. **2012,** *3*(6), 581–596. https://doi.org/10.1002/WCC.195.

Samoilys, M.; Pabari, M.; Andrew, T.; Maina, W. G.; Church, J.; Momanyi, A.; Mibei, B.; Monjane, M.; Shah, A.; Menomussanga, M.; Mutta, D. *Resilience of Coastal Systems and their Human Partners in the Western Indian Ocean.* Fitzgerald, M. A. (Ed.), IUCN ESARO, WIOMSA, and UNEP. **2015.** https://www.iucn.org/resources/publication/resilience-coastal-systems-and-their-human-partners.

Shemsanga, C.; Omambia, A. N.; Gu, Y. The cost of climate change in Tanzania: Impacts and adaptations. *J. Amer. Sci.* **2010,** *6*(63), 182–196.

Shongwe, M. E.; van Oldenborgh, G. J.; van den Hurk, B.; van Aalst, M. Projected changes in mean and extreme precipitation in Africa under global warming. Part II: East Africa. *J. Climate*, **2011,** *24*(14), 3718–3733. https://doi.org/10.1175/2010JCLI2883.1.

Sintayehu, D. W. Impact of climate change on biodiversity and associated key ecosystem services in Africa: A systematic review. *Ecosyst. Health Sustain.* **2018,** *4*(9), 225–239. https://doi.org/10.1080/20964129.2018.1530054.

Wanyama, I.; Pelster, D. E.; Butterbach-Bahl, K.; Verchot, L. V.; Martius, C.; Rufino, M. C. Soil carbon dioxide and methane fluxes from forests and other land use types in an African tropical montane region. *Biogeochemistry* **2019,** *143*(2), 171–190. https://doi.org/10.1007/S10533-019-00555-8/FULLTEXT.HTML.

Wassmann, R.; Jagadish, S.; Sumfleth, K.; Pathak, H.; Howell, G.; Ismail, A.; Serraj, R.; Redona, E.; Singh, R.; Heuer, S. Regional vulnerability of climate change impacts on Asian rice production and scope for adaptation. *Adv. Agron.* **2009,** *102*, 91–133.

Zhao, Y.; Wang, C.; Wang, S.; Tibig, L. V. Impacts of present and future climate variability on agriculture and forestry in the humid and sub-humid tropics. *Clim. Change* **2005,** *70,* 73–116.

INDEX

J

K

L

M

N